THE
ELEPHANT
IN THE ROOM

THE
ELEPHANT
IN THE ROOM

How to Stop Making Ourselves
and Other Animals Sick

LIZ KALAUGHER

The University of Chicago Press

The University of Chicago Press, Chicago 60637
© 2025 by Liz Kalaugher
Published 2025
Printed in the United States of America

34 33 32 31 30 29 28 27 26 25   1 2 3 4 5

ISBN-13: 978-0-226-84090-1 (cloth)
ISBN-13: 978-0-226-84091-8 (e-book)
DOI: https://doi.org/10.7208/chicago/9780226840918.001.0001

Published in the United Kingdom in 2025 by Icon Books.

Library of Congress Control Number: 2024041097

♾ This paper meets the requirements of ANSI/NISO
Z39.48-1992 (Permanence of Paper).

To Sue, Patrick, Mags, Catherine, Justin, Dan, Tom, Josh and Krystal

Contents

Color illustrations follow page 176

Preface

One evening in early 2020, as I walked back to my cabin at a hotel near the base of Mount Kenya, creatures screeched from high in the trees. It was too dark to see but they sounded vicious and large. Once my feet had rejoined the ground, a voice in my head said, 'Blimey, archaeopteryx. I thought flying dinosaurs had gone extinct.' The next day, I learned the screechers were tree hyraxes – nocturnal mammals related to elephants but a little larger than guinea pigs. Over the next few evenings they bawled and rasped outside the cabin's ground-floor bathroom like murderers with chainsaws, and howled from the chimney-top so that the sound reverberated from the stone walls as I tried to sleep.

These small animals' bark was far more scary than their bite. Far away in China, however, a toothless but much more deadly danger had quietly emerged from the wild. As I travelled across Kenya in search of the Grevy's zebra, a rare and threatened species that copes well with harsh, dry conditions, the SARS-CoV-2 virus, which causes Covid-19, spread around the globe. And pretty much everyone's focus turned to zoonotic diseases – illnesses that jump from wildlife to humans.

In recent years, more and more of these diseases have emerged from the wild and made people ill – HIV, SARS-CoV-1 and SARS-CoV-2, MERS, swine flu, Zika, Ebola, mpox and more. But we've paid much less attention to the ways that humans have made other animals sick. We've done this for millennia, pretty much since we first left Africa and encountered Neanderthals.

As soon as the first anatomically modern humans left Africa, we took bugs with us in our bodies, spreading disease when we ourselves were sick. Later we grew more sophisticated in the ways we impart diseases. Some 12,000 years ago, when we started farming other animals, we began living closely alongside poultry and livestock and putting these animals in cramped conditions. That enabled bugs to spread more easily from non-human animal to non-human animal, from other animals to us, and from us to other animals. Our farming made the animals we'd domesticated sick; not just livestock like cows, sheep, goats, pigs, camels and llamas, but also working and companion animals like dogs and cats. Many of those animals have close relatives still living in the wild – goats, antelope, wild boar, wolves, big cats – that had similar biologies to their domesticated and newly disease-rich cousins. Domestic dogs, we think, have spread canine distemper virus to lions and African wild dogs. Nearly 90 per cent of mammals threatened by parasites, including viruses and bacteria, are from the orders Artiodactyla, especially the swine family and the bovid family of cloven-hoofed ruminants such as cattle, bison, buffalo and antelope, and Carnivora, particularly the dog and cat families.[1] That doesn't seem like a coincidence.

When we developed arable farming, our crop stores attracted rodents that shared pathogens – disease-causing agents – with us too. Since rodents have a live-fast-die-young approach to immune systems and don't invest much effort in fighting germs, they were largely the major gift-givers. In turn, we shared those gifts worldwide, with other animals and other humans.

Then, of course, there's trade. Humans started moving non-human animals around the world as we explored. We took domestic animals with us, whether by design, in the case of livestock, or accident, in the case of rats, and we began to trade wild animals – live or as carcasses – from the places we'd inhabited or invaded. We introduced invasive species to islands and continents where they didn't exist before, taking their bugs and diseases with them. This killed off, as just one example, many species of the finch-like birds known as Hawaiian honeycreepers, leaving the lower slopes of Hawaii's mountains desolate. More recently, trade has spread chytrid fungus around the world, devastating frog populations. The wildlife trade for pets, food, trophies, souvenirs and medicine continues today, some of it legal, much of it not.

Trade isn't last on the list of the ways we've spread diseases to wildlife. We've also logged forests and destroyed habitat, forcing wild animals to live more closely alongside each other and us. What's more, damaged, less biodiverse ecosystems are less resilient and more likely to harbour pathogens. We've warmed the climate too. And we've begun farming wildlife, and farming livestock on an industrial-scale, cramming non-human animals into massive sheds. High-density poultry farms kick-started a more dangerous strain of bird flu that spilled back into the wild, and today threatens humans with a new pandemic. Even the wildlife conservation programmes that have tried to help have in some cases inadvertently introduced disease, for example by passing pathogens from one species to another during captive breeding programmes, then letting the germs loose in a new area when they let that species free. As we'll see later, that's how the chytrid fungus, deadly to many amphibians, reached the island of Mallorca.

Our harms broadly fall into four categories – moving pathogens to new places, encroaching on the lives of other animals, harming or destroying ecosystems, and influencing climate change. The elephant in the room is us. And it's an elephant with many faces – the story for each species and pathogen unfolds in its own unique way.

PATHOGENS RISING

The result of all these changes? We've increased the threat of new diseases emerging in other animals. Like those in humans, wildlife diseases have emerged more frequently in recent decades. Wildlife biologists only recently became aware of just how big a role disease played in extinctions both ancient and modern. Disease has been the neglected problem child of the conservation world. Instead, attention has focused on its four ugly sisters: the 'evil quartet'[2] of habitat destruction, introduced species, overkill – humans slaying too many individuals – and secondary extinctions – species wipe-outs that result from other extinctions, for example the two species of louse that lived only on the passenger pigeon shot to oblivion in North America in the nineteenth century.[3]

American scientist and environmental historian Jared Diamond coined the term evil quartet in the early 1980s, before global warming showed its true face; many would now add climate change to the ugly sisterhood.[4] The same goes for diseases. 'Normally infectious disease is left off that list,' says infectious disease expert Hamish McCallum of Griffith University, Australia. 'My view is that's probably a mistake.' Even back in 1933 environmentalist Aldo Leopold stated that 'the role of disease in wildlife conservation has probably been radically underestimated'.[5,6] Illness is often the final straw that breaks the camel's, or Tasmanian tiger's, back. Today, as we'll see, the African wild dog, Tasmanian devil and North America's black-footed ferret are all at risk of extinction from disease.

FINDING THE ELEPHANT

To begin, *The Elephant in the Room* will introduce human and wild-life diseases, as well as the Grevy's zebra – a species threatened by many of those evil step-sisters, particularly habitat loss, overkill and, increasingly, climate change. As an animal that lives alone or in small groups and has a loose social structure, the Grevy's is, thankfully, relatively unaffected by infections. But, with conditions becoming more extreme, early signs indicate that even the Grevy's is suffering more strongly from disease. Yet this equid's story shows us what can happen when people change their behaviour in order to protect wildlife; the Grevy's zebra is now a rare success story in the field of conservation. Its stripes form a lens for us to examine our actions and choose a path forward.

Next, we will head back to the dawn of modern man and explore how early humans' germs may have killed off Europe's Neanderthals. From then on we will journey forwards in time, through the potential role of disease in the extinction of North American woolly mammoths and giant sloths, the triple import whammy that harmed Hawaii's native birds, the accidental introduction of cattle plague to Africa, and the demise of the Tasmanian tiger. Each chapter in this book will focus on a single wildlife case study that illustrates the animal's unique value, the manmade factors contributing to its disease woes, and, where possible, how we can stem the emergence and spread of such diseases. I've largely avoided detailed descriptions of sick animals. As an animal-lover, I didn't want to write them and they don't make for a fun read.

As this book moves through time and the diseases come thick and fast, the manmade factors change in parallel with our lifestyles. We begin to farm. We stop walking between continents in favour of travel by ship and then planes. We chop down rainforests for ranches and plantations. We burn fossil fuels. We build industrial-scale farms. Case by case, the full picture of our planetary damage reveals itself.

As we reach the 1980s and 1990s, when wildlife diseases emerge faster, we turn to the case of the black-footed ferret, which introduces the risks associated with conservation techniques; the African wild dogs and seals that share related viruses thanks to close contact between humans and other animals; foxes that highlight the trials and tribulations of city living; frogs killed by the increasing globalisation of trade; and American crows that reveal how climate change allows diseases to take new ground.

Here in the present day, *The Elephant in the Room* looks at the results for monkeys, apes, bats and wild birds as people have fragmented habitats, cleared more land for agriculture and commercial plantations, come into ever closer contact with wildlife, and set up industrial-scale sheds housing chickens and pigs: outbreaks of yellow fever, Ebola, flu and more. As I wrote the chapter about bird flu, the first transmission of a bird flu virus from a cow to a farmer took place in the US; many fear the H5N1 virus is adapting to spread more easily between mammals. That prospect is terrifying.

UNIVERSAL HEALTH

The same human activities that have spread diseases to wildlife have come back to bite us too – they've also heightened the risk of diseases spreading from wildlife and livestock to humans, whether virally, as with the 1918 influenza or Covid-19 pandemic, or by other means, like with the chronic illness Lyme disease, which spread to us from wild mammals via ticks (partly due to habitat disruption). The pet trade introduced mpox – the pox formerly known as monkeypox that's generally found in rodents and non-human primates in western and central Africa – to the US in 2003. That made 81 people ill. By May 2022 the disease, which had caused sporadic human cases in Africa for decades, had become better at transmitting from human to human and broke out in humans in several countries, including the UK, the US and Australia. In 2024, as this book neared completion, the World Health Organization declared a new strain of mpox a public health emergency of international concern, its strongest form of global alert, after an outbreak in the Democratic Republic of Congo and other nations nearby.[7]

We've made ourselves as well as other animals more likely to get sick. In 2008 prestigious science journal *Nature* published a paper that showed that 60 per cent of infectious diseases in humans originate in non-human animals, with more than 70 per cent of these zoonotic diseases coming from wildlife, and increasing over time.[8]

Wildlife diseases and human pandemics are two sides of the same coin. Yet we're mainly looking only at 'our' side, a selfish approach that is unlikely to serve us well. We're animals too and our health depends on the health of our ecosystems and of other animal species, whether livestock or wild. The burgeoning academic discipline of One Health acknowledges that human, non-human animal and ecosystem health are closely linked. Indigenous cultures have aligned with this concept for tens of thousands of years.[9] In the West, although ancient Greek philosopher Hippocrates was an early proponent,[10] One Health only became a mainstream topic from around 2004, when scientists at a meeting in New York organised by the Wildlife Conservation Society set up twelve Manhattan Principles on 'One World, One Health'.[11] Even in this field we're still neglecting wildlife; currently most people working in One Health focus on disease surveillance in livestock and humans, and rarely consider how the environment feeds in to emerging health threats.[12]

'The stress that humanity is placing on nature is leading many [non-human] animals to become more susceptible to infection, and to be more likely to shed virus when they are infected, even if they don't get sick themselves,' says practising medical doctor and epidemiologist Neil Vora of Conservation International. 'These are all factors that come back to hurt our own health because the pathogens that cause most new infectious diseases originate in animals and then jump into people. We're all connected.'

When I set out to write the final chapter of this book, I planned to look at ways of guarding wildlife health then move on to protecting humans from pandemics. But arranging the material this way was like trying to untangle a thicket of brambles – everything was too closely entwined to tease apart without pain. Instead, I decided to ignore the artificial borders we currently imagine between the health of humans, other animals and the environment, and proceed as if humans are part of nature, which we are, no matter how divorced we feel from the wild as we tramp along the hard surfaces of city pavements. Our bodies and immune systems know it even if our minds don't. They've proved it by catching diseases from other animals. Harming wildlife and the environment is a form of self-harm, though you might not guess it based on how we're treating the natural world, ourselves included.

TAMING THE ELEPHANT

The remedies for this harm range wide. We could regulate trade and farming, protect forests and other habitats, turn away from fossil fuels. This would likely cost less than another pandemic. As yet, our efforts to stop new infectious diseases emerging for both wildlife and people have been minimal. The chances of another global pandemic are high and wildlife diseases and extinctions continue apace. Unless more people understand why and call for government and international action, we'll see more and more wildlife loss and more and more pandemics.

Many of these changes would have a hefty side-benefit – they would also save wildlife from population declines and extinctions not linked to disease. In turn, keeping ecosystems healthy in this way would protect all living beings, by making diseases even less likely to break out. It's a win-win situation. Let's choose panacea not pandemic. Until we mend our ways and our ecosystems, us, the Grevy's, the tree hyrax and all other species face an uncertain future. Let's focus on, then usher out, this elephant in the room.

Liz Kalaugher
Bristol, UK, 2024

1

Wild Horse Chase: Grevy's Zebra, Extinctions and Wildlife Diseases

'Our wildlife is our river of life; if we finish the wildlife it would be like the river we fetch water from drying up.'

Lmantros Lenangetai, Grevy's Zebra
Trust Ambassador

LATE JANUARY 2020, SAMBURU COUNTY, NORTHERN KENYA

It's dusk. I'm 'powdering my nose' in the room behind my tent at the Grevy's Zebra Trust HQ. I've climbed the six wooden steps from the stone floor of the room with the sink, then turned right. Just above my head, chicken wire frames the sky in the handspan gap between the rendered mud wall and the plant-stem roof. In the distance, a bird serenades the evening and a goat bleats. The light is fading fast, it's not far to the equator. Better get on. As I sit, something zooms past my right ear. It comes from near the window, turns 90 degrees to head down the stairs and swoops out into the sky through the open roof of the cubicle for bucket showers. My heart beats almost as fast as the flutter of its sharp-cornered wings – it's a bat.

I like bats. I own a bat detector. But generally, I go out to look for them in a field or park or garden. I wasn't expecting a bat to join me in the bathroom. *It must have come through the window*, I think. *There must be a gap in the wire.*

It's only when it comes up in conversation at dinner that I realise the bat didn't fly through the window. It sped out of the long-drop toilet just before I blocked the exit. Over the next few nights, I become savvier. If I hear the bats flapping as they circle the exit from their cavernous roost, I move politely to one side until they've left. I don't want to excrete on anybody. With the aid of my head torch, I strike a balance between being in the way and getting a good view; it's dark down there, though I'm grateful I can't see the bottom. One evening, I glimpse two bats flying several circuits of the long-drop before they venture out. Gradually I grow more confident. I move less far away, give the bats less space. Another evening, when I hear the flutter of a bat inside the long-drop, I shuffle forward with my head turned to look behind. The bat flaps out and up. I'm a shock to it. We're face to face, the bat so close it's a blur, a fuzzy shape growing larger fast. I gasp and freeze. The bat veers away over my shoulder. All it leaves is a memory of a small pink face with a prominent snout, edged by brown fur. I don't know if that's what I saw or what I imagine I saw. The image doesn't arrive until the bat has gone, like a flash of nostalgia. *That was close*, I think. *Good work, bat.*

We were in each other's faces, we breathed the same air. Should I be worried? Later, I learn that bats host a multitude of viruses; I'm fully aware of that now it's affected my life and those around me. Bats can live symptom-free with infections that would – and do – kill us.[1] These animals may be protected against becoming ill by their evolution of flight, whose stringent requirements mean that, almost as a side-effect, they developed an immune system that counteracts viruses while limiting inflammatory responses that can damage the bat itself.[2] Bats make us sick with all sorts of diseases – they've passed us Ebola, Marburg virus, SARS-CoV-1, MERS, Nipah, Hendra and now, chances are, SARS-CoV-2, which causes Covid-19. Though humans have also made bats sick, notably by transporting across the Atlantic the fungus that causes white-nose syndrome in North American bats.

When bats make humans ill, it's possible that the fever our bodies create to kill these viruses only makes the invaders feel more at home, by bringing our body temperature closer to that of a bat. I don't, as far as I'm aware, catch anything from my close encounter. No fever, no cough, no loss of taste or smell. The bat's faeces are safely inside the long-drop and the Covid-19 pandemic is a distant shadow, a flutter on the edge of my consciousness. We aren't yet face to face.

Already in our faces, even though we're ignoring it, is the flipside of this – how much humans have inflicted harm on wildlife by spreading diseases throughout history. Over time, our numbers have soared, from 1 billion in 1800 to 8.2 billion in 2024.[3] The human population curve mimics the rise in cases at the start of an epidemic, though on a longer timescale. For the luckiest of us, lifestyles grew lavish – houses, central heating, aircon, cars, washing machines, plastic toys, televisions, laptops, mobile phones, and food, pets and lab animals flown in from other countries. So we 'needed' more resources to sustain these lifestyles, more land to grow timber and crops, to mine metals and minerals, and to set up factories for goods and livestock.

Often the devastating consequences, for us and other animals, of our cosseted Western lives happen far away, out of sight and out of mind, easy to ignore until they come back to hit us in the lungs. This hunger for land, for homes for our growing numbers and to meet our growing needs, has seen us encroach on wildlife territories. More people have moved deeper into areas that were recently wild, into forests and savannah. That's disrupted ecosystems, sent species into population decline and brought us into closer contact with wildlife. As we should have learned over the centuries of our encounters with livestock, contact with other species means opportunities for pathogens to jump the species divide.

Before the road network expanded, chances were that when a deadly virus that could transmit from human to human spilled over from a rainforest, only a few people from that first victim's village and perhaps a few neighbouring villages would die before the outbreak sputtered out. Nowadays, almost everywhere has access to a road and more people travel to, trade with, and live in towns and cities. A virus can range further, faster, the flames of its devastation spreading like wildfire across the globe rather than flickering in isolated patches that burn themselves out. International air travel means pathogens that do make it into a human can be in another continent the very next day, almost before their victim has time to cough. Or they can hide out in a wild or farmed animal – or a wild farmed animal in China – that we ship or fly around the world dead or alive. Chances are West Nile virus reached the US, where it kills birds and people, by plane or ship from Israel in 1999. Today it's spread across the continent. Our industrial farms bring non-human animals into closer contact, chicken crammed in next to chicken near sheds packed full with pigs, while wild birds peck the dirt outside or feed in the ponds where the chicken waste falls. It's enough to give a person chills. In fact, it does – it lets flu strains mix and match from pig to bird to human. That's dangerous.

SHARE ALIKE

We've found many ways to worsen other animals' health. Sometimes we're almost literally hands-on and give animals that are biologically similar to us diseases directly, for example by visiting chimps and gorillas in nature reserves and passing on colds and other human respiratory viruses that may kill other apes. Some call diseases that jump from human to non-human animal zooanthroponoses, while some choose the term reverse zoonoses. Diseases that jump species in the other direction, from non-human to human, as SARS and likely Covid-19 did, are known as zoonoses or, if you have time on your hands, anthropozoonoses. Here, we'll take the easy route and call everything a zoonosis. Often we won't need to use the word, as the way we cause wildlife diseases isn't always direct. Sometimes our livestock and companion animals make wildlife sick, and sometimes our harm is less direct even than that.

As biologists look back at past extinctions, they find the hand of disease caused by man in many more cases than they realised, or rather, the suspected fingerprints of the hand of disease caused by man – it's hard to diagnose diseases when physical evidence has long decayed. It wasn't until the seventeenth century that we first saw bacteria in early optical microscopes, and not until the nineteenth century that we linked these bacteria to disease. Virus discovery was even slower – filtering sap from a diseased plant proved the existence of viruses in the late 1800s and it wasn't until the 1930s and the invention of the electron microscope that we could look at them in detail. Even then, it was a specialist job and not available to those on the ground seeing wildlife, livestock and their relations die.

Whether the cause was known at the time or not, the march of creatures disappearing to disease goes on, from Christmas Island's native rats to the Hawaiian birds lost in the nineteenth and twentieth centuries, the demise of the Tasmanian tiger in the 1930s, and more recently, some 90 species of amphibians and counting, including Costa Rica's golden toad. All gone. Vanished into the mists of the past with a helping shove from man.

The other stresses we're imposing make matters worse. Between 1970 and 2016, global populations of vertebrates – backboned species like mammals, reptiles, birds, fish, amphibians – crashed by roughly two-thirds, according to a report by the WWF.[4] Nearly a million of Earth's species are at risk of extinction, say UN figures, and the current rate of extinctions is tens to hundreds of times higher than the average for the past 10 million years.[5] In 2010 the UN Convention on Biological

Diversity set 20 biodiversity targets for 2020. We haven't fully achieved a single one.[6] Habitat loss is a major factor but there are many others alongside emerging diseases – competition for resources, invasive species transplanted by man, climate change, pollution, hunting, poaching and human–wildlife conflict.

The problem is, diseases and drops in headcount are like the currents in a whirlpool sucking wildlife towards oblivion. Their forces combine and accelerate. Once numbers of an animal decline because of habitat loss or climate change or for whatever manmade reason, that species is much more likely to be wiped out by disease. A small population that lacks genetic diversity, is stressed, has lost much of its habitat so is crowded together, perhaps alongside new competition from invasive species, and in a climate that it's not well adapted to, is not best placed to withstand an onslaught of infection. Animals become more suscep-tible to the pathogens that bring diseases, which could cause so many individuals to die that there wouldn't be enough left to build numbers back. Disease can pile pressure onto animals already under threat from habitat loss or exploitation or poaching or pollution.

These losses create an even stronger push towards the centre of the whirlpool – both population declines and extinctions disrupt ecosystems, which in turn makes diseases more likely to break out. Generally, the larger, longer-lived animals go extinct first. And they tend to have bet-ter immune systems and be more disease-resistant. Take these animals out and you have an ecosystem where disease can take hold and spread more easily. This 'dilution effect' theory, that ecosystems with more biodiversity are more resilient against disease, has been controversial over the years but the evidence has finally tipped towards it.

Each species lost leaves a gap in the structure of an ecosystem that risks it crumbling into disarray. If nature goes, we'll suffer too – we're part of it. No bees to pollinate our crops, no detritivores to clean up rotting vegetation. No protection for our health. As we destroy nature further, both wildlife and humans will suffer more diseases; the rates for both have already increased. In the 1980s and 1990s, seals died in their thousands in the North Sea thanks to an outbreak of phocine distemper virus that attacked their lungs and brains. Canine distemper virus harmed African wild dogs and lions and nearly wiped out the black-footed ferret in the US. My home city of Bristol's foxes fell to mange. Tasmanian devils caught an infectious cancer. Frogs disap-peared around the globe. North American birds perished in the wild and in zoos.

WILD HORSES

The point of my travels, the Grevy's zebra, is unusually resistant to disease, even for a relatively large and long-lived animal. Unlike plains zebras, Grevy's don't live in harems – groups of females and their young headed by a male. Instead, Grevy's have a more flexible social structure known as fission-fusion, because the animals split apart and come together depending on their needs, an approach that helps the Grevy's cope with dry conditions where resources are patchy. About one-tenth of adult males are territorial and guard a patch of land with grass and water that attract females.

This loose social structure also makes the Grevy's zebra relatively hardy to disease. Under normal circumstances, the Grevy's suffers less from parasitic worms than the more sociable plains zebra, for example. Though the Grevy's can, if humans relocate it, pass on a virus it carries symptom-free. In 2007, a Grevy's killed a polar bear at a zoo in San Diego, by giving it equine herpesvirus-9. And nowadays we've made even the relatively hardy Grevy's zebra sick. Recently two Grevy's at a nature reserve were so stressed by food shortage due to a drought most likely caused by climate change that they fell to intestinal parasites, an infestation they'd normally be able to live with.

The Grevy's I see in Kenya, however, all look healthy. Like my encounter with the bats, my first sighting of these animals was also a surprise. The story of the Grevy's zebra stands out in a conservation landscape that's otherwise bleak. After twelve years writing about climate change, I was desperate for good news. On my very first afternoon in Kenya I found myself on an upholstered bench on the veranda of a former colonial ranch house in the central highlands, at Mpala Research Centre. On the horizon, far beyond a wooded valley, lay the foothills of Mount Kenya; the mountain itself was shrouded in cloud. Superb starlings with petrol-blue chests and oil-sheen backs flitted onto the lawn in front of me while lesser-striped swallows swooped their copper heads and sparkling navy wings into a nest in the eaves above. *This isn't real*, I thought. *This is paradise.* From a tree, a red-chested cuckoo cried as if saying 'it will rain' in three mournful notes descending in pitch, and rock hyraxes grazed like oversized guinea pigs or scuttled up the veranda steps. One peered at me from the gutter above, beady-eyed. At night, I'd soon discover, they and the vervet monkeys would scamper across my bedroom roof, the scraping of their claws like stones bumping down corrugated iron.

Seven hours after leaving Nairobi, I was in a scene that could have been in *Out of Africa*, lazing on a veranda, deep in white privilege. It was an accident because I had booked late and the working accommodation for researchers was full. So I was a special guest. About three-quarters of an hour after taking my seat, I looked away from the lawn and the trees and the hyraxes and hornbills and superb starlings, and glanced at the hill on the other side of the valley. Two Grevy's zebras grazed, heads down, near a termite mound in a patch of open ground edged by thickets and small thorned trees. Distracted by the birds, I hadn't seen them arrive. They weren't there and then they were, fawn shapes against the dry grass highlighted only by the white of their ears and bellies. In places, on a neck, on a rump, I could just about make out their thin black and white stripes. I imagined the sound of grazing, a gentle munching, the ripping of grass stems, the swish of a tail, the buzz of flies. *This is why I'm here*, I thought. *This seems too easy.* The animal I'd travelled more than 6,000 kilometres, or 4,000 miles, to see had wandered into my path within an hour of my arrival.

The zebra on the left faced away from me and was swiping the brush of dark brown hairs at the end of its tail from side to side to ward off flies. The patch of white on this animal's hindquarters gleamed in the sunshine; the Samburu name for these animals is Loiborkurum, or white-rumped. Some have used other names. The Romans paraded 'imperial zebras' into colosseums before gladiators battled, prizing them for their upright bearing and regal stripes. Centuries later, in 1882, Abyssinian emperor Menelik II presented one to French president Jules Grévy. When a French naturalist realised this gift was a separate species, he gave it a new name, after its new host.

Until recent times, Grevy's roamed only the Horn of Africa, from the coast of the Arabian Gulf down into northern Kenya. Like the water in a reservoir during a drought, Grevy's homelands have shrunk into small, isolated patches. Today these zebras make the leader board of African mammals whose range has dwindled the most. Grevy's no longer graze the arid lands of Eritrea, Djibouti or Somalia, and only a couple of hundred linger in the deserts of Ethiopia, mainly in the Aledeghi Nature Reserve. The rest dwell in dry and semi-arid northern and central Kenya, chiefly in the north's Samburu County, where the Grevy's Zebra Trust has its field HQ, and in the Laikipia region around Mpala Research Centre.[7] Mpala is further south than Grevy's traditional habitat, where the growth in the number of humans pushed them out.

In 2023 estimates put the Grevy's population at some 3,000,[8] a drop of more than four-fifths from the near 15,000 alive in the late 1970s.[9] In that decade zebra hides were bang on trend for Western interiors and accessories, and trade in this 'commodity' flourished. Even after Kenya banned commercial hunting in 1977, the population continued to fall. By then, the nation's human population had boomed – from nearly 8 million in 1960 to more than 56 million in 2024 – and pastoral herders had stopped moving around the landscape so much, increasingly settling near schools and health clinics. This degraded the land and reduced the availability of vegetation, forcing wildlife and livestock into smaller spaces. Like a million of Earth's other species, the Grevy's zebra is on a slow slide to extinction; in 1983 Andy Warhol featured a Grevy's, brightly coloured but dejected, in a screen-print for his *Endangered Species* series. The International Union for the Conservation of Nature (IUCN) classifies this species as endangered – very likely to become extinct in the near future.

REVERSE REVERSE

As I watch these Grevy's graze in the sunshine, Covid-19 is spreading west into Europe. In December 2019, cases of unexplained pneumonia spiked in Wuhan, a city of some 11 million people in Hubei province, eastern China. By the end of the year, doctors had implicated the previously unknown virus 2019-nCoV – the letters stand for novel coronavirus – as the culprit for the disease they dubbed Covid-19. By 11 January, Chinese scientists had released the genetic code of the virus; the next day, given its similarity to the SARS virus that caused a pandemic in 2002 and 2003, the World Health Organization renamed 2019-nCoV as SARS-CoV-2. And by 23 January, four days before I encountered the bat at Grevy's HQ, Wuhan was in lockdown following 540 official cases of Covid-19 and 17 deaths. The lockdown didn't stop the spread of the virus around the globe. All most likely because of a bat, though people, as we'll see, must share the blame. Chances are that encroachment into wildlife territory, close contact between humans and wildlife, trade and international travel were all factors.

There's much we don't yet know about SARS-CoV-2. What we do know for sure is its genetic code and that it's a coronavirus. At its heart it has strands of RNA, which is a nucleic acid like the DNA that forms our genes, but curls in single-stranded spirals, not the twisting coupled ladders of DNA's famous double helix. Sheathing these strands of RNA is a coat embellished with the protein spikes that give the family its name; they fan out around the virus like a crown. SARS-CoV-2 basically consists of RNA, proteins and a double layer of lipid molecules that link the proteins in its coat. That's it – genetic code in a bag. A very small bag, roughly one thousandth of the diameter of a human hair. Under a powerful microscope SARS-CoV-2 looks innocuous, like a child's drawing or a fluffy burr. It's what it does that harms.

Despite their simple structures and small size, viruses kill. In humans, SARS-CoV-2's club-shaped protein spikes chiefly latch on to the ACE2 (Angiotensin-converting enzyme 2) protein receptors coating the cells that line our airways – nose, throat, lungs. The virus tricks its way inside a cell via these protein receptors. Once there, it subverts the cell's DNA-reading machinery to read viral RNA instead. So the cell replicates the viral proteins that the RNA codes for instead of making more of its own proteins. 'Don't copy you,' the virus whispers inside the cell. 'Copy me.' And, bewitched or bewildered, the cell does, over and over again. It helps its own invader multiply. These home-grown copies of the virus sneak back out through the cell wall and infect other cells, where they replicate again in an ever-expanding onslaught. Some people become infected but don't show symptoms, others cough or lose their sense of taste or smell or burn a fever as their immune system fights back. An unlucky few per cent, a number that soon becomes hundreds of thousands of individuals when millions are infected, struggle to breathe when their lungs fill with fluid, and the unluckiest of these people die. Millions of people around the world lost their lives to SARS-CoV-2; my heart goes out to everyone who didn't make it and to those they left behind, especially those forced to grieve alone.

SARS-CoV-2 is a killer. Yet not all coronaviruses are this lethal. There's diversity here too; researchers estimate there are around 3,000 to 5,000 coronaviruses, all of them likely to have originated in bats. The virus doesn't care which animal's cells it enters, as long as its proteins hold the key to getting inside and it can take over the cell's workings to replicate. Coronaviruses have probably existed alongside us and other animals for thousands, if not millions of years. Science identified the first only in 1931, on a farm in the US state of North Dakota, where it caused avian infectious bronchitis in chickens and killed up to 90 per cent of infected chicks.[10] In the 1960s, researchers discovered that some coronaviruses give humans colds, although most colds are caused by rhinoviruses, which are three times smaller than SARS-CoV-2. In people and birds, coronaviruses mainly affect the respiratory system. Other species react differently. In mice, coronaviruses cause brain inflammation and hepatitis, while in cows and pigs they bring diarrhoea. In bats too, coronaviruses tend to thrive in the gut; bats typically shed particles of these viruses in their faeces. Given what we now know, the traditional practice in Chinese medicine of treating diseases by sprinkling powdered bat faeces into people's eyes doesn't seem a great idea; some practitioners have called for a ban.

In November 2002 we discovered coronaviruses could give us something more deadly than a cold when Severe Acute Respiratory Syndrome, or the SARS virus, made itself known in Guangdong province in southern China. This virus infected more than 8,000 people around the world,[11] and killed around one in ten,[12] a rate some ten times higher than SARS-CoV-2. The SARS virus hasn't infected a human since 2004 but that doesn't mean it couldn't jump species to us again. It's likely that SARS came from horseshoe bats in a cave in China's Yunnan province, also in the south. When scientists tested these bats' faeces and swabbed their insides, they found that although none were infected with SARS, they did harbour a number of different bat coronaviruses.[13] Together these viruses contained all the genetic building blocks for making the SARS virus. Jumbling – or recombination – of these building blocks when a bat had two or more viruses at once could over time, the team believes, have created a SARS-like virus. Spread of this SARS-like virus to an intermediate animal such as the cat-sized masked palm civet,[14] may have let the virus evolve into a form that could spread between humans. When one of these intermediate hosts ended up at a wet market, a forum for selling unrefrigerated vegetables and meat, sometimes freshly killed, it passed the virus to the first human. And SARS was ready to run rife from one person to the next.

There are probably many more spillovers of non-human animal diseases into humans each year than we ever realise. In the events that disappear without trace, the crucial difference is that the person who catches the disease doesn't pass it on; the pathogen isn't well enough adapted to us to use our bodies to transmit itself. The virus or bacterium cadging a ride inside us doesn't have the right ticket to transfer vehicles, if you like, and the human it enters becomes a terminus, a dead end. Perhaps literally. But that can change, as it did with SARS, probably with the help of a civet. Ten years after SARS, in 2012, Middle East Respiratory Syndrome coronavirus, MERS-CoV for short, was detected in Jeddah, Saudi Arabia. MERS transmitted, scientists think, from an African bat to a North African camel then moved to humans in the Middle East via the camel trade.[15] Also known as camel flu, MERS has killed roughly one-third of the 2,000 or so people it's infected to date,[16] mainly through contact with diseased camels or in hospitals.[17]

When Covid-19 came on the scene and scientists learned that the virus that causes this disease is genetically 80 per cent similar to the original SARS virus, both viruses got new names. Original SARS unofficially became SARS-CoV-1 and 2019-nCoV became SARS-CoV-2.[18] The two share a common ancestor; SARS-CoV-2 is not a direct descendant of SARS-CoV-1 but a cousin, perhaps a distant one. Together with similar viruses in bats, the pair are known as severe acute respiratory syndrome-related coronaviruses, or SARSr-CoV. This is not a field of snappy titles. Since different viruses often mix and match – or recombine – their genetic material, virus family history tends to be murky and hard to trace. To find out where SARS-CoV-2 came from, researchers strip out its recombinant regions – the mixed-in parts – and look at them individually,[19] as if they were tracing the history of a car that's had new doors, a new engine and a change of upholstery piece by piece. Such analysis indicates that, like its relative, SARS-CoV-2 also has its origins among horseshoe bats; it's more than 96 per cent similar to several different bat coronaviruses found in South-west China,[20] as well as northern Laos.[21] It's the viruses further from Wuhan, in Laos, which have genes related to binding to host cells – the make-or-break part when it comes to the ability to infect humans – that are more similar to those of SARS-CoV-2.[22] These viruses last had a common ancestor with – or in other words, diverged from – SARS-CoV-2 only a couple of years before it emerged in humans in Wuhan.[23] 'Given the extent of recombination going on, we're unlikely to find an individual bat hosting the SARS-CoV-2 virus,' says virologist David Robertson of the University of Glasgow, UK. But it does mean that horseshoe bats host other viruses that could infect humans too.

These bats don't usually migrate,[24] and are unlikely to have dispersed fast enough to transport the virus hundreds of kilometres on this timescale. So it looks like humans helped, moving the virus to Hubei province inside an as-yet unknown intermediate species via the trade in wild and farmed animals.[25]

Once there, SARS-CoV-2 jumped to a human at a wet market in Wuhan.[26, 27] 'There's certainly no evidence for any lab involvement, while there's lots of evidence for a spillover event very much like SARS,' says Robertson. 'The evidence that SARS-CoV-2 got into humans through an intermediate animal host at the Huanan market in Wuhan is now very compelling.' It's notable that some of those who called most strongly for further inquiries into whether the virus leaked from the Wuhan Institute of Virology were politicians, who may have been lobbied by those with vested interests in our continued exploitation of the environment. If the virus escaped from a lab, the lobbyists can argue, then the myriad ways we're harming nature simply aren't relevant, and politicians are in no way to blame for failing to regulate this. Yet the lab leak theory itself creates an anti-science agenda that leaves us more vulnerable to future pandemics.[28]

Some have touted pangolins, also known as scaly anteaters, as the intermediary where the virus developed the ability to transmit itself from human to human. But Robertson doesn't believe this happened. Only a few of the pangolin viruses studied have the same human-ready receptor-binding sequence as SARS-CoV-2. And no one has found evidence of a pangolin carrying SARS-CoV-2 at the Huanan market.[29] The pangolin may be a wild-goose chase, along with the snakes, civets, ferrets and members of the cat family who were on the early suspect lists as intermediaries. On my travels, someone joked that the link to pangolins was a rumour spread by conservationists – these animals are severely endangered, chiefly because of the demand for their body parts from traditional Chinese medicine. Though, since these animals can carry SARSr-CoV-like viruses, they're still a possible culprit and remain a potential source of future spillovers.[30]

Whether or not pangolins contributed, SARS-CoV-2 isn't a lone phenomenon, a single chance happening with incredibly low odds. This is no black swan. At least 60 per cent of the 335 infectious human diseases that emerged between 1940 and 2004 came from other animals, according to ecologist Kate Jones of University College London and colleagues.[31] Since 2015 the World Health Organization has kept a list, updated each year, of up to ten of the most serious emerging infectious diseases, those 'likely to cause severe outbreaks in the near future' and for which there are no good treatments, cures or vaccines. Every single disease on that first list was zoonotic, found in a non-human animal with the potential to jump to humans: SARS and MERS, which came from bats; Crimean Congo haemorrhagic fever, carried from livestock via ticks; Ebola and Marburg virus diseases from bats; Lassa Fever spread by rodents in West Africa; Nipah virus and henipaviral disease, which jumped to us from fruit bats in Asia via pigs; and Rift Valley Fever, transmitted by mosquitoes from livestock in North and East Africa, including Kenya.[32]

Given that it's never spread in the country where I live, it's not rational, but it's Ebola that scares me the most – up to 90 per cent of its hosts die. Named after a river in the Democratic Republic of Congo, this snake-shaped filovirus has killed thousands since its discovery in 1976, in outbreaks in Sudan and across West and Central Africa, most notably in West Africa in 2014–16. Even when we've stamped out one human outbreak of the virus, it lingers in chimp, gorilla and bat 'reservoirs', ready to emerge from the forest and flare up into the next human epidemic. It's only a matter of time. The harm caused by our lifestyles is catching up with us.

SAVING OUR STRIPES

So how do we save wildlife and ourselves? Here the Grevy's provides signs of hope. In the early twenty-first century, the Grevy's fortunes turned around when Samburu livestock herders monitored the zebras' behaviour with the aid of GPS receivers. Once they realised chasing Grevy's away from water sources and the best grass harmed them, particularly the youngsters, they changed their ways. Grevy's numbers stabilised in 2015, with infants and juveniles making up nearly 28 per cent of the group, a promising sign for population growth. Behaviour change by a few humans made a big difference to wildlife survival. This zebra has, for now, ceased to slip towards the void, though outside forces still loom.

A couple of nights after the bat and I come face to face at the Grevy's Trust field HQ, the Trust staff and their visitors eat around a campfire in the dry riverbed. Only the glow of the flames lights the night; there are no streetlamps and the buildings – the kitchen and the gathering area – lie behind the trees. I look up at the stars, bright spots in a sea of navy. Most of the constellations are new to me, though Orion, who hangs outside my home on winter nights, is here too. There are so many stars, this universe has depth. Something falls into my eye. It stings and I blink as tears form. My first thought is that it's a bat taking its revenge by weeing on me. More likely, it's ash from the fire.

The fates of humans and wildlife are closely linked and a stinging eye is the least of our worries. Now that we've seen and felt the human, social and economic damage of a pandemic close to home, will we be wise enough to jolt into action to prevent the next? To stop destroying habitat, exploiting wildlife, wiping out species, cramming farm animals into industrial-scale sheds and shipping them around the globe? Or will we put our heads back in the sand and distract ourselves by rebuilding the economy? Or become waylaid by techno-fixes – cataloguing viruses from the wild, developing universal vaccines – that deal with symptoms of the problem, not the cause? Although they could be useful, focusing on techno-fixes is like plugging a hole in a leaking boat from the outside while a man on the inside hacks at the hull with an axe, splintering it to pieces. It can't work long-term. And we need to fix this now, for good, for all animals including ourselves.

CROSS COUNTRY

As I travel across Kenya, to the national parks in the south and the Indian Ocean, then back to Nairobi by train with a seat-side view of elephants and antelope, I track progress of the coronavirus from afar. It lurches across Asia and Europe. At a hostel in Kilifi, near the coast, in early March, I chat to a couple of German tourists. Germany has far more cases of Covid-19 than the UK but fewer deaths. We work out the percentages in our heads and look at each other confused. Why is the UK death rate so much worse? At the poolside a Kenyan man sneezes loudly and effusively. 'Corona,' he says and laughs. Then he looks at me and says, 'Sorry, I shouldn't joke.' Kenya is not yet reporting any cases and the comments beneath the story on a national news website say that black people can't catch the virus. I wish this was true. Kenya doesn't have many hospitals.

On my last full day in Kenya, Friday 13 March, I visit the elephant orphanage in Nairobi. It's shutting, they announce while I'm there, as Kenya had its first confirmed case the day before, in a woman who flew in from the US via London, and the government has banned public gatherings. On the way back to the hostel I ask the taxi driver to stop at a shopping mall in the suburbs so I can buy eye drops. One of my eyes is itchy and red. Instead of seeking out powdered bat faeces I opt for a Western-style pharmacy. Inside, a white woman has put three giant bottles of hand sanitizer – they must contain a litre each – in her shopping trolley. The transparent gel gleams under the fluorescent lights. Trade, as we'll see, was part of the problem that brought us to this point. Now it's providing this woman a solution, of sorts.

On the flight home, the sky is clear as we pass above Mount Kenya; her jagged peaks stand proud of the low cloud at her base. After flying by Lake Turkana, home to some of the very first hominins, we cross miles and miles of rippled umber sand and the Nile leads us north to the sudden stop of the Egyptian coastline. Greek islands adorn the blue sea like jewels before they are cut. Who knows what pathogens I'm bringing with me, inside my body or on my clothes? And what their risk is to other people, and to wildlife that is already struggling in the face of numerous troubles? By the time we reach mainland Europe clouds have gathered. The view is gone.

2

A Mammoth Problem: Early Travel, Disappearing Neanderthals and Vanishing Megafauna

'Trust not the horse, O Trojans.
Be it what it may, I fear the Grecians even when they offer gifts.'
Virgil, *Aeneid* (*c.* 29–19 BCE), II, 4
Translated by J. W. Mackail

MARCH 2020, BRISTOL BUS STATION

Arriving home is a big relief. I'd worried, at one point, that transport would stop and I'd have to wheel my suitcase down the side of a spookily traffic-free motorway. I'd worried too that I might bring Covid back with me, hidden inside. The taxi driver when I arrive at Bristol coach station from Heathrow is wearing a thick coat. For a moment, I'm confused. Then I realise it's winter and the air is cold. It takes my body a second to catch up, and then I'm cold too. It's a shock. I shrink from the raw air, huddling into the jumper I've just put on as the chill grasps at my bones. And what of the very first people to leave the African continent, our early forebears? How did they fare? No fleeces for them, just furs. And no Covid either, though likely plenty of other bugs.

Setting off at least 120,000 years ago, these early travellers would have experienced very different journeys to mine. The biggest challenge I faced was finding a boy already sitting in my seat. He'd misread the row numbers so I sat in his seat instead and waved at the air stewards every time

I saw a vegetarian meal heading his way. Back then, the journey probably lasted generations, not one day, and meals weren't provided. The first anatomically modern humans – humans for short, from now on – travelled north-east to leave Africa on foot, walked on into the Levant then east to Asia and, later, west into Europe. The lands they reached were already inhabited, not by other humans but by Neanderthals.

This group of hominins had evolved some 300,000 years earlier from a branch of the human family tree that left Africa a million years before. They were smaller than today's humans; at a little over 1.68 metres (5 feet 6 inches), I'm taller than the average Neanderthal man. Neanderthals were also stocky, which helped protect against the cold, as did their large noses that warmed and humidified freezing air before it entered their lungs.

Some Neanderthals made it to southern England, where I live today. As a human from outside Africa, chances are that my DNA is between 1 and 4 per cent Neanderthal.[1] I have blue eyes, one of my sisters has red hair and some days I only want to grunt at things, especially in the mornings. So I reckon I'm at the higher end of this range. But that's just a hunch, so, like a true scientist, I order a DNA test. Having questioned previous customers about their traits, the company will be able to reveal what my genetic heritage might mean for my ability to match musical pitch, how often mosquitoes bite me and my favourite flavour of ice cream. On the webpage where I checked which tests include Neanderthality (a word that should definitely exist), an image displays a group of Neanderthals warming their hands around a fire. Furs drape their torsos and the men wear some kind of trousers, but the women and children have bare legs and everyone has bare, surprisingly hair-free arms. Woolly mammoths plod across a snow-clad ridge in the background and I can't help thinking the Neanderthals would have benefited from inventing sleeves. Nevertheless, they were the first hominins to survive a glacial ecosystem.

WHY DID WE WIN?

The Neanderthal man sculpted by the Natural History Museum in London has dark eyes that twinkle beneath his strong eyebrow ridge, a wide nose above a gentle smile and a Charles Darwin-style full beard.[2] Yet the Neanderthal of legend is hairy, grunty, violent, stupid and short. The species was simply inferior to humans, according to some of the first explanations for why Neanderthals died out and humans didn't – the last Neanderthal we know of lived 38,000 years ago, some 7,000 years after humans reached Europe.

There's evidence for short – Neanderthal skeletons show they were around 1.5–1.6 metres tall (4 feet 11 inches to 5 feet 3 inches). Hairy, grunty, violent and stupid are more debatable, I learn. Since they possessed the FOXP2 gene that brought humans language, Neanderthals may have been able to speak, while ancient pollen reveals that they may have placed flowers on the graves of their dead, from species with medicinal powers, though others believe that bees may have transported at least some of the pollen.[3] In any case, there's good evidence that Neanderthals cared for individuals who could no longer look after themselves,[4] and may have decorated shells and painted the walls of their caves.[5]

Over the years, we've discovered that humans weren't any cleverer or faster than Neanderthals; we didn't have better tools, our hunting skills weren't sharper and Neanderthals may have communicated just as we did. When today's archaeologists find stone tools on a dig site, they can only tell whether they belonged to Neanderthals or humans if they discover the toolmakers' bones alongside their creations. The cultures of Neanderthals and early humans were equally advanced. If anything, humans were less strong and less well adapted to living in the cold. Put a Neanderthal and a human in one of those hypothetical 'which would win a fight?' pub arguments and the answer wouldn't be clear. Yet Neanderthals were the ones who vanished, not us.

In 2019, Gili Greenbaum, an expert in computational genomics at the Hebrew University of Jerusalem in Israel, came up with a new theory to explain the Neanderthals' disappearance.[6] 'Thinking has changed quite dramatically in the last couple of decades,' he tells me carefully. 'Neanderthals had the brain capacity to do what modern humans did.' It's clear Greenbaum is passionate about his subject and he wants to get this right. And he doesn't think humans persisted because we were 'better' than Neanderthals.

When evidence is scarce, there's room for imagination. You can play with theories and models, think about what could have happened, Greenbaum explains, whereas if you have an experiment, you have to try to explain that particular result. This flexibility let many an idea flourish over the years. Perhaps, researchers pondered once they'd realised humans were no better than Neanderthals, climate change wiped out these other hominins, or they split into groups too small to survive catastrophes.

Or maybe it was something else. When I get home from Kenya, I'm sick. I have a bad stomach and I'm feverish and fatigued. I cancel all my plans – meet-ups with friends I haven't seen for two months – and lie in bed. I've brought something back with me. I hoicked it from one continent to another in less than a day. I just hope it was something I ate rather than a bug I've given to others. Is that what happened to the Neanderthals? Did disease lead to their downfall?

Perhaps the humans brought a disease with them, inside their bodies, in one of the very first cases of pathogen pollution. Arriving from Africa, they were likely riddled with pathogens that had made the easy leap from apes and monkeys as humans butchered their biologically similar cousins for meat. 'There's more diversity in the tropics,' says Greenbaum. 'Trees, animals, plants, bacteria and pathogens – there's just more types.' Some individuals lucked out and had immune systems better at fighting off this myriad of germs. More of these individuals survived to pass on their disease-resistant genes. In this way, over the generations, the human species gained at least some immunity to the pathogens it had picked up from other primates and elsewhere. In a battle long-fought and hard-won, human immune systems grew wise to the old germs' tricks.

But the Neanderthals' ancestors had left Africa and its plethora of germs hundreds of thousands of years before. Chances are the Neanderthals had lost their immunity to these diseases. When it comes to pathogens, the problem is often novelty. A germ new to your immune system – because it's jumped from another species or been 'forgotten' over the millennia since your forebears encountered it – is bad news. Your body is immunologically naive – it has few defences and is much more likely to succumb. The humans could well have had an 'African advantage' over the Neanderthals on the pathogen front. Similarly, by the fifteenth century, when Europeans invaded North America, they had a 'Eurasian advantage' due to their long association with domesticated animals and their diseases. Pigs, cows and sheep were the new non-human primates. The result for many Native Americans was horrific. This fifteenth-century decimation of the human population may have led species such as the passenger pigeon to soar in numbers, now that humans were no longer competing with it for tree nuts, acorns and maize. But when European colonialists shot all the passenger pigeons, sometimes in hundreds of thousands at a time, the birds no longer guzzled acorns and the resulting excess of food may have led to growing numbers of deer mice, which are the main reservoir for Lyme disease.[7] (A reservoir population keeps the pathogen causing a disease in circulation and may pass it on to another type of host, often one that's less tolerant to the infection and suffers more harm.)

LEVANTINE LOCAL

Coming from Israel, Greenbaum has a unique perspective on the 7,000 years it took humans to spread across Europe and replace their distant relatives. He got the Neanderthal bug – though hopefully no diseases – while leading youth groups on hikes into the Carmel mountains. 'One of the sites that I really liked is near Haifa and has a lot of Neanderthal findings,' he says. 'It's like magic to go into a cave and imagine Neanderthals lived there and you can see where the people sat and where they made fire.'

By all accounts, pine, olive, laurel and oak trees scatter the landscape of the Carmel mountains today. When Greenbaum talks of the mountains, I imagine the tang of oils evaporating from hot bark, buzzing insects and caves peering like small dark eyes from cliffs of crumpled rock. Bones from 100,000 years ago tell us that humans dwelt in some of these caves at roughly the same time as Neanderthals. For tens of thousands of years, Neanderthals and humans lived alongside each other in the Levant – the lands around the eastern end of the Mediterranean, including modern-day Syria, Lebanon, Jordan, Israel, Palestine and most of Turkey. Then, around 45,000 years ago, humans spread further afield, into Asia and Europe, and soon the Neanderthals died out.

Greenbaum feels that researchers who've concentrated on the last few thousand years of Neanderthals' struggles to survive in Europe may be missing out by ignoring the hundred thousand years of interactions beforehand. Though this isn't purely Eurocentrism. 'It makes sense to focus on this very dramatic process,' Greenbaum says. 'But maybe looking at this more messy interaction in the Levant could give us hints.' Was the key to what happened in the pause before the drama? If you take this longer view, it's striking that it took humans some 55,000 years to spread into Europe and start replacing Neanderthals. No matter how tired they were after their journey from Africa, that's a long recovery time.

After completing a PhD on the Asiatic wild ass, which conservationists reintroduced to southern Israel in the 1980s, Greenbaum turned back to prehistory. One day, he put his head together with Orem Kolodny, a friend who is an expert in ecological modelling. Could ideas from ecology explain the disappearance of the Neanderthals? To find out, Greenbaum wanted to make a model, and for that he needed data. Fortunately, a study in the UK, where badgers have been implicated in spreading bovine tuberculosis (bTB), even though most transmission occurs between cattle,[8] provided just what was needed. Researchers had put location-tracking loggers on cattle and badgers to find out how often these species were meeting. The badgers, it turned out, used cattle pasture but tended to avoid cattle, indicating that the disease may spread between the two species not by direct contact, but through contamination of the environment.[9] Measures such as improving biosecurity, for example by preventing badger access to cattle feed troughs and sheds,[10] increasing testing of cattle, and vaccinating cattle and, perhaps, badgers could all reduce transmission.[11] There is no clear evidence that culling badgers (as was government policy in England, but not Scotland and Wales, at time of writing) reduces the spread of bTB in cattle.[12]

Thinking about the data from the UK badger study and other models of inter-species diseases inspired Greenbaum to simulate how the Neanderthals interacted with early humans. The plan was to explain the archaeological story by thinking about the two groups as two species that met, transmitted diseases and evolved. Two species that were from different branches of the same family tree. When humans arrived in Israel and met Neanderthals, it was like a family get-together. The two species were still so similar that pathogens saw them as the same, Greenbaum says. 'Reunited, you get diseases.' The same thing happened with squirrels in the UK. When Victorians imported grey squirrels (*Sciurus carolinensis*) from North America in the late nineteenth century, the incomers gave squirrelpox to their long-lost relation the native red squirrel (*Sciurus vulgaris*). New to the virus, red squirrels have much less tolerance and almost always die; it's part of the reason, along with habitat loss and competition with greys for resources, that the grey squirrel, which often carries the virus without symptoms, is pushing the red out of the UK.

The humans had pathogens from biodiverse tropics, and other diseases had no doubt evolved alongside the Neanderthals. So each population had germs that were novel for the other and in all probability caused epidemics to rampage. What Greenbaum doesn't know is which diseases the two groups swapped, though ancient DNA could provide clues. The last two years have seen an explosion in ancient DNA sequencing, and scientists have now analysed more than 10,000 human genomes, the earliest dating back to roughly 45,000 years ago. And we've also examined the genome of Neanderthals from 400,000 years ago. The past sends us messages inside these tiny packages of information. A prevalence of genetic variants that provide resistance to a particular disease indicates that disease was rife at the time. 'It's very exciting in that we're getting a lot of evidence,' Greenbaum says. 'It will probably open up more questions than answers, but it's fun.' One day, we may even gain an insight into the DNA of ancient pathogens. So far, researchers have investigated DNA from leprosy bacteria from a skeleton in England from the fifth or sixth century,[13] and of plague bacteria from the teeth of humans in Asia and Europe from up to 5,000 years ago.[14] And in 2024, just as this book was going to press, researchers discovered the DNA of three viruses in Neanderthal bones from 50,000 years ago:[15] the adenovirus that causes modern humans colds; the herpesvirus responsible for today's cold sores and the papillomavirus responsible for genital warts and cancer.[16]

Whatever pathogens the two groups gave each other where their territories overlapped, the exchange would have slashed the headcount of both humans and Neanderthals. With fewer people and Neanderthals remaining, they'd have met less frequently and been less likely to make each other sick. According to Greenbaum's model, numbers would have dropped to a 'sweet spot where they were suppressed' but could sustain themselves. 'There were low population sizes and they were still interacting a bit, but even this tiny bit of interaction transmitted diseases,' Greenbaum says. It was as if Neanderthals and humans were playing volleyball and each team member had their own ball that killed whoever on the other team touched it, as well as that opponent's closest teammates. Add more players to a team, and the likelihood of dying went up for both sides – the opposing team got hit by more balls and the bigger team grew more crowded and more likely to make contact with a ball lobbed over by the other side. The 'team size' reached a natural limit and stayed that way for generations, accounting for the 55,000 years that humans remained in the Levant before they ventured into Europe and Asia.[17] They didn't just like the area, they needed time to learn to live with Neanderthal diseases.

DNA DISCOVERIES

Neanderthals turned out to be part of the solution as well as the problem. Interacting with a Neanderthal meant risking an encounter with a novel disease. But DNA testing proves that humans and Neanderthals definitely did interact. Today, the genomes of many people outside sub-Saharan Africa contain around 1–4 per cent Neanderthal DNA.[18] These percentages are misleading at first glance, as human and Neanderthal genomes differ by only a few tenths of a per cent. The differences arise in small regions of the genome that are highly variable in modern humans, and it's in these small regions that up to 4 per cent of the differences in the genetic code – also known as variants – may have been inherited from Neanderthals. (The differences in DNA between individual humans are even tinier – only 0.1 per cent of my DNA differs from yours.)

At the time, some 100,000 years ago or less, the offspring of a Neanderthal and a human would have had 50 per cent Neanderthal DNA, 50 per cent human. Some of these genes were the ones that provided immunity from disease – each group gave each other its diseases but also their cures, as Greenbaum puts it. Over time, this interbreeding changed the status quo – the volleyballs became less lethal. The Neanderthal-derived genes that are most common in modern humans today are immune-related, providing evidence that disease exchange was prevalent at the time, Greenbaum says. Evolution would have selected out genes that were less useful to humans, in a process known as adaptive introgression. Humans who lucked out and happened to receive a Neanderthal gene that contributed to their immune system probably gained protection against Neanderthal diseases, whereas a Neanderthal gene for eye colour wouldn't have helped you very much.

Before I get my DIY DNA test results, I put myself on a DIY DNA refresher course. Humans have some 20,000 genes, the collections of DNA molecules that provide a code our bodies copy to make proteins. Most of these genes sit inside 23 pairs of linear chromosomes within the nucleus of all human cells. The exceptions are red blood cells, which don't have nuclei, and egg or sperm cells, which have a single set of 23 chromosomes and must literally find their other half to create new life. Outside the nucleus lie 37 more genes, in a circular chromosome inside the 'powerhouses' of the cell – the mitochondria. Unlike the DNA in the nucleus, where one chromosome in each pair comes from each of your parents, you inherit this mitochondrial DNA only from your mother. But no one has yet found any similarity between mitochondrial DNA in humans and Neanderthals, so my Neanderthal rating will be based only on the DNA inside the nuclei of my saliva cells, which I've inherited from both my parents, not just the female line. When I tell my mum

I'm doing a test, she says that any Neanderthal heritage must have come from my dad.

My family have a lot of views on the test. When my DNA sample gets stuck in the postal strike on its way to the lab, my non-ginger sister points out that if they can take DNA from a million-year-old mammoth, a two-week wait should be OK. And she turns out to be right. When the results finally come in, I'm less than 2 per cent Neanderthal and a little disappointed. Only 222 of the 7,462 genetic variants the company tested for me have Neanderthal origins (though my mum is in the frame after all, as some of these variants came from both parents). Just 24 per cent of their customers have less Neanderthal DNA than me.

My hunch was wrong. My sister's hair colour turns out to be a red herring – recent evidence indicates that Neanderthals weren't typically ginger after all.[19] Research has also linked Neanderthal genes in northern latitudes to a propensity to be a night owl, meaning I'm no closer to an explanation of my own aversion to dark winter mornings.[20] And before getting the test results, I'd thought that a strong Neanderthal heritage might explain why one of my cousins needed help with breathing when he got Covid-19; one of the biggest genetic risk factors for suffering a serious case – a cluster of genes on chromosome 3 carried by roughly half of people in South Asia and 16 per cent of people in Europe – came from Neanderthals and may lead the immune system to go into overdrive.[21] Though other Neanderthal genes protect against the disease.[22] The Neanderthal genes that survive inside many of us today, influencing our immune systems, may have helped those earlier humans spread outside the Levant. Greenbaum believes humans slowly gained immunity to Neanderthal diseases, both by acquiring Neanderthal genes and by evolving their own genetic resistance. Eventually, they had enough disease-resistance for their population to grow. They were free to head out from the Levant and deeper into Neanderthal territory. Even though the Neanderthals had gained at least some immunity to human diseases the same way, the African advantage skewed the fairness of the 'game'. When you allow for the human team lobbing more balls at the Neanderthals, Greenbaum's model fits the history. Even if humans carried only a few more diseases, the human population could have bounced back long before Neanderthals adapted. Our Neanderthal cousins, who cared for their sick and injured kinfolk, had the disease odds stacked against them.

That said, it can be difficult for diseases to make a species go extinct. If it spreads in only one type of animal and transmits by contact between individuals, the disease generally fizzles out once fewer individuals remain and their encounters grow rarer. Typically, such a density-dependent specialist disease goes extinct before it can kill off the species. Sometimes, however, disease transmission depends not on the density of the individuals but on how often they interact, whether that's through fighting or sex or being bitten by a tick, mosquito or other disease-transmitting 'vector' that's previously fed from one of their diseased brethren. These frequency-dependent diseases can spread until they've killed every single member of the species. It looked, for a while, as if the Tasmanian devil – in which a transmissible cancer spreads when individuals bite each other – might go that way and the animal is still not totally safe. Or, for a density-dependent disease that infects two or more species, if one species happens by chance to be less susceptible, its members may act as a 'natural reservoir' that lets the disease flow even as the susceptible species grows rarer, then disappears. Domestic dogs are a reservoir for the canine distemper virus that threatens African wild dogs and black-footed ferrets, for example. And humans perhaps did this for Neanderthals.

Once humans ventured beyond the Levant, the Neanderthals they met were even worse off. Like the Neanderthals in the Levant 55,000 years earlier, they'd never interacted with humans and so had no immunity to human diseases. Their populations may have crashed, ultimately leading to Neanderthals' 'replacement' by humans now largely immune to their diseases. 'We were able to provide a very detailed, although theoretical, explanation of what could have happened in terms only of diseases,' Greenbaum says. 'That doesn't mean that there's no other processes involved, but diseases are a sufficient explanation.' Greenbaum's model accounts for both the mysterious hiatus in the human journey from Africa to Europe and the Neanderthals' disappearance; it fits the evidence on the ground.

What's more, this down-to-earth theory is testable. When humans arrived, the number of Neanderthals in the Levant should have dropped compared to previous years. And comparing numbers of Neanderthals in the Levant to areas of similar size elsewhere should also reveal a Neanderthal population drop accompanying the advent of humans. Following release from the suppression phase – once human bodies had gained the ability to deal with Neanderthal diseases – the population density of humans should have risen. 'That's what I liked about the paper, that it's not just coming up with an idea,' Greenbaum says. 'That's a prediction that now we can study and test.' It does require more digging, however. Israel may have the depth of archaeological research needed, but other parts of the Levant don't; Greenbaum is keen for someone to take up the gauntlet.

MISSING MEGAFAUNA

Eventually, humans and their new-found immunity to Neanderthal diseases spread even further around the globe. The dates on the human tour T-shirt would list Africa, the Levant, Europe and Asia, Australia, the Americas then selected islands such as Madagascar, Polynesia and New Zealand. But it's the Americas that interest us here. Opinions – and the quality of the evidence – vary on when humans first arrived in North America then headed further south. The current consensus is that it was somewhere between 30,000 and 20,000 years ago and that the early settlers walked in from Siberia over the Bering land bridge. I feel I have an idea of the landscape that faced them, though my visit was thousands of years later and hundreds of kilometres further east. In 2008, I travelled to the far North of Canada to spend time with climate change researchers on an Arctic icebreaker in Franklin Bay. The journey from the UK lasted more than two days; I jetted across the Atlantic to Toronto, transferred to Edmonton for a night in a hotel, then flew to Yellowknife, then Norman Wells then Inuvik, a settlement near the northern coast of the Northwest Territories. Nothing as arduous as walking from Siberia, even if the jet lag made it feel like a slog.

For the journey up to Norman Wells, I was the only woman on the plane apart from the air stewards. Some 40 men surrounded us; they looked like they worked in oil, sporting suede or leather jackets, jeans and boots, tanned outdoorsy skin, and an air of wealth and machismo. My white cardboard coffee cup bore a blue sketch of an Inuit person, their face hidden by a hood, driving a sled drawn by a team of dogs. I still have it, on a shelf next to my collection of wildlife books. After another night in Inuvik I flew to the coast on a Twin Otter. As we headed north, the landscape changed. The thin, hardy-looking conifers of the boreal forest, and their downward-sloping branches to keep off the snow, disappeared, and the ground turned the colour of dusty olives and grew crazed with the squares and irregular hexagons of pools formed by melting permafrost. Thermokarst. Flat and bleak with not a single tree. Not a landscape I'd want to walk across, even back when it had more grasses and sedges. The taxi I took in Inuvik had a crack almost the whole way across the windscreen. In spring, the driver said, the frozen ground melts and fills the roads with ruts and dips – one of them threw up a rock that smashed the glass. By June, when I was there, the dips had evened out but ruts an axle-width apart remained. No asphalt here.

No smooth roads for the early settlers, either. Having trudged into North America through difficult terrain, they may have moved south along the Rocky Mountains, according to one theory. This is where the researcher who was one of the first to single out disease as a potential culprit for the extinction of the megafauna spent a couple of summers doing fieldwork, while he was a student at the University of Alberta, Canada. Ross MacPhee, now a specialist in mammals and their extinctions at the American Museum of Natural History, searched caves near the eastern slope of the Rocky Mountains for signs of the first human arrivals. Invited to take part in an archaeological dig 'very far back in the last century', at the time he knew absolutely nothing and 'was just gobsmacked by the idea that there were all these huge megaherbivores that existed at the same time as the first Indians'.

Those early humans would have encountered woolly mammoths. Herds of these early elephants plodded trunk to tail across the frozen ground of the northern 'mammoth steppe', feeding on grasses, sedges and herbs as they roamed. Their feet, the size of dinner plates, no doubt crunched in the frost; the warm air from their trunks and mouths would have condensed into puffs of cloud as they bellowed and breathed, cloud too cold to melt the ice that crusted the lashes around their small brown eyes. From the top of their domed heads, their shaggy-furred backs sloped down towards thin tails and their tusks circled outwards in a wide arc before curving back up and in. I imagine their family groups to be like the herd of elephants I watched at Kenya's Ol Pejeta Nature Conservancy. Quietly magnificent, they stood and rested in the late afternoon, a harmonious collection of individuals milling about as the sun crept lower in the sky. Some of the smallest youngsters suckled while one, a quarter the size of his mother, kept close beside her tree-trunk legs, rubbing his eye with his trunk like a human baby ready for sleep. I've never wanted to hug an elephant so much.

Unlike elephants, who are endangered but currently still with us, woolly mammoths no longer exist. By some 10,000 years ago, according to all but the most recent evidence, these giants had disappeared from mainland North America, though they survived on an island off Siberia until around 2000 BCE.[23] After that, the woolly mammoth remained on Earth only as giant bones scattered in the dirt, a few bodies frozen in permafrost, and as likenesses scratched by early humans onto cave walls and into its own tusks.

Starting about 50,000 years ago, huge numbers of other megafauna – defined as animals that weigh more than 44.5 kilograms (about 100 lbs), roughly the same as a small sheep – went missing around the globe. The species we lost include the giant ground sloth, the sabre-toothed cat and mastodons – mammoth-like creatures named after the Greek for 'breast tooth' by a naturalist who thought the bumps on their teeth looked like nipples. Altogether MacPhee estimates that between 750 and 1,000 land-living vertebrate species – mammals, birds and reptiles – went extinct. And most of them were large. To this day we don't know precisely why, but there were most likely different reasons for different species and in different places.[24]

While Africa and southern Asia largely escaped the worst of the extinctions, and the losses in northern Eurasia weren't too extreme, other places fared less well. Australia's big animals mostly went extinct around 40,000 years ago. In North and South America most of the megafauna disappeared 30,000 years later. Islands such as Madagascar and New Zealand lost their megafauna less than a thousand years ago.[25] 'This is why people continue to argue that people had something to do with it,' says MacPhee, citing the 'close tracking of the human diaspora over the planet'. After people left Africa and entered Europe, there's 'a similarity in respect of the spread of human populations and the downfall of fauna everywhere, particularly on islands'. Did we kill off the megafauna the same way we may have killed off the Neanderthals?

MacPhee always says that working out what happened is like investigating Napoleon's disastrous march on Russia in 1812 without any written records – no newspaper reports, no military dispatches, no letters home. If all you had was whatever might show up in an archaeological record, you'd be hard pushed to make conclusions, he explains. You might discover the graves of French soldiers along the route of the march, but this evidence would be patchy and it would be hard to tell if they'd died in the same timeframe. You'd be unlikely to deduce that Napoleon's goal was a blitzkrieg to bring Russia to collapse – you just wouldn't have the evidence. 'That's where we are with all kinds of questions in historical biology,' MacPhee says. 'Yes, we have fossils. Yes, we can, to a degree, determine whether individual fossils are contemporaneous by radiocarbon dating, but we can tell very little about whether or not losses were coordinated.' It's all an educated guess.

Like Greenbaum, MacPhee went on to study something else, in his case biological anthropology. But when he joined the museum in

New York City in 1988, he was able to investigate a broader range of species than humans and their ancestors. 'I went back to my interest in trying to sort out what might have happened [to the megafauna],' he says. 'Completely unsuccessfully just like everybody else, but it was a good exercise.' Suspects included overhunting and climate change, despite evidence that the megafauna had survived the waxing and waning of several ice ages before people arrived. MacPhee feels it's unlikely that the small groups of humans living on the continent in those early days could have hunted even one species to extinction, let alone dozens of species all at roughly the same time. 'To me that was just hopeless as an argument,' he says. But MacPhee does think it feasible that humans brought disaster with them. 'This is typically what we do,' he says. 'Mostly it's seen when humans start messing up environments in which they're resident, but one of the things that they do frequently to each other in the course of human history, is to introduce diseases into naive populations.' With horrific results.

In the mid-1990s, together with virologist Preston Marx of Tulane University in New Orleans, MacPhee came up with the 'hyperdisease theory' for the megafauna extinction.[26] The newly arrived humans, or perhaps the dogs accompanying later incomers, may have been a Trojan horse, hiding a disease inside themselves. 'If there were these killer plagues released by early people coming into north America, and they were very good at jumping from one species to another,' MacPhee says, 'then you could have had this Armageddon that could have gone to completion because there was no need to have humans running about doing all the work – it was just [non-human] animal to animal contact.' To be this apocalyptic, the hyperdisease would have needed a mortality rate of some 80–90 per cent, and to be non-lethal to humans so they could act as a reservoir. Social herbivores such as horses, mammoths and mastodons might have been particularly vulnerable because they lived in groups. And once they'd gone, the megacarnivores would have starved.

On the wall of MacPhee's office is an illustration of two creatures that look like giant meerkats, their heads protruding from the treetops. When I ask, MacPhee tells me that they're *Amblyrhiza inundata*, rodents that weighed in at 90–150 kilograms (198–330 lbs), at least one and a half times my weight. (That's my own estimate, MacPhee is far too polite to make comparisons.) Closer up, these giants look like hunchbacked kangaroos. Mouse-like pale pink ears sit above their deep-cheeked brown faces, and their fine-boned front legs drape from powerful brown shoulders highlighted by the cream of their bellies. They're massive. Nothing against rodents, but I'm almost glad they're extinct. MacPhee worked out that this species, which lived on the West Indian islands of Anguilla and St Martin and was the largest island rodent ever known, died out 120,000 years ago when sea levels rose, long before humans arrived.[27] For this species, which is also known as the blunt-toothed giant hutia, we were not to blame. 'I take a very dim view of humans in general, but there are some things we didn't do,' he says.

Humans might not have killed off this giant hutia, but we did, MacPhee learned during the course of his megafauna research, kill a more modern rodent by introducing disease. At the end of the nineteenth century, British doctor Herbert Durham was sent to Christmas Island, a small island south of Sumatra in the eastern Indian Ocean to investigate why hundreds of workers in the phosphate mines there had died from beriberi. People had only recently begun to live on the island and Durham searched in vain for an environmental cause. He didn't know, as we do today, that beriberi is due to a shortage of B vitamins – the workers were Chinese and their rice-heavy diet lacked these essentials. In the downtime around his fruitless quest, perhaps to take his mind off his failure, Durham studied the island's wildlife. In those days, MacPhee says, physicians were expected to show an interest in natural history as part of their training, and naval doctors and surgeons kept themselves occupied between battles with natural history collections and describing new species – think naval surgeon Stephen Maturin in *Master and Commander*.

One day, Durham noticed one of the island's nocturnal rats stumbling around on pathways in the middle of the day, putting it at risk from predatory birds. The island had two native species, the Maclear's rat and the bulldog rat. In a Beatrix Potter-style illustration from 1887, the Maclear's rat has long hairs standing out from its rump and a bald tail that's dark at the body end and white from halfway down to the tip. It looks like it is frowning and has urgent matters to attend to. In its own picture, the bulldog rat seems milder mannered and has a head that's astonishingly small, perhaps because its back has a thick layer of fat. It's not clear whether Durham had a favourite or which rat he observed acting strangely during the day. But he did, even though he had no way to prove it, surmise that the native rats had picked up trypanosome parasites from fleas that had fed on black rats accidentally shipped in by humans. These stowaways had stowaways on their stowaways. And that, for the native rats, was the end of their journey here on Earth. A few years later a naturalist wrote that they were 'morbid' and neither species was seen after 1908. Christmas Island's two species of rat were extinct within a decade of the opening of the mines.

Some hundred years later, MacPhee was investigating whether a hyperdisease killed off the megafauna and looking for recent examples with more evidence. 'I never would have looked twice at Christmas Island rodents had it not been for that decision [of Durham's], just because he was interested in making natural history observations and saying, "I don't know, but it could be",' MacPhee explains. In 2008, together with his then postdoctoral associate Alex Greenwood, a specialist in ancient DNA, MacPhee sampled bone from specimens of the Christmas Island rats in some of the only known collections, at the UK's University of Cambridge, University of Oxford and Natural History Museum in London. Their findings proved that the physician guessed correctly. Sure enough, native rats collected after black rats arrived contained DNA sequences for a trypanosome, while those collected earlier did not. It was the first verified introduction of a novel pathogen before extinction of a mammal.[28] The unlucky rats received the award posthumously.

'Because no trypanosomes had infected these rats before, it was very, very mortal,' MacPhee says, 'like sleeping sickness can be in humans.' Also caused by a trypanosome, a protozoan (single-celled organism) with a tiny beating tail that lives in the bloodstream of vertebrates and the guts of invertebrates, sleeping sickness was a particular problem for Europeans invading Africa in the nineteenth century. It was new to, or at least forgotten by, these humans' bodies too. The rats native to Christmas Island, who hadn't even left home, suffered a mini hyperdisease situation. 'These poor rodents got infected from interlopers from the outside,' says MacPhee. At the same time as the black rats arrived, humans were changing the environment in all sorts of other ways, including by mining phosphates. 'I'm sure the rodents couldn't possibly have been very happy about how life was for them at that point,' MacPhee adds. 'But the big deal was the introduction of pathogens.'

Is that what happened to the megafauna too? MacPhee's hyperdisease theory received much attention on its first release, both positive and negative. Among the dissenters were disease physiologists who said there's no disease in nature like that. MacPhee argued that's not the same as knowing what went on in the past, 'but progressively, as we tried various options to plump up the argument, it never seemed to get very far'.

In 2021, more evidence came in. It's not of a pathogen, but it changes our thinking about their role. Tyler Murchie and Hendrik Poinar of

Canada's McMaster University and their team had been drilling samples from permafrost in the Yukon and looking for DNA shed into the environment by passing animals long ago. Biologists also use this environmental DNA technique to find evidence of today's wildlife, including, to take an example from the UK, checking if great-crested newts live in a particular pond by sampling the water. You don't look for the animal itself but for the traces it has left behind. Among the particles of ancient sediment in their permafrost cores, the team found the frozen DNA of horses and mammoths that lived 6,000 years ago – around 4,000 years after they were supposed to have died out.[29]

The horses were a particular surprise, as researchers already knew that some mammoths had survived later, on islands rather than on the mainland. Although horses evolved in the Americas, they only survived the extinction of the megafauna in Eurasia. Yet the oral traditions of indigenous North American peoples from at least seven nations across the US say that they always had the horse, and the animal features in these nations' creation stories. So when Europeans brought horses to the continent in the fifteenth century, it was a reintroduction, not an introduction, according to the indigenous peoples.

Palaeontologists had denied the possibility that horses ever existed at the same time as people on the continent – even the earliest human arrivals couldn't have seen horses and spoken of them in their histories, the experts said. Looks like the indigenous peoples may have known best. The findings could ultimately save the wild mustangs currently persecuted by ranchers in the western US as an apparent threat to their cattle's food supply. If horses survived in North America into modern times, they would be categorised as a native species rather than an invasive one, and receive more protection as a result.

Ranchers want the mustangs gone, but others love them. 'There's a battle going on,' MacPhee says. 'I want mustangs to survive because they're among the very last megafauna that we have in this country.'

As well as offering the prospect of salvation for today's wild horses, Murchie and Poinar's discovery shifts our thinking on how the megafauna went extinct. Another 4,000 years of archaeological record for horses and mammoths makes all the difference. In the 1960s, when it looked like people first arrived on the continent 14,000 years ago, it made sense for overhunting to be the prime suspect for extinctions that were largely complete 4,000 years later. The argument back then was, according to MacPhee, that 'the big animals didn't understand that humans were a danger and by the time they did, they were mostly dead'. Today there's plausible evidence that people in fact lived in North America 30,000 years ago or perhaps even further back, and mammoths and horses survived for longer than we'd thought. With humans in place at least 20,000 years before the extinctions, that overhunting argument seems less feasible.

In recent years the climate change argument has taken on a lot more oomph because there are massive amounts of new data. It no longer looks like humans were indirectly to blame. But things aren't totally cast in stone, even the fossils. Some people do still believe that humans were complicit in the megafauna extinctions at the end of the last ice age, and perhaps even the dominant factor. MacPhee himself thinks that disease may have wiped out one or two large species and could well have contributed to the demise of others. He cites the example of today's monk seals. 'That's already a group that's in very bad shape,' he says. 'The West Indian monk seal disappeared in the 1950s and the Mediterranean monk seal never seems to make any kind of comeback. It keeps getting hit by the seal plague.' Only some 700 Mediterranean monk seals remain, mainly around the coasts of Greece and Turkey. And a small population is much less likely to recover; any hardship is a more substantial danger than it would be for lots of individuals spread widely across the ocean or landscape. 'Infectious disease does make a difference,' MacPhee says. 'That's the significant thing.'

The only way to be more sure about what happened would be finding evidence of a pathogen. MacPhee and colleagues tried this in the mid-1990s using the ancient DNA techniques of the time, but it proved too tough. Now that science has moved on and DNA analysis is automated, MacPhee thinks the chances of finding DNA from really ancient diseases are outstanding. When MacPhee approached ancient DNA expert Poinar, the McMaster University researcher basically said, 'Ross, you find me the money and we'll do the work'. The dream would be finding disease in a group of individuals from the same species who died at the same time at a single site; that would help us know for definite why the megafauna's journey was cut short.

DISEASES DIG IN

Like MacPhee, biologist Matthew Moran of Hendrix College, US, thinks diseases could have raced ahead of the human settlers entering North America. 'Imagine you have an elephant population that has not seen humans yet and this wave of disease comes through first and kills a lot of them off, then humans arrive and overhunt them,' he says. 'That's like the final blow.' Especially in combination with climate change. 'Those things together could have been enough to drive these animals to extinction,' Moran adds. 'I think the disease angle has been overlooked or at least understated.'

Having examined a range of diseases, Moran reckons anthrax and TB are the most likely suspects for killing off the megafauna.[30] Some mastodon and mammoth fossils appear to show tuberculosis damage to the bones, he says. 'It's very likely that was brought into North America by humans.' Caused by bacteria, TB has a long incubation period after which it may kill. Anthrax also originated outside the continent – these bacteria have a high genetic diversity in Asia, suggesting the species evolved there. Once it's taken hold, anthrax is almost always deadly, though it's 'not super easy' for humans to catch. Early settlers could have imported anthrax spores into North America on their furs. When these spores infiltrated the soil, grazing animals would have eaten or inhaled them and grown ill.

It's also plausible, according to Moran, that a disease killed the megafauna then died out with the last of them, leaving no trace of itself today. Moran agrees with MacPhee that we need more evidence: we must scour the fossil record and analyse animals frozen into the tundra to find snippets of DNA and RNA from ancient bacteria and viruses. Looking at the timing could also help. Did we find, for instance, TB in preserved specimens at a certain point, but not before that, Moran asks, and does that match up with when humans arrived?

ENGINEERING HEAVYWEIGHTS

In 2008, a pair of beavers appeared on the River Otter, a few kilometres north of the pebbled beach at Budleigh Salterton, a small town in South-west England featured in a famous Victorian painting of Elizabethan seafarer Walter Raleigh as a boy. Born in 1552, Raleigh probably just missed seeing beavers; we'd hunted them to extinction in the UK by the mid-sixteenth century for their thick, waterproof fur (which we turned into hats), their meat and the castoreum from their scent glands, which we used in perfumes, as a flavouring and in medicines.

Nobody knows where these newly returned beavers came from, or even who saw them first. Certainly, when a local ecologist spotted signs of gnawing on riverside trees, he put up a camera trap. That very first night, he filmed a beaver in action, sitting on its haunches, its webbed hind feet providing a stable base as it grasps the trunk of a sapling with its sturdily clawed front paws. The trunk, which is about a handspan wide, sports a V-shaped notch on each side. At one point the beaver stops gnawing and runs a paw across the narrowest point. Then it gnaws some more. As soon as the notches almost meet, the beaver shoves the trunk and it jumps neatly to one side and stands upright next to its stump. The beaver returns to all fours and lumbers round to look at the camera side-on, its eye shining in the dark and its scaly 'scoop' tail trailing behind like a rounded leather bookmark.

Beavers like the water around their lodges and burrows to be at least 70 centimetres deep, to protect them from foxes and the otters who are rumoured, although not proven, to attack their kits. The main part of the River Otter is mostly deep enough to keep beavers happy. But on the river's shallower tributaries the animals have to engineer the conditions they like by building dams. These have created new areas of pools and wetland, forming an ideal habitat not just for beavers but also for amphibians, wildfowl, water voles, fish, aquatic insects and other invertebrates. The engineering works have even benefited humans. Beaver dams store water then release it slowly; once a pair of beavers built a dam on the River Otter upstream from a village where up to 60 buildings had flooded in past deluges, the river saw lower peak flows after heavy rain.

Over time, the beavers have grown their population to around 15 family groups, which are active along the more than 65-kilometre-long river and some of its tributaries.[31] August of 2020 brought good news; the government said that beavers could stay on the River Otter, following the completion of a five-year trial of their impact on the landscape and local people. To even start that trial, the beavers had to pass another test – they had to be the Eurasian species, not the North American, and they had to be free of diseases, including the *Echinococcus multilocularis* tapeworm that lives in Europe but not the UK and spreads to dogs, foxes and, sometimes, humans. The beavers proved disease-free and, following the trial, became the first legally sanctioned reintroduction of an extinct native animal anywhere in England. These ecosystem engineers can remain, restoring the landscape to what it once was.

At around 30 kilograms (66 lbs) and the size of a Labrador, beavers fall shy of the official weight category for megafauna. Though they are the second-largest living rodent after South America's capybaras, the biggest of which weigh more than double the 44.5-kilogram (100 lb) megafauna pass mark. Beavers may not quite be megafauna, but their tree felling and other engineering has a massive impact on their habitat and other wildlife. Imagine, then, how much impact the disappearance of so many bigger animals, the megafauna, had on the ecosystems of the time. And there's a twist in the tale – those extinctions hundreds of centuries ago may affect human diseases today.[32] When I talk to the researcher who made this link, Chris Doughty, an ecoinformatics professor at Northern Arizona University, US, his ideas – and words – come thick and fast and my pen struggles to keep up. As Doughty explains, we 'lopped off the upper half of the animal size distribution pretty much globally, outside of Africa'. Finding out how this devastation altered ecosystems, he reckoned, would be vital not only for our knowledge of the past, but also because many of the big animals that did survive are going extinct now. Disrupting ecosystems is rarely a good idea – they function with the intricacy of a Victorian pocket watch. The flow of energy, water and carbon through nature relies on animals and plants like the motion of the hands around the clockface relies on the cogs and springs that make up the watch. All must be in place and working in balance.

The ancient megafauna acted as the watchmakers who set everything in motion and kept all the components running smoothly. In many cases the ecosystem 'watch' stopped, as Doughty found out, when these giant animals disappeared. With great size comes great responsibility – large animals are an ecosystem's engineers. They eat a lot, move a long way each day, and have long guts. As they go, they control the structure of the habitat – where and how much plants grow – and transport nutrients, seeds and more via their bodies and droppings.

The ancient megafauna were also giving free rides to pathogens and to parasites, like ticks and fleas, that might have contained pathogens. As with much engineering, this transport of diseases wasn't glamorous work. And that's what Doughty looked at. He modelled how far the megafauna would have transported common gut bacteria such as *E.Coli*, some of whose strains cause diarrhoea and vomiting. A wolf, say, could have picked up the bacteria by sniffing another animal's droppings before later depositing them in droppings of its own. He also investigated how far a generalist tick would be likely to travel on board a member of the megafauna. Some parasites – the generalists – aren't too fussy about who they feed on, attaching themselves to their nearest warm-blooded meal then falling off later.

When the non-human megafauna disappeared, the free rides stopped. The giant engineers were no longer 'building bridges' between microbes living in different regions by carrying them from place to place. According to Doughty's models, when the average body size of animals dropped by an order of magnitude almost worldwide, the distance a pathogen would typically move shrank by a factor of seven. For a gut microbe, the mean distance travelled fell from around three kilometres a day to just half a kilometre, while a pathogen onboard a parasite such as a tick, flea or louse moved just over one kilometre each day instead of nearly eight.[33] The megafauna extinction was like the sudden appearance of a mountain – it separated populations of disease-causing organisms by removing the creatures that spread them far and wide, stopping them from mixing. Separated populations may eventually evolve to form separate species, like the Galapagos finches that Charles Darwin realised had developed their own style of beak tailored to the local food on different islands.

The pathogens which the megafauna had previously transported long distances were now confined to specific areas. Over millennia, this allowed them to evolve and split into new species, perhaps with the ability to prey more effectively on the animal species that remained in those areas. When these evolved pathogens encountered new megafauna – our domestic animals such as cows, sheep, goats and camels, bred for fast growth rather than strong immune systems and often with similar biologies to extinct megaherbivores – they suddenly hit the jackpot. From our domestic animals it's but a short hop to us – Doughty's calculations show that the ancient loss of the megafauna may have increased the emergence of infectious diseases in humans today by some 5–8 per cent, or roughly 10 diseases.[34] That doesn't seem like something we're paying enough attention to. Extinctions, huge losses in and of themselves, removing creatures forged by evolution over many generations, also brought us as many as 10 new diseases. They killed humans as well as other animals. Those long-lost giants were keeping us healthy. Yet we're still killing the few species of really large megafauna that survive to this day; elephants, rhinos, tigers and bison are all in trouble. At Mpala Research Centre in Kenya, where I encountered Grevy's zebras, a collection of wire enclosures like mesh-walled tennis courts with roofs excludes large herbivores. Experiments a few years ago showed that keeping giant grazers like elephants and giraffes at bay changed the vegetation inside these enclosures. The signs are still visible today, the trees denser, the scrub scrubbier. At the time, rodent numbers doubled.[35] And rodents spread diseases, an ominous sign for the future if we let today's big herbivores die out.

The megafauna extinctions could well have harmed other wildlife too. Most generalist parasites probably survived and focused on smaller animals, passing on the diseases they were already used to. Some parasites, however, had likely co-evolved to live exclusively on certain species of megafauna. 'Those that were specialists, did they all go extinct or did some of them jump to a new species causing new wildlife outbreaks?' Doughty asks. 'That's something we have no idea about. Likely wildlife suffered as well.' Doughty and his student Tomos Prys-Jones are currently analysing the DNA of bacteria in fossilised dung from caves in northern Arizona to find out which gut microbes lurked inside megafauna such as Columbian mammoths, Shasta ground sloths and ancient bison.

The species Doughty would most like to see come back to life is the ground sloth or woolly mammoth – one of the really massive animals. In his view there's no point bringing back a species with the same role in the ecosystem as something alive today, like a deer or a bison. Doughty has been talking to Colossal, a US-based company that aims to create an approximation of the woolly mammoth and return it to Siberia. Colossal has another of this mammoth's engineering services in mind. By destroying trees, the giants could restore grasslands to the Arctic tundra, potentially keeping more carbon in the soil and protecting us against climate change. Thousands of years ago, when mammoths and other grazing megafauna disappeared, climate in these northern latitudes warmed by up to 0.2°C on average, and perhaps even as much as 1°C regionally, according to Doughty's calculations.[36] Left uneaten, young birch trees grew into forests whose darker colour absorbed more of the Sun's heat than the grasslands they replaced. Colossal plans to add genes from frozen woolly mammoths to cells from Asian elephants. In 2021, the company said that it expected its first mammoth calves within six years. 'It could happen,' Doughty says of this de-extinction. Though species brought back from extinction could harm ecosystems or no longer find today's environment suitable.[37]

So humans could be involved in both destroying the woolly mammoth and its re-creation. Perhaps, one day, a form of mammoth will again crunch its way across the northern steppes as its warm breath clouds the air. With the mammoth's journey restarted, its ticks, fleas, lice and gut bacteria will once more hitch a long-distance lift, bringing mixing within an ecosystem that could stop new diseases from evolving further down the line. Nobody knows how such a creature will fare in the face of today's pathogens. What is certain is that something in its past wiped out the woolly mammoth, perhaps a disease imported by humans on some of their earliest journeys, in their bodies, on their possessions or with their dogs, maybe in combination with other ills. Neanderthals may have gone that way too. The same human-wrought pathogen pollution and ecosystem disruption that are triggering the rise of extinctions and infectious diseases today had already reared their mammoth heads tens of thousands of years ago. And humans hadn't yet started to trade or farm animals; so far we'd created disease havoc using only our own bodies.

3

The Canary in the Hawaiian Chain: Honeycreepers, Colonial Ships, Accidental Imports, Avian Pox and Avian Malaria

'Sweet is the voice of the land shell
Happy voice of the damp forest
Their voices are like that of the 'i'iwi bird
My sweet lei mamo'

'Sweet Lei Mamo',[1] Words by Huelani
Music by Charles Hopkins

More than 3,000 kilometres (2,000 miles) west of North America, a string of volcanic islands juts from the Pacific, as if a giant moving south-east let a trail of jagged pebbles fall from her pocket. Strewn with lush vegetation, the slopes of dark-rocked mountains tumble to the shore. There they meet palm trees and platinum sands that shade into seas of azure blue stretching to the horizon and beyond – as well as being far from any continent, this archipelago is 4,000 kilometres (2,500 miles) from any other island. The only mammals able to swim or fly for long enough to reach Hawaii are seals and bats; until 800 CE, the islands were free from humans. Birds ruled supreme, some flightless, some taking to the skies. Their calls, chirrups and whistles would have rung out from the coast to the edge of the highest forests, where 'ōhi'a lehua trees bloomed flowers spiked with fiery red stamens like erupting lava. Festooned with moss, branches soared above tree ferns on the forest floor, whose trunks were shrouded by past years' brown, bedraggled fronds beneath an umbrella of fresh green. Here rang out the high-pitched upwards whistle of the 'amakihi (pronounced ah-mah-KEE-hee), its look-at-me-hee note swooping skywards over and over. The 'i'iwi (ee-EE-vee) trilled in squeaks and squawks reminiscent of a string of rising 'mm-hmms' that drop to a cynical 'yeah'. And the 'apapane (ah-pah-PAH-neh) yodelled a few seconds of melody followed by peeping and a series of clicks like a budgie's. It was a bird paradise.

Apart from visiting migrants such as sea- and shore-birds, most of the archipelago's birds lived only in Hawaii; the islands supported at least 113 of these endemic species.[2] Yet today the calls of many native birds ring out only in Hawaii's highest forests on just a handful of the most mountainous islands. In 1987, from the branch of a tree at high altitude, a male Kaua'i 'ō'ō (Kuh-WAI oh-OH) gaped his long beak wide as he sang. A dark-coloured bird about the size of a starling, this 'ō'ō had yellow eyes that matched a small patch of colour at the top of each leg. His fluting calls echoed around the trees.[3] After his first few whistles, he paused to let his mate sing her own phrase. When she didn't respond, the male resumed his haunting tune, waiting time after time for his mate-for-life to join the duet.[4] But she was no longer there. This male was the last Kaua'i 'ō'ō ever recorded, and probably the very last survivor of all four species in the 'ō'ō group. Other Hawaiian birds also perished. Some 71 of the islands' original 113 bird species have been silenced.[5] Given the lack of recent sight or sound of several species, it's likely that only a third of Hawaii's endemic bird species survived the arrival of humans. These days, with most of the native birds surviving only in the hills, the birdsong that resounds in the lowlands is largely foreign, chiefly from species that were imported.

What created this alien soundscape wasn't easy to discern. Researchers didn't manage to home in on the full cause of the disappearances and relocations until at least the late 1960s, centuries after the birds' problems began and too late to save many of the species. Even now that we know, it's not clear how much we can do, although one technique does show promise. At the time it was all too obvious that the incoming humans destroyed natural habitat and brought in animals that preyed on birds. Against this backdrop of blatant damage it was hard to pick out the more subtle notes of destruction. For key to the precipitous decline in Hawaii's native birds were a trio of much less visible arrivals, only one of which made any sound at all.

BEAK VARIATIONS

Today ʻōʻō are commemorated by the title of a 2009 jazz album by John Zorn. Every track is named after a threatened or extinct bird; third in the list is Poʻoʻuli (po-UHL-ee), in honour of a species of Hawaiian honeycreeper. Discovered in 1973 on the island of Maui,[6] which lies just to the north-west of Hawaii Island, the largest and furthest south of the archipelago's land masses, this bird was smartly turned out in subdued tones – grey cap, black head, cream body and brown wings. It faded into the darkness when the last known individuals died in 2004. The same fate met more than two-thirds of the original 55 or so known species of Hawaiian honeycreeper, while nine of the surviving species are listed as endangered by the International Union for Conservation of Nature (IUCN).[7] Before people entered Hawaii's horizons, these birds had thrived. Over time, a type of finch that reached the archipelago 5 million years ago adapted into tens of different forest-dwelling honeycreeper species.[8] The smallest Hawaiian honeycreeper was about 10 centimetres long, including the tail, and the largest, the Hawaii mamo and greater koa finch, some 23 centimetres. Like Charles Darwin's Galapagos finches, the bill of each species was uniquely suited to its food. Also like the finches, some of the honeycreepers only lived on a single island. The ʻiʻiwi, or scarlet honeycreeper, stuck to the job description and drank 'honey' from the curved flowers of Hawaiian lobelioids, using its magnificent salmon-pink downwards-curving beak to delve for nectar. The ʻakialoas (ah-kee-ah-LOH-ahs) were – four of the seven species have gone extinct – also nectar feeders and probers for insects, some with bills so long, at roughly two inches, that they were one-third the length of their ochre-yellow and buff bodies. Many honeycreepers changed tack and specialised in insects, fruit or seeds. The canary-yellow and olive-green ʻakiapōlāʻau (ah-kee-ah-pohh-LAHH-AU) adopted the same ecological niche as the woodpecker and developed a bill with a straight underside for chiselling into bark and a curved upper for picking out insects. Others gleaned insects from surfaces with short straight beaks. Seed-eaters, harking back to their finch origins, grew short, wide bills.

These birds varied in the colour of their feathers too. From the reds of the ʻiʻiwi, ʻākepa (ahh-KEH-pah) and ʻapapane to the yellows of the Laysan Finch, Nihoa Finch, palila and ʻakiapolaʻau, through the pale greens of the ʻamakihi, Maui ʻalauahio, ʻanianiau, ʻakekeʻe and Hawaii creeper, to the striking black and gold of the now extinct Hawaii mamo, the forests must have been a sight as well as a sound. Colours would have flashed in all directions as the birds busied themselves feeding, grooming, finding a mate, nesting and raising young. Despite their often striking appearance, the birds' plumage had a heavy, clinging odour like that of old canvas tents,[9] or steam rising from an asphalt road.[10] But for tens of thousands of years no human was around to smell it.

Until, that is, roughly 1,200 years ago, when people breached this paradise and the birds' troubles started. Polynesians paddled in by canoe, bringing dogs, chickens and pigs with them and accidentally importing rats. The dogs and chickens probably didn't cause Hawaii's native birds too much trouble – apart from the flightless and ground-nesting birds – but the pigs chomped on native vegetation and the rats were murderous. The Polynesians themselves felled trees to clear farmland and hunted birds for food and feathers. Over the years a tradition developed; the new arrivals turned feathers or 'hulu' into ceremonial capes and head-dresses. The cloak of Kamehameha the Great, a ruler who united all the Hawaiian islands in 1810, incorporates the feathers of some 80,000 Hawaii mamo and looks like it's spun from gold.[11] Whether harmed by the hunting, the rats or the loss of their habitat, 48 species of native birds, including most of the flightless waterfowl, ibises and rails, went extinct.[12] Yet the Polynesians probably harmed birds less than later human arrivals, who also hunted, brought in unwelcome visitors and destroyed habitat.[13]

In 1778, while sailing across the Pacific, British explorer James Cook happened on Kaua'i, the most northerly of Hawaii's larger islands. When Cook landed, islanders bestowed him with at least seven feathered cloaks and capes.[14] But the 'gifts' Cook's sailors gave in return were vile, and this chance visit to Kaua'i proved unlucky, for Hawaii's people, birds and even for Cook himself who, relations having soured, was killed on his third visit. Although Cook tried to keep his crew safely on board the ship, the sailors passed on their sexually transmitted diseases to local women. These, along with the cholera, influenza, mumps, measles, whooping cough, smallpox, leprosy and other maladies that Westerners imported later,[15] were devastating. The population of native Hawaiians plummeted by as much as 90 per cent in half a century – a near decimation.[16]

Bird life fared similarly badly. At the time, forests stretched from the mountains to the coasts. 'There's always these accounts of the birdsong being deafening as sailors approached the shore,' says ecologist Dennis LaPointe of the US Geological Survey. 'You get a hint of this sometimes if you're up in the high rainforest and it's the dawn chorus. It's quite an experience.' Today, LaPointe can watch two surviving species of honeycreeper six feet from his office window at Kilauea Field Station, near the edge of the rain shadow, in Hawaii Volcanoes National Park on Hawaii Island. He can also hear their song from the 'ohi'a and koa trees that rise above an understory of dense tree ferns. As LaPointe describes the birds, his face lights up. 'I'm fortunate in that we still have healthy populations of 'apapane and 'amakihi,' he says. 'Sometimes I get to see a nice bright 'amakihi male going off to make a nest, or an 'apapane coming down and probing in blossoms.' The 'apapane is slightly smaller than a house sparrow and mostly red with black wings, a black tail and a white rump, while the Hawaii 'amakihi is the size of a wren with an olive-green body and delicate grey legs. In the breeding season, the male's breast turns bright yellow. Both species have hooked bills. Seeing one of the native honeycreepers still out there doing what they've done forever is a good feeling, LaPointe says.

Early naturalists must have felt good too – many were impressed by the diversity and density of Hawaii's bird life. But by 1854 this was no longer the case. That year, George Washington Bates wrote in his book about the islands that 'a traveler is more impressed with what there is not, than what he sees ... the almost universal absence of singing birds.'[17]

SILENCE SPRINGS

In August 2023 I went on a 'wildlife discovery' meditation retreat in South-west England. On an early morning bird walk I realised that to me, birdsong had just been background noise. When I followed the expert's lead and tuned in to the individual calls, the cheeps and twitters and snippets of song, I could, with his help, not only tell which birds were around me but what they were doing, what they wanted and what was bothering them. The birds had been telling me their stories, but I hadn't listened. In Hawaii in the mid-nineteenth century, people were listening but the birds' stories had come to end.

In 1902, nearly five decades after Bates noted the absence of singing birds, naturalist Henry Wetherbee Henshaw wrote that within the six years he'd lived in Hawaii 'large areas of forest, which are yet scarcely touched by the axe save on the edges and except for a few trails, have become almost absolute solitude. One may spend hours in them and not hear the note of a single native bird.'[18] A few years earlier, in contrast, 'the notes of the oo, amakihi, iiwi, akakani [also known as the 'apapane], omao, elepaio and others might have been heard on all sides'. Roughly 120 years after Europeans invaded the islands of Hawaii, much of the birdlife had disappeared. Henshaw was mystified. 'The ohia blossoms as freely as it used to and secretes abundant nectar for the iiwi, akakani and amakihi,' he wrote. 'The ieie still fruits and offers its crimson spike of seeds, as of old, to the ou. So far as the human eye can see, their old home offers to the birds practically all that it used to, but the birds themselves are no longer there.' Of the birds Henshaw mentions, the 'ō'ō and 'ō'ū, a fruit-eating honeycreeper with a yellow head, dark green back and olive-green belly, are now extinct; numbers of the others, a mix of honeycreepers, flycatchers and thrushes, are greatly reduced.

In areas fully 'touched by the axe' and converted to farmland and plantations for sugarcane and pineapples, the birds' disappearance was no mystery. Their preferred habitat had vanished and they followed suit. As well as large-scale agriculture, the colonial invaders brought cattle, goats, sheep, a different type of pig, mongoose, cats and two more species of rats. The livestock trampled and ate native vegetation, while rats, cats and mongoose ate birds and their eggs. But this didn't explain why birds disappeared from lowland forests 'scarcely touched by the axe' in such huge numbers, especially when the same species were thriving in the unaxed forests higher up the mountains. It would take until the late 1960s, nearly 70 years later, to fully solve the conundrum. In the meantime, people continued to unwittingly make birds' problems worse.

Alongside more mammals, the incomers brought new birds. As well as poultry, some of the first Westerners imported domestic rock pigeons that escaped into the wild and went feral. Early human immigrants from Asia brought cage birds such as spotted doves (a pigeon with a striking black and white spotted collar), the melodious laughingthrush (a brown bird with a white eye stripe reminiscent of Cleopatra's kohl) and the red avadavat (in which the males are red with white spots, like a toadstool) as pets or for food. Some of these escaped too. From 1865, introductions to the wild became more deliberate – the Hawaii Board of Agriculture and Forestry imported new bird species to control crop-guzzling pests, starting with Asia's common myna to battle outbreaks of armyworms. New birds were everywhere.[19]

A POX ON YOUR HOUSE

At least one of these new arrivals unwittingly introduced another visitor. In the late 1800s, naturalists on Hawaii Island, the largest and most south-easterly of Hawaii's islands, found dead and dying birds with tumour-like growths on their legs, feet, around their eyes and near their beaks – basically in areas where they didn't have feathers.[20] These lesions went through phases – sometimes they were dry and grey, at others wet and red or crusty pink/yellow.[21] Bird-lovers were mystified until 1902, when Henshaw gave an 'ākepa, a honeycreeper with orange-red plumage that likes to nest in holes, to the Bureau of Animal Industry in Washington DC, US.[22] Bureau staff diagnosed avian pox. Spread by a double-stranded DNA virus that's relatively large at around 200 microns in diameter,[23] this disease can cause lesions too, on the mouth, windpipe and gullet. Although some birds recover, perhaps bearing scars such as missing claws or toe joints, many grow thin and die. Avipoxviruses spread by close contact, for example if fluids from a weeping lesion on one bird contaminate water, food or dust, or touch a patch of grazed skin on another bird. And avian pox can also spread with assistance – more on that later.

But how did the disease reach Hawaii in the first place? In domestic chickens avian pox had been diagnosed on the island of O'ahu the year before, in 1901. But the two forms of avipoxvirus found in Hawaii's native forest birds resemble canarypox; the thirteen species of avipoxvirus each generally infects a range of birds from the same genus or family.[24] So the finger points to an introduced canary-like bird as the initial culprit for infecting Hawaii's wild birds. Where that canary-like bird came from isn't clear. Avian pox affects more than 200 species of birds worldwide; the virus lurks everywhere apart from the Arctic and Antarctica, and people had introduced birds to Hawaii from many corners of the globe. (In Britain, you're most likely to see wood pigeon, dunnock and great tit suffering from avian pox.) What is clear, as genetic testing of avipoxviruses from the bodies of native birds in museums has shown, is that a pox-bearing canary-like bird arrived well before Henshaw discovered the cause of the tumour-like growths on Hawaii's native birds; by 1902, the disease was already well established. Avian pox was probably responsible for the first major wave of extinctions of Hawaiian forest birds in the late 1800s.[25] At about the same time that bird numbers began to fall in intact forests, naturalists noticed that signs of pox had become prominent, LaPointe says. 'Pox undoubtedly played an important role in the early extinction of forest birds.' It looked like Henshaw had worked out what had killed Hawaii's native birds in the now silent lowland forests. But why were birds in forests higher up the mountains still thriving? To explain that, we need to turn to another accidental introduction, one for which we know a more definite arrival date.

Back in 1826, a new sound joined the birds' chorus. After a stop on the west coast of Mexico, the whaling ship *Wellington* reprovisioned in the town of Lahaina on Maui. According to one account, members of the crew took the ship's water casks and 'drained dregs alive with wrigglers into a pure stream'.[26] The wrigglers were the larvae of the 'night mosquito' (*Culex quinquefasciatus*) and would grow into adults about 4 millimetres long with a white-banded abdomen. Common in the tropics and sub-tropics, this mosquito is thought to hail from Africa or Asia and to have travelled around the world on ships serving the Atlantic slave trade, the Old China trade and the American whale oil industry. Often known as the southern house mosquito,[27] this insect likes to live near people and their mess – in polluted water, heavily organic water, human sewage or livestock sewage. The southern house mosquito thrives where it looks like nothing should.

SONG WITH A STING

Once this import had multiplied, native Hawaiians gathered in the evenings to listen to its sound. Soon people near Lahaina were reporting 'a hitherto unknown kind of itch, inflicted by a new kind of nalo [fly] described as "singing in the ear"'.[28] The insect spread to most of Hawaii's islands; it can fly at least 22 kilometres (14 miles), creating that high-pitched 'song' as it flaps its wings hundreds of times a second. At some point this people-loving mosquito entered the forests, far from humans. Here, however, it received help from another recent import – the pig. Pig varieties originating from Europe bred with descendants of the pigs that accompanied the Polynesians, creating a large feral population living free in the forests. These feral pigs knock over tree ferns, strip away the outer woody layers and guzzle down their starchy pith – it's one of their favourite foods. Then they move on, leaving a small wooden bowl where the tree fern stood. These bowls, dotted throughout the forest, harbour water and make perfect habitat for mosquito larvae, as do the pools that form where pigs have wallowed, stagnant pools by streams and pools where trees have fallen.

In turn, these pig-assisted mosquitoes do their bit to help out avian pox. When a female mosquito bites an infected bird (only the females need blood), she harbours the avipoxvirus on her mouthparts for more than a month,[29] and injects it into the next birds she bites. Scientists can only speculate how much this mosquito boosted transmission of avian pox, but as the insect was in place in lowland forests in, or soon after, 1826, LaPointe imagines that mosquito-borne pox was a major issue. The two unwelcome visitors, the pox and the mosquito, may have arrived around the same time.

Native birds proved particularly susceptible, having only recently encountered the disease.[30] Introduced species such as the warbling white-eye, a straight-billed, lime-green and beige bird with a striking white ring around its eye, have had generations to evolve resistance and usually avoid infection. Part of the native birds' problem may have been their lack of night-time defences. 'A daytime-biting mosquito isn't going to have a chance near a bird because the bird's going to peck it,' LaPointe says. 'Basically no animal wants to lose any blood.' But *Culex quinquefasciatus* seeks out blood at night. On the mainland, many birds fluff up their feathers before sleep, hunker down so that their feet aren't exposed, and tuck their bills, and perhaps as much of their heads as possible, under a wing. As well as keeping the bird warm, this posture protects most of its exposed skin and the areas mosquitoes generally bite – the feet, around the eyelids and the base of the bill. But sleeping native Hawaiian birds don't cover their skin, they just perch. When LaPointe introduced mosquitoes to sleeping birds in the lab, the insects managed to feed on native birds much more than on non-native species.[31]

In the lowland forests, native birds were suffering from the newly introduced avian pox. Half the explanation was in place. But why were birds unscathed in 'the more profound solitudes', as Henshaw put it, of the forests at higher altitudes? The *Cambridge Dictionary* definition of solitude – 'the situation of being alone without other people' – provides an answer after just a small adaptation to 'the situation of being alone without mosquitoes'. These insects couldn't survive the colder temperatures of the higher forests and so didn't carry avian pox from bird to bird; at altitude the virus lost one of its transmission routes and more birds survived.

Mystery solved. But the problem of the absent lowland birds remained and people missed their song – the squeaks and squawks of introduced birds such as sparrows and mynas weren't nearly so popular with the humans who'd recently arrived in the lowlands of Hawaii. The absent birds were missed. When I was working on this chapter, I listened to a talk about wellbeing that suggested listening to bird-song while you write. The next morning, I put on the RSPB's Birdsong radio. The second I heard birds twitter and flute, my mood soared. So I empathise with these settlers pining for birdsong. But chances are their actions introduced the problem that caused the second wave of Hawaii's bird extinctions, from the 1960s onwards. What did these settlers do? Understandably enough, they imported even more birds.

From children to gardeners, not to mention the game board, pretty much everyone was bringing new birds to Hawaii. In the late 1920s the Buy-a-Bird campaign saw school pupils raise funds to introduce more species, and a ladies' gardening club set up the Hui Manu society, which imported songbirds until 1968. Hunters wanted game birds to shoot so government agencies obliged, with a particularly big push on imports in the 1930s. The ring-necked pheasant (native to Asia), warbling white-eye (native to East Asia) and northern cardinal (a crested North American bird in which the males are bright red) proved popular additions – as many as four different organisations released these species. And in the late 1950s, the Hawaii Board of Agriculture and Forestry imported the barn owl to prey on rats and mice in sugar cane fields, and the cattle egret to eat flies.

Most of the introduced birds were passerines (songbirds and perching birds, species similar to Hawaii's native honeycreepers, flycatchers, warblers and thrushes), large ground-feeding birds like chickens and pheasants, or doves. So far, humans have imported more than 170 species of bird to Hawaii and 54 of them have established themselves in the wild. Today, the birds in the archipelago's lowlands are pretty much exclusively from imported species. From 1945 onwards, restrictions on importing birds to Hawaii for deliberate release became tighter; most new arrivals in the wild are now pet birds released either accidentally or intentionally.[32]

AN ILL WIND

Sometimes birdsong on Hawaii is drowned out by storms. In winter native birds would sometimes move from the hills down to the coast to find calmer skies. In 1920, Kaua'i resident James Clapper saw a flock of honeycreepers avoid a storm by congregating in a hedge of hibiscus – another introduction – that was blooming at the edge of the road in the lowland area around Waimea in the west of the island (now dubbed 'Hawaii's *Original* Visitor Destination' thanks to Cook's arrival there).[33] The flock contained at least 30 birds, some red, some yellow, some with bills that were 'long and decurved, like that of the curlew, and at least two inches long'.

Clapper noticed the birds in the afternoon; they stayed a few hours and were gone the following morning. If they'd lingered until nightfall, even this brief visit to the lowlands may have put them at risk of yet another danger arguably worse than the storms. A danger they couldn't see and so couldn't avoid. About this time, most likely in the 1920s or 1930s, the *Plasmodium relictum* parasite that causes avian malaria arrived. The parasite occurs worldwide apart from in Antarctica,[34] and affects more than 400 species of birds.[35] Humans and other non-bird vertebrates have their own *Plasmodium* species, although the human versions are not a problem in Hawaii, as the *Anopheles* mosquitoes that transmit them haven't reached the islands.[36] Like avian pox, it looks like this unwelcome visitor arrived in Hawaii alongside a songbird, this time in its bloodstream or liver. *Plasmodium relictum* is 'typically thought of as a parasite of songbirds only', according to LaPointe. One of the birds imported to replace lost birdsong likely led to the loss of even more birds and even quieter forests.

Nobody knows which species carried in the parasite, although recent genetic detective work indicates it was a bird from the Old World (Africa, Asia and Europe).[37] Some think the GRW4 strain of the parasite that entered Hawaii originated from islands in the Indian Ocean off Africa.[38] The first known Hawaiian victims, found in the 1940s at Hawaii Volcanoes National Park, where LaPointe is based, were two introduced birds, a warbling white-eye and a red-billed leiothrix, a colourful bird with a bright yellow chin, pale-rimmed eyes, a bluish-grey back, and russet, yellow and black patches on its wings.[39] To catch avian malaria, these birds must have been bitten by a mosquito. Unlike the avian pox virus, which also spreads via close contact, the *P. relictum* parasite needs a mosquito to transfer it to another host. And the mosquito doesn't just provide its mouthparts as an injection needle, its whole body enables the parasite to complete its life cycle. Had a bird imported the parasite before those sailors poured mosquito larvae into a Kaua'i river, avian malaria wouldn't have been able to spread.

But now it could. As soon as *Plasmodium relictum* arrived in an infected bird's blood, a female mosquito no doubt guzzled it up. Inside her gut, the male and female forms of the parasite then got together and reproduced. After a couple more stages of development and reproduction, this time asexual, a spore-like version of the parasite headed to her salivary glands, ready for injection into the next bird she fed from, along with plenty of anti-coagulant to keep the blood flowing. Once inside that new bird, the parasites matured and spread from the bloodstream to internal organs such as the liver, spleen, lungs and brain. There they multiplied before infecting the red blood cells, where they multiplied so much that they burst the cells and gave the bird anaemia. The same process repeats each time an infected mosquito bites a new bird. 'As you would imagine, the birds become very weak and listless, they pretty much sit in the cage,' says LaPointe, who studied avian malaria for his PhD. 'They have a food source right next to them but they don't feed, and they eventually succumb to this weakened condition.'

Not all birds die from avian malaria; some species are immune or can live with the disease. But again, Hawaii's native birds proved highly susceptible to a new disease. Especially Hawaiian honeycreepers who, together with penguins, perhaps because both groups evolved in the absence of the parasite, are the bird group most susceptible to the infection.[40] The *P. relictum* parasite is on the IUCN-backed list of 100 worst invasive species, where it's blamed for the extinction of at least ten of Hawaii's native bird species to date.[41] For some experts the feather of suspicion points to avian malaria as the culprit behind the second wave of extinctions of Hawaii's native forest birds, which began in 1960.[42]

As with avian pox, birds in mountainside forests too cold for mosquitoes escaped the ravages of malaria. But on the islands of Oʻahu, Molokaʻi, Lānaʻi and Kahoʻolawe, where there is no land higher than 1500 metres, native bird species have almost completely vanished, obliterated by the arrivals of a virus, a parasite and a mosquito. Most of Hawaii's native forest birds, including all endangered species, now live on the islands of Kauaʻi, Maui and Hawaii above this 'mosquito line' at 1500 metres.[43] Hawaii's native forest birds have largely retreated to areas higher than Britain's highest mountain – Ben Nevis peaks at 1,345 metres, or to altitudes 3.5 times the height of the Empire State Building. 'The hope was,' says LaPointe, 'that this "mosquito line" would remain consistent so that birds that hadn't been lost would remain. But we all know that things don't ever stay the way they are.'

CHANGING CLIMATE

Hawaii's birds face a new invisible threat caused by humans, climate change having arrived with a vengeance. Temperatures on the slopes of Hawaii's ancient volcanoes are rising and cases of avian malaria are appearing at higher altitudes than ever before. Early modelling by LaPointe's colleague Tracy Benning, who is now at the University of

San Francisco, US, indicated that bird populations on Kaua'i would decline before those on Maui and Hawaii because the island's peaks are lower.[44] And that model proved to be correct.

Over the last decade or so in Kaua'i, numbers of two honeycreepers, the 'akikiki (ah-kee-KEE-kee), a near-sparrow-sized insect-eater with a white face that shades into its delicate pink beak and forms a splash of light against its charcoal back and pale-grey belly, and the slightly smaller greenish-yellow, chunky-billed 'akeke'e (ah-keh-KEH-eh) have declined rapidly. When LaPointe worked in the island's Alaka'i swamp, a plateau at an elevation of roughly 1,300 metres, in the late 1990s, he'd regularly see almost all the surviving species of native forest bird, including the 'akikiki and 'akeke'e. But by 2007, numbers of 'akikiki and 'akeke'e were shrinking, and so were their ranges. Taking blood samples from the birds revealed that avian malaria was up to ten times more prevalent than it had been in the late 1990s. When LaPointe revisited his site in 2016, he found that 'those species that were rare when we were observing them were now completely absent, and other species had become extremely rare in those areas in which they had been common'.[45]

The 'akikiki is now for all practical purposes extinct in the wild. Only five individuals live free in the forest; the remainder have been taken into captivity, as the species was so close to lost. The hope for these 'back-up' birds is that conditions in their natural habitat will improve. Yet even in captivity they're not completely safe. Some of the birds were moved to a facility on Maui, to spread the risk. In 2023 these 'akikiki had a close encounter with the island's wildfires when conservation workers successfully fought off flames that were nearing the buildings holding the birds.[46] The 'akeke'e, in contrast, is not yet completely on the edge. At the last count, in 2012, there were roughly a thousand left.[47] But this species doesn't do as well in captivity.

The next drops in bird numbers, if reality continues to follow the science, will be on Maui followed by Hawaii Island, which has the highest-level forest. It looks like that's becoming true too. On Maui the numbers of kiwikiu (Maui Parrotbill) and 'ākohekohe (ahh-KOH-heh-KOH-heh) are declining fast; the species could both be gone within ten years. It's not just rising temperature that's a problem; the islands have dried as well as warmed over the last few decades. Streams that were once permanent are now intermittent, and when they're not flowing, potholes in the streambed provide new homes for mosquito larvae. As temperatures hot up and the islands dry, Hawaii's forest birds may have nowhere mosquito-free to go.

Thousands of kilometres away in London, avian malaria – and climate change – may also be harming the humble sparrow.[48] (Although it's resistant to avian pox, the sparrow does succumb to avian malaria.) Numbers of this bird in the capital have declined 70 per cent in the last 20 years. In one study, nearly three-quarters of the birds tested carried *Plasmodium relictum*, the highest proportion found in any population of wild birds in northern Europe. And colonies that had experienced the biggest drops in population tended to be home to birds, especially juveniles, infected with the highest numbers of parasites. It could be that the disease is stopping juveniles from surviving the winter. Climate change may be exacerbating the problem – the warmer and wetter conditions of recent years expose sparrows to more mosquitoes. Habitat loss may also play a role; sparrows don't tend to roam far from where they hatched, so populations in cities that have lost a lot of sparrow habitat may be isolated and less genetically diverse, reducing their ability to survive malaria.

RISING RESISTANCE?

Protecting habitat alone can't save birds from *P. relictum*. But some birds in protected areas of Hawaii have started to save themselves. Inside the confines of the Hawaii Volcanoes National Park the landscape must be striking. Mauna Loa, the largest shield volcano in the world, has a peak capped by winter snow and gently bulging slopes built up over centuries by the ooze of highly fluid lava. Neighbouring volcano Kilauea is younger and more active, sending lava into the sea where it sparks, steams and spews out scalding jets of water. It's near Kilauea that LaPointe has his office with a view of 'apapane and 'amakihi. If he ventures up into the mountains he'll also see 'i'iwi, a honeycreeper a little larger than the 'apapane but similar in appearance – red with a curved beak and black markings on its wings and tail. Both the 'i'iwi and 'apapane prefer the nectar of the red-flowered 'Ohi'a tree, which grows from the highest forests all the way down to the coast, although it is now under threat from Rapid 'Ohi'a Death, a fungal disease that was first detected here after an outbreak in 2010. Caused by two species of fungi, one that likely originated in the eastern US and one from Asia, the disease yellows the tree's leaves. It can be spread by wood-boring beetles, the wind and humans who unwittingly transport its spores.[49] In search of healthy 'Ohi'a trees that are in flower, the 'i'iwi and 'apapane migrate up and down the mountains. At times, that means they encounter many mosquitoes. Hawaii 'amakihi, on the other hand, don't stray far from home. These birds feed more extensively on insects as well as nectar so they're less tied to seeking 'Ohi'a flowers.

In the early 2000s, LaPointe and his colleagues discovered that something strange was going on. More 'amakihi were living at low elevations, where mosquitoes and avian malaria are rife, than in the disease-free heights.[50] In the south-east of Hawaii Island, the lowland population was even increasing. Some 90 per cent of these low-living 'amakihi were chronically infected with avian malaria – they had a level of *Plasmodium* parasites in their blood low enough for them to survive but high enough to pass on to the mosquitoes that bit them. Any newly hatched 'amakihi would be exposed to avian malaria in an intense evolutionary baptism by fire. And it looks like that baptism gave the birds a blessing in disguise – the lowland 'amakihi population rapidly evolved a genetic tolerance for avian malaria.[51] These days the populations of 'amakihi at low and high elevations are genetically distinct. In one study, when exposed to avian malaria, at least 90 per cent of the lowland 'amakihi survived, whereas only 20–30 per cent of the 'amakihi that had lived higher up made it through. In the face of continual exposure to avian malaria, the lowland birds have adapted. It looks like the 'amakihi could have a future. But since 'apapane and 'i'iwi move up and down the mountains, it's unlikely that birds that happen to hatch at lower levels will pass on any greater tolerance for malaria to their offspring as they often mix, and mate, with birds from higher up. And, following the arrival of Rapid 'Ohi'a Death fungus, the forest that supported the malaria-tolerant 'amakihi is disappearing.[52]

MESSING WITH MOSQUITOES

So what could we do to help keep the 'apapane and 'i'iwi singing? With global efforts to halt climate change currently woefully inadequate, Hawaii's native birds need assistance on other fronts. There are several options. Firstly, we could declare war on rats. The black rats that came over on European sailing ships climbed the rigging faster than you could shake their tails. So nipping up a tree to take eggs and chicks from a Hawaiian bird's nest is rat child's play. Several projects are controlling or have controlled rats to help critically low bird populations survive, including on the island of O'ahu to protect the O'ahu 'elepaio, and the

Kauai Forest Bird Recovery Program in the Alak'i Swamp to protect the 'akikiki (although unfortunately this latter project failed as avian malaria decimated the 'akikiki).[53, 54]

Secondly, we could declare war on mosquitoes. These insects transmit the avian pox and avian malaria that are among the most crucial factors harming Hawaii's native birds. But eradicating mosquitoes is no mean feat. Fencing off areas of forest to keep out feral pigs stops the animals creating tree-fern cavities where larvae flourish and has the added advantage of preventing plants in the understory from becoming a pig's dinner. But larvae also thrive in pools alongside streams or where trees have fallen down. Killing these 'wrigglers' with a non-specific pesticide is off-limits in a natural area – there'd be too much harm to other insects and invertebrates. The biological pesticide Bti is more precise, killing only the larvae of mosquitoes and a few other insects. Based on toxins produced by the bacterium *Bacillus thuringiensis* serotype *israelensis*, Bti is used in the wild around the globe. And trials are underway in Hawaii.

The latest push to silence Hawaii's mosquitoes will use another type of bacteria, from the *Wolbachia* genus. These tiny organisms live alongside many species of insect, including mosquitoes, and even arthropods such as crustaceans. The bacteria concentrate in these creatures' reproductive organs, where they play an unusual role. If two mosquitoes mate and one carries *Wolbachia* bacteria but the other doesn't, the eggs the female lays won't develop properly. The same lack of progeny results if the pair carry different strains of *Wolbachia*. By manipulating which, if any, strains of the bacteria male and female mosquitoes carry, scientists are able to control how successfully mosquitoes reproduce. This technique has been used in Asia, the Americas and Oceania to reduce populations of mosquitoes that spread human diseases like dengue and Zika virus.

So how might it work in Hawaii? Microbiologist, and long-time colleague and friend of LaPointe's, Carter Atkinson noted that southern house mosquitoes in Hawaii mostly carry only one strain of *Wolbachia*. Now, along with other members of the 'Birds, Not Mosquitoes' consortium, the plan is to release male southern house mosquitoes that carry a different strain.[55] Over time, by stopping many of the islands' female mosquitoes from producing eggs that hatch, this 'incompatible insect technique' should drastically reduce the mosquito population. A deliberate introduction that could, all being well, counteract the harms of the accidental introductions of mosquitoes, avian pox and avian malaria. Though it's not without its controversies – many Hawaiian residents

are understandably wary of yet another import that will alter their ecosystems, even if its aim is to restore them.

After small-scale pilot releases of the incompatible male mosquitoes
on Maui, trials took place on private, state and federal lands in East Maui
in the autumn of 2023, using helicopters to access Hawaii's remote landscapes. By June of 2024, the project had released more than 10 million
male mosquitoes.[56] After Maui and Kaua'i, the plan is to expand
throughout the main Hawaiian islands. If these *Wolbachia* manipulations suppress mosquitoes in Kaua'i's prime 'akikiki habitat, the captive
birds could be returned to the wild. 'I hope it works,' LaPointe says.
'I'm nearing the end of my career and I'd like to see something come
of it. I don't enjoy basically watching another bird go extinct.' Perhaps,
one day, the tweets, calls, whistles, peeps and trills of Hawaii's surviving
native birds will once more ring out from the mountains to the shores.

Even if the *Wolbachia* project succeeds in wiping out the southern
house mosquito, there are still other dangers. Another *Culex* mosquito
that's capable of spreading avian malaria may arrive in the islands; Hawaii
now has six or seven species of biting mosquitoes, all brought here by
humans. Or a different pathogen may turn up. We no longer only arrive
by ship; from time to time, the southern house mosquito is intercepted
on plane cabins at Hawaii's airports. When West Nile virus swept across
the US in 1999, LaPointe was convinced it was a matter of time before
an infected mosquito arrived in Hawaii from California, spelling trouble
for both birds and people.[57] Yet if humans get their act together, we could
learn from the past and bring in stricter regulations on the introduction of
any insect, plant or animal that could harm the islands' vulnerable ecosystems, whether on its own by competing with or eating other species, by
transmitting pathogens as mosquitos and other insects do, or by importing a disease inside its own body. We could also aim to develop tools that
knock back these invasive species if they do arrive, as well as ways to deal
with the unwelcome visitors that are already well established.

As globalisation continues apace, by sea and air, who knows what
else we're accidentally bringing in alongside our deliberate introductions? It's time to pay more attention to our imports so that birds can
continue to lift our hearts with their calls. In John Zorn's album track
Po'o'uli, human music fills the pauses between the honeycreeper's song,
piano notes bubbling like a river beneath the bird's trills. The last note
is digital and hangs in the air unresolved.

4

Measly Migration: Close Contact, War, and Rinderpest's Deadly Jump from Cattle to African Wildlife

'Why is the land not plowed in Mota and Qaranyo?
Why is the land not plowed in Dambaçça and Dabra Warq?
I came from there all the way to here without seeing an ox ...'
                                        Line from Ethiopian poem, 1890s[1]

Like Hawaii's native birds, which have been devastated by two imported diseases and an introduced disease carrier in the shape of the mosquito, wildlife in Kenya has also suffered from the accidental arrival of a pathogen. One day, after a short drive from a campsite near Laisamis in northern Kenya, we reached the ranger station. In front of one of the low buildings, a lesser kudu – a near-threatened antelope – stood in a patch of long grass. Widely spaced white lines drizzled her fawn flanks, and her large diamond-shaped ears pricked towards us. As we walked over, she gazed at us calmly from round brown eyes; once up close we could see the long eyelashes fringing her upper eyelid and the white stripe that swept forward from the inner corner. Tame from being fed, one of the 80,000 lesser kudus that remain let us stroke her. I expected wiriness, an echo of the sand and thorns she lived among; instead, the hide on her neck was velvet soft.

This kudu seemed healthy – her hide was in good condition, she wasn't too thin and her eyes were bright. But in 1994, British veterinarian Richard Kock, then working for the Kenya Wildlife Service, found that many lesser kudus in Tsavo National Park had gone blind. And it seemed to be due to an infection.

'The question was, which one?' Kock tells me. 'It didn't fit much of the literature around known diseases at the time.' To check if his suspicions about an infection were correct, Kock arranged electron microscopy, a technique that could magnify any pathogens much more than an optical microscope, of samples from the kudus. Sure enough, he was right – the tissues showed the presence of particles that resembled a member of the morbillivirus genus of RNA viruses, a group that includes the pathogens that cause measles, canine distemper and, crucially in this case, rinderpest, or cattle plague.

If the morbillivirus in question was pinned down as rinderpest virus, this was bad news; efforts to eradicate this virus from Africa were thought to be so near completion that the champagne was practically on ice. Treading his way carefully through the politics of this potential disappointment, Kock persuaded an official from the African Union – Interafrican Bureau for Animal Resources – on his way from Nairobi to visit family on the coast for Christmas, to stop off and look at a sick buffalo in the Chyulu Hills to the north of Tsavo West National Park. This official, who had diplomatic immunity and high enough political status to persuade the Kenya government and veterinary services to take action, had worked with rinderpest before and recognised the disease. Further tests confirmed it; rinderpest was definitely rife.

'This disease was not meant to be where we found it,' says Kock. 'It was meant to be in a tiny enclave in Sudan. Essentially they had the whole thing under control, so that [discovery] blew it wide open. There was a big panic – the worry was that this disease would spread again across a wider area of Africa.'

DOWN IN HISTORY

A relative newcomer to Africa, rinderpest is an age-old disease in Europe and Asia. Nobody's certain where or how rinderpest emerged. Some believe the virus evolved with the ancestors of cattle and antelope in the Ice Age.[2] Kock speculates that its forebear circulated among a gregarious animal such as Saiga antelope in central Asia. When humans domesticated animals, the virus gained new victims; animals kept by humans are often in more crowded conditions, more stressed and more genetically similar than in the wild, making them more susceptible to diseases and enabling those diseases to spread more quickly. Pathogens could jump more easily from domestic animal to domestic animal, from domestic animal to us, and from us to domestic animal. Basically, we gave rinderpest an opportunity to become dominant and the virus leapt at it. 'In balanced ecologies where you have natural processes, often these microorganisms just behave a bit like the species that they infect,' Kock says. 'They just get on with things without causing too much harm. But it seems that in the modern age, everything's so disrupted that we get these imbalances ecologically, and then you end up with pathogens. Not always, but a large number of pathogens are really a product of agriculture, a product of humanity and its behaviour, and the disruption of ecological systems.'

The first mammals that humans lived right up close with were sheep, which we kept in large numbers on the Eurasian steppe.[3] Around 11,000 BCE it was the turn of cattle. At some point, perhaps as early as the sixth century BCE,[4] after humans began living in large communities with their cattle near rivers, a virus that had by now evolved into something more like rinderpest jumped from cattle to humans, becoming an early form of measles. It's likely it made this jump before in a smaller group of humans but the outbreak fizzled out. Measles needs at least a quarter to half a million people to stay in circulation,[5] otherwise the disease can't find enough susceptible victims to transmit to and peters out. It's hard to know exactly what happened several thousand years ago, especially when the virus that causes rinderpest was only formally identified in 1902. We do know that, in the fifth century BCE, the Romans reported epidemics that affected both cattle and humans,[6] as did others in the early medieval period, during the sixth to tenth centuries CE. It's possible these outbreaks were due to an immediate ancestor of both the measles and rinderpest viruses that caused disease in both species. And today we know that rinderpest viruses are typically spherical and around 100–300 nanometres in diameter. They're sensitive to their environment and can be destroyed by heat, light and drying out. As RNA viruses, they tend to mutate easily – we'll encounter this family again.

By domesticating animals, humans shifted the disease landscape. Our farming made the animals we were tending sick, not just livestock like cows, sheep, goats, pigs, camels and llamas, but also working animals and companions like dogs and cats. Many of those animals have close relatives in the wild – antelope, wild goats, wild boar, wolves, big cats – that have similar biologies to their domesticated disease-rich cousins. And similar bodies mean it's easy to share bugs. Today nearly 90 per cent of wild mammals threatened by diseases and other parasites are from the order Carnivora, particularly the dog and cat families, and the order Artiodactyla, especially the swine family and bovids such as cattle, bison, buffalo and antelope. The health of wildlife, domesticated animals, companion animals such as cats and dogs, people and the environment are tightly linked.

Like Kock, others believe that rinderpest originated on the Central Eurasian steppe around the time that we domesticated cattle, with cattle becoming the primary host. The disease may first have spread west from Central Asia with the aid of nomads from Iran known as the Scythians in the ninth century BCE. Hundreds of years later, in the fourth century CE, Huns from east of Europe may have kick-started a chain of epizootics – the non-human animal equivalent of an epidemic, a sudden widespread outbreak of disease – that Charlemagne's armies continued to spread in the ninth century.[7] Conflict involved moving cattle – to provision troops, haul guns or as spoils of war – and rinderpest often travelled to new areas inside the bodies of the cattle. Conflict may also have spread the disease both west and east from the Central Asian steppe when marauding armies, including those of Genghis Khan in the thirteenth century, brought Grey Steppe oxen with them. These animals didn't show clinical signs of rinderpest but infected other cattle and buffalo, leaving the invaded population without transport or a means to plough their fields.[8]

Rinderpest spread so fast and killed so many animals that it's one of the epizootics for which we have the most history; people recorded their devastating losses.[9] The disease may have first reached Britain thanks to invading Saxons or Danes and returned to the nation with a vengeance in 1745, killing half a million cattle. Across that same century, rinderpest wiped out more than 200 million cattle in Europe. And this triggered a fightback. In 1711, Pope Clement XI instructed human physician Dr Giovanni Maria Lancisi to study rinderpest, and Lancisi wrote a book detailing eleven tactics, including quarantines, to combat the disease.[10] These sanitary measures and the new veterinary profession that had emerged, initially in France, to fight rinderpest eventually quelled the disease into submission in Europe.[11] By the end of the nineteenth century the West of the continent was rinderpest-free.[12]

Even today, the veterinary profession still focuses on livestock and companion animals more than wildlife. 'No disrespect to my profession but they trained for dogs and cats and cows and sheep and so on,' says Kock. 'And that training doesn't extend to ecology or to wildlife species.' Around the world, Kock estimates there are some 1,500 competent wildlife vets compared to more than 10 million competent human doctors.

Kock himself grew up in Zimbabwe, where his experience was truly with nature, not 'with David Attenborough'. Some of his experiences as an ecologist in the field have been mind-blowing. 'One's in awe, really,' he says. 'I think the body and the mind and the spirit are destroyed by not being exposed to nature. I wish we weren't in such a dire situation because there are a lot of human beings and nature is very much under pressure now.'

In Western Europe, rinderpest resurged shortly after the First World War.[13] During the Second World War, there was fear in the United States and Canada that the Japanese would launch aerial attacks using a pulverised form of rinderpest that would devastate the American and Canadian cattle populations.[14] The Americas had only had one short-lived brush with rinderpest in 1920, when a shipment of infected zebu cattle from India arrived in Brazil. Both Brazil and Belgium, a stopping-off point en route, suffered brief outbreaks of rinderpest. The shipment led to the creation in 1924 of the Office International des Epizooties (OIE) to coordinate scientific knowledge on rinderpest and other diseases.[15] Today the office is known as the World Organisation for Animal Health (WOAH).

INTO AFRICA

Conflict likely caused the spread of rinderpest across Africa too. By the late 1880s, the disease was already present on the continent in Egypt and some other parts of North Africa, but had never transmitted beyond the Sahara.[16] At that time, having occupied the Red Sea coast at Eritrea, Italy was preparing to invade Abyssinia/Ethiopia, according to Thaddeus Sunseri, a historian at Colorado State University, US, who became fascinated by rinderpest because of its colossal impact on African history. Italy began to provision its armies with cattle from India or the Middle East that were infected with rinderpest, and this time the disease spread with a vengeance.

By killing 95 per cent of African cattle, rinderpest made European invaders' lives easier. 'For much of Eastern, Southern as well as Western Africa, rinderpest was devastating and undermined a lot of African societies right at the moment of colonial conquest,' Sunseri says. Once it arrived in Eritrea and Ethiopia, rinderpest spread down through eastern Africa into southern Africa, taking about nine years to complete the journey.

Cattle infected with rinderpest, unlike greater and lesser kudu and giraffe, don't generally go blind.[17] They, and other particularly susceptible animals, exhibit the 'famous 4Ds',[18] as the World Organisation for Animal Health puts it – Depression, Discharges, Diarrhoea and Death due to another D, dehydration.[19] They may also have 'a loathsome stench' and suffer fever and mouth sores.[20] The disease is quick to kill – in the 1890s whole herds could disappear seemingly overnight.[21] The rinderpest virus spread via aerosol droplets when animals got close or via discharges onto pasture land or into water sources that another animal encountered soon afterwards. Herding cattle to the best pastures each season, trade by sea, and military raids probably all helped rinderpest spread.

Having reached the Horn of Africa, the disease slowly spread south, and west to the Atlantic coast. Writing from Uganda in East Africa in June 1891, British explorer Frederick Lugard called it a 'terrible plague which has spread inwards through East Africa [that] has carried off millions of cattle, and has inflicted a terrible blow on the pastoral tribes. No such epidemic has visited Africa within the memory of man.'[22] Up to 100 per cent of cattle died, and the humans who'd relied on them for milk, blood, meat or to pull their ploughs, suffered famine and outbreaks of smallpox thanks to the social disruption. According to one elder from the Maasai tribe of Kenya and Tanzania, recalling his youth several decades later, the corpses of cattle and people littered the landscape 'so many and so close together that the vultures had forgotten how to fly'.[23] Two-thirds of the Maasai people in Tanzania may have died.[24] The disease broke up traditional nomadic, herding lifestyles such as those of the Maasai, the Wahima from the Great Lakes region and the Oromo of Ethiopia by destroying the cattle on which they'd built their societies, often forcing women into prostitution or work on plantations and men to become soldiers, mercenaries or policemen for the colonial powers of Britain and Germany.[25] The price of the few remaining cattle soared.[26] Many men couldn't afford cows to give to their prospective bride's family and couldn't marry.

In 1891 southern Ethiopia was hit by rinderpest followed by drought, outbreaks of locusts and caterpillars and a smallpox epidemic.[27] The resulting crisis, the worst famine in Ethiopia that anyone at the time could remember,[28] became known as ciinna, 'the end of everything'.[29] Communities broke up, people scattered in search of food, and some fell victim to predators who had lost their normal wildlife or livestock prey.[30] Other people pawned their children in exchange for food or broke taboos and ate donkeys, elephants, warthogs, horses or even carrion.[31] It's rumoured that some turned to cannibalism.[32] Two-thirds of Oromo people died; Ethiopia as a whole lost one-third of its population.[33]

In addition to pastoralists, farmers fared badly as they relied on oxen to plough up the hard ground. Today it's hard for us to picture how catastrophic the disease was.

WILDLIFE WELL?

During this African outbreak, unlike in Europe or Asia in preceding centuries, large numbers of wildlife died too. Sometimes before cattle did, sometimes alongside them. 'Never before in the memory of man, or by the voice of tradition, have the cattle died in such vast numbers, never before has the wild game suffered,' Lugard wrote in 1893. 'Nearly all the buffalo and eland are gone. The giraffe has suffered, and many of the small antelope, the bush-buck, and reedbuck – I believe – especially ... The pig (wart-hog) seem to have nearly all died ... It is noticeable that the animals nearest akin to the cattle have died – viz.: the buffalo, and the most bovine of antelope, the eland.'

Rinderpest affected artiodactyls, a group of mammals whose members often have cloven hooves and are known as even-toed ungulates, though some members are whales and dolphins, which is frankly unhelpful. Around 40 species of even-toed ungulate were susceptible to rinderpest, although buffalo, bovine antelope such as eland, bushbuck and lesser kudu, giraffe and warthog suffered the most severely.[34] Often warthog died before antelope had shown any signs of illness, though up to 90 per cent of buffalo and bovine antelope succumbed.[35]

'Buffalo, which used to roam in hundreds over this country [Ethiopia], are now almost extinct, nor did I see any sign of hartebeest which are also said to have been common: both have been carried off by rinderpest,' wrote British civil servant J.L. Baird in 1900.[36] 'Indeed the only game seen up to this point was reedbuck.'

Wildebeest, according to the Maasai, were the last to be infected and only succumbed after most cattle had died. As they migrated across the Serengeti-Mara plains in Tanzania and Kenya, wildebeest may have helped rinderpest to spread. Some populations of the western giant eland (*Tragelaphus derbianus ssp. derbianus*), an antelope with spiralling horns that point backwards at the sky and with flanks drizzled, like the kudu's, with thin white stripes, never fully recovered. Rinderpest, in combination with overhunting, wiped out the population in the Gambia,[37] making the giant eland locally extinct in the early 1900s.[38] 'In Africa, the virus, as it came through, hammered certain species so much that the distribution changed,' says Kock. 'It was that serious. It literally caused extinction of certain species in certain countries.' Rinderpest was, in part, the reason for the creation of conservation reserves that after the Second World War would become national parks and exclude people from their grazing lands.[39]

So why didn't wildlife in Europe seem to succumb to rinderpest? The only reports were of a single wild boar dying in a German wood in the mid-eighteenth century, and of a few captive wild animals dying in parks and zoos.[40] The deer family is not particularly susceptible to rinderpest,[41] though it's possible that wildlife was dying unnoticed in forests and other habitat outside farmed areas – the continent was more fenced off than Africa and cattle didn't share territory with wildlife in the same way. In Africa the disease arrived during a series of droughts that forced cattle and wildlife to cluster together at the same water sources and on the same struggling grassland, enabling the disease to spread rapidly.[42] The droughts also led wildlife and cattle to move long distances in search of water, taking the virus with them. And Africa has large herds of gregarious cow-like animals, in the form of Cape buffalo, that Europe lacks.

HIDDEN HURT

When rinderpest hit Africa, this absence of visible wildlife deaths from the disease in Europe muddied the waters. Many medical professionals (who were sometimes human doctors, as vets were in short supply) decided that the outbreak couldn't be rinderpest as that disease didn't kill wildlife. The name of the disease didn't help. Until the outbreak reached South Africa, English-speakers called rinderpest 'cattle plague'.[43] And many different diseases plague cattle, often with similar symptoms. (German-speakers weren't much better off – rinderpest means 'cattle plague' in German). Was that animal sick with lungsickness (contagious bovine pleuro-pneumonia), East Coast Fever, bovine sleeping sickness, Texas Fever, foot and mouth disease, or anthrax? There wasn't a proper diagnostic test for rinderpest at this stage; the only option was to chop the animal up and look at its insides once it had died. As a result of this confusion, the outbreak didn't receive an official diagnosis until 1896, from a British vet who happened to be working as a telegraph operator in Southern Rhodesia.[44] By this time it had spread all the way down the continent.

'In the 1890s, the south was the only area where there was significant European involvement,' says Kock. 'Rinderpest wiped out all these people's cattle. It was life-threatening to the settlers, and that's what really made it a big drama – rinderpest now faced a battle with humans determined to eradicate it.' Only when it began to harm large numbers of European settlers in addition to pastoral herders did the disease face the measures that could contain it. The colonial authorities, having acknowledged what was causing the plague, followed the strict approaches that had exterminated rinderpest in Western Europe by the 1880s.

'To a certain degree, I think that European borders were established because of the awareness that rinderpest was flowing in from the east,' says Sunseri. 'And so they could begin to monitor borders, they could begin to monitor railway traffic, they can disinfect railway carriages, they can quarantine cattle in port cities. But in Africa you couldn't do that because there were no real border controls.'

In southern Africa, the authorities shot cattle that had been exposed to the virus, buried cattle that had died from the disease rather than leaving them to rot, and fenced in areas to stop cattle mixing. The decades of overhunting that meant the area was low in wildlife that could act as a reservoir for the disease probably helped. Along with early inoculations – a technique invented in 1897, based on blood serum or bile from infected animals and developed with the help of famous German bacteriologist Robert Koch, employed by the South African government – the fearsome anti-rinderpest regime managed to stamp out the disease in southern Africa by the turn of the century, a few years after it had arrived. But the process enraged people by confiscating their cattle, fencing livestock in with barbed wire, dynamiting mountain passes, and forcibly disinfecting train passengers (a measure that, with zero scientific basis, was applied more thoroughly to those with black skin). Uprisings and rebellions ensued; the human cost was high.[45]

By this time, even to the north the Great African Panzootic (the non-human animal equivalent of a pandemic) that began in the late 1880s had burned itself out. Most of the susceptible animals – the eland, bushbuck, kudu, buffalo, cattle, warthog, giraffe, wildebeest and more – had either died or survived and earned lifelong immunity. The South was pretty much devoid of cattle, and attempts to restock using livestock from Madagascar and East Africa brought in East Coast Fever and the tick that carries the protozoan parasites that cause it, introducing another epizootic of cattle disease around the time of the Anglo-Boer War of 1899–1902. In some areas, vegetation ran wild as the few remaining herbivores no longer kept it in check. The Serengeti began to experience intense annual wildfires after its wildebeest population stopped reducing the amount of dry grass.[46] And the tsetse fly, which likes to live in woods, could spread and its trypanasome parasites could bring more sleeping sickness to humans and cattle.[47] 'That virus certainly demonstrated that one single disease could lead to a complete tipping point and change the whole vegetation and therefore species composition, structure, everything – insects, you name it,' says Kock.

KILLING ZONE

When a disease feared to be rinderpest broke out in Kenya in 1910, the authorities in German East Africa, which lay to the south, slaughtered wildlife in a 50 kilometre (30 mile)-wide belt running for about 200 kilometres (125 miles) parallel to the border in an attempt to create a cordon sanitaire that would stop the disease in its tracks.[48] 'Hundreds of thousands of cartridges were issued to slaughter wildlife to prevent them from bringing in rinderpest,' says Sunseri. 'It was a very controversial policy. A lot of wildlife conservationists were really appalled at the slaughter.' But as the governor of German East Africa would point out, protecting the cattle of German East Africa also protected the wildlife. Reports came in of around 5,000 wildlife deaths and the killing lasted for a month, stopping only when vets declared the outbreak wasn't rinderpest after all. Following several more scares, rinderpest resurged in German East Africa in 1912.[49]

By 1918 in East Africa, where before the panzootic European settlers hadn't hunted or displaced much wildlife, numbers of all species had almost completely recovered.[50] Apart, that is, from the roan antelope, which has striking black and white patches on its face and muzzle that mask the position of its eyes, and the greater kudu, which almost died out completely – populations took some ten more years to recover. In Kenya, the eland was thought near-exterminated and was a rare sight fifteen years after the panzootic, but by 1928 was as plentiful as in 'the old days'.[51] Some have even suggested that the Great African Panzootic set the geographic distribution of most artiodactyl species in Africa today.[52]

Cattle populations too bounced back fast. Even though 95 per cent of cattle in East Africa were killed between about 1891 and 1893, in about a decade numbers began to recover.[53] But whereas southern Africa had eradicated the rinderpest virus, in the north it was outstaying its welcome. The virus had become enzootic – the non-human animal equivalent of endemic, meaning it was persisting in the region at a fairly low, consistent level. Rinderpest broke out periodically,[54] often in association with drought, when herds were weakened by the lack of water and poor food.[55] Chances are the disease repeatedly spilled over from cattle to wildlife, and vice versa, particularly for species such as wildebeest and buffalo that often share waterholes and pasture with cattle. Each year in the Serengeti-Mara region many wildebeest calves between seven and twelve months old died of rinderpest, also known as 'yearling disease'.[56] Younger calves received rinderpest antibodies in the first milk from their mothers, but this protection lasted only around five to six months,[57] or even less.[58] And the next generation up was immune, as only rinderpest survivors remained. This 'disease tax' on the young kept the wildebeest population much lower than it otherwise would have been, and changed the landscape by enhancing tree growth.[59]

ENTER ERADICATION

In Africa, rinderpest held out for decades longer than in Europe. 'The impact of the introduction of rinderpest to Africa was pretty dramatic, particularly in the early years,' says Kock. 'But even right to the end, it was costing a lot of money and causing a lot of aggro and loss of economic opportunity, et cetera.' To stem these economic losses, by the middle of the twentieth century, 22 countries were co-ordinating their vaccination of cattle in West, Central and East Africa in the hope of eradicating the disease,[60] vaccines having become more sophisticated.[61] But there was a snag – it's hard to vaccinate or quarantine wildlife, though vaccinating cattle did seem to protect wildlife too. By 1976 only two nations were reporting rinderpest. Then conflict stepped in again. Hidden pockets of rinderpest in conflict-ridden southern Sudan, where access for vaccination was difficult, and in the Niger let the virus rampage across Africa once more. Much of the vaccination work was undone. From 1979 to 1983, Africa saw an upsurge of rinderpest in which one million cattle died. Numbers of giant eland (*Tragelaphus derbianus*) in the Central African region crashed by 60–80 per cent, although they later recovered.[62]

FINAL HUNT

This wasn't to be the only time the fight against rinderpest would suffer a setback. In 1987 the Pan African Rinderpest Campaign (PARC), another cattle vaccination programme, kicked off. After years of effort, this had apparently beaten back rinderpest to a final hold-out in Sudan – until Kock discovered the disease inside a Kenyan kudu in 1994.[63] Analysing the genetic make-up of the virus in Kenyan wildlife showed it had come from the unvaccinated cattle of refugees who'd crossed Kenya from Somalia to reach northern Tanzania.[64] 'It was spreading in wildlife,' Kock says. 'The virus, after a hundred years or so, was pretty mild in [East African] cattle. Viruses often adapt and become less problematic; it's not in their interest to kill animals.' A virus that kills its host can no longer spread via coughs, sneezes, vomit, diarrhoea and so on. Because the virus wasn't causing cattle symptoms, it was hard to detect. But this version of rinderpest was still a big problem for wildlife. More than 29,000 Cape buffalo, 3,000 eland and 7,000 giraffes died.[65] 'We lost something like 60 per cent of the buffalo in the main population of Kenya,' Kock says. 'That's pretty significant and a concern to the conservation community.' Kock pushed the Kenyan government to continue to support eradication. 'If we hadn't picked it up, it probably would have spread again subtly until it got into a very naive, susceptible population,' he says. 'There would have been another big outbreak somewhere – Zambia, or who knows? That's the nature of the virus. Once it was in the slightly cryptic [hidden] form, it could move long distances.' Pakistan also suffered an outbreak of rinderpest in 1994, in yaks and cattle.[66]

By the end of the year, the Food and Agriculture Organization (FAO) of the UN had launched the Global Rinderpest Eradication Programme (GREP). Four years later, in 1998, the African Union hired Kock to lead the wildlife side of rinderpest eradication across the whole of Africa. Having hampered eradication by acting as a hidden reservoir for the virus, wildlife now redeemed itself by providing information about its whereabouts and history. The worry was that the virus still lurked even in areas thought to be rinderpest-free, so Kock led investigations in Kenya and roughly ten countries on its borders or further afield in West and Central Africa, including Burkina Faso, Central African Republic, Chad, DRC, Ethiopia, Senegal, Sudan, Tanzania and Uganda. If you found rinderpest antibodies in a cow's bloodstream, you couldn't be sure if the animal had been exposed to the disease or merely vaccinated. But antibodies in wildlife, none of which had been vaccinated, showed you not only that rinderpest had been in an area, but also roughly when. Antibodies against rinderpest last a lifetime, so by testing animals of different ages you could learn the date when a wave of the virus had passed through. If no animals younger than fifteen, say, had antibodies, you knew that the last outbreak was fifteen years ago.

'You could tie up the history of the elimination of the virus and confirm that yes, it looks like this country's clear,' says Kock. 'It was a lot of difficult work in very remote areas but we had sufficient resources to get that done. We gradually worked our way back into East Africa.' Kock, his colleagues and the people they'd trained in about ten countries were able to prove that the only place in the whole world that rinderpest remained was the Somali ecosystem,[67] including southern Somalia and north-eastern Kenya. The team also linked antibodies in wildlife to spillover from outbreaks in livestock. So unvaccinated cattle gave rinderpest to wildlife, where the disease could also circulate freely in unvaccinated animals, which, chances are, later returned it to unvaccinated cattle. The obvious solution was to ramp up cattle vaccination in the Somalia ecosystem, quite a feat for somewhere this remote. But the programme stepped up. About four years after Kock started his role at the African Union, 'everybody was saying, "We've done it, we've achieved the final goal",' he says. 'People were getting pretty excited and relieved that no more money had to be spent.' The last phase of the project had cost something like 60 million dollars and it looked like the world was finally rinderpest-free. Kock's discovery had helped stop rinderpest re-emerging and spreading back across the whole continent.

WILDEBEEST WOES

Then Kock took some students to Kenya's Meru National Park, which is pretty much in the middle of the country, as part of the Envirovet programme to expose participants from the US and Europe to African wildlife health. 'By this stage I had a bit of a reputation, not of being a troublemaker, but of bringing out or exposing issues,' he says. Sure enough, issues ensued. At Meru, Kock saw some youngish buffalo that didn't look quite right. 'When you work a lot with species in a natural state, you get sensitive to these things,' he says. 'It's a bit like you get sensitive to your dog being off-colour, not eating so well. I got to know buffalo so well that if a buffalo wasn't in top form, I knew it.' These buffalo weren't exhibiting the classic signs of rinderpest – fluid seeping from their noses, depression, dehydration, diarrhoea or being dead. But as the sick animals were too young to have survived the last rinderpest outbreak, Kock decided to take a look. He immobilised and tested two or three buffalo that were about eighteen months old, using samples of a new type of test for rinderpest that he had been given to try out in the field and happened to have in his back pocket. A little like a lateral flow test for Covid-19, these kits incorporated monoclonal antibodies and detected proteins on the surface of the virus. Without further ado, coloured lines appeared – these buffalo had rinderpest. Kock's gut feeling had been right.

By this time, Kock says, sceptics in the funding agencies were advocating giving up on eradication, but fortunately additional money did come through and more cattle in Kenya and Somalia received the rinderpest vaccination. By June 2011 the job was done. The UN FAO declared rinderpest eradicated from the planet, the first non-human animal virus ever to be eliminated, and the second virus of any kind after the human virus smallpox, snuffed out in 1980. Kock had diagnosed the very last case of rinderpest in 2003. 'That I have some pride in,' he says. 'I remember that case extremely well because it was a young buffalo, and after waking it from the immobilising drug, it chased me up a tree. I was stuck up this tree shouting to my colleagues, "For goodness sake, chase this buffalo away, it's too embarrassing".'

These days, rather than being greeted as a potential troublemaker, Kock receives awards, and a bronze statue of a buffalo at Meru National Park marks the eradication of rinderpest. A few labs held on to samples in case the virus somehow re-emerged in the wild and scientists needed to develop a new vaccine. (Should rinderpest re-appear, the first confirmed case will be treated as a global emergency and kick off action plans to re-eradicate it.)[68] In 2019, researchers at the Pirbright Institute in the UK noted the exact proteins along the length of the virus then destroyed their remaining physical copies.[69] Should they need to, they'll be able to recreate rinderpest from their digital copy. The team hopes that the remaining labs with stocks of rinderpest will do the same and then rinderpest will be truly extinct.

ANOTHER PETIT PROBLÈME

Rinderpest was no more. But unintended consequences can be a menace. Another virus in the morbillivirus genus causes Peste des Petits Ruminants (PPR), which is French for 'small ruminant plague'. This disease had been enzootic in sheep and goats across the Sahel in northern Africa, from Senegal in the west to western Ethiopia, since the

1940s. In a curious twist of fate, Kock became involved with this second morbillivirus, which some people believe diverged from the rinderpest virus around 6,000 years ago.[70] By 2015 he was working at the UK's Royal Veterinary College and was called on to investigate a mass die-off of Saiga antelope (*Saiga tatarica*) in Central Kazakhstan, where 200,000 of the nation's roughly 600,000 Saiga had succumbed.[71] Today the near-threatened Saiga live only in Kazakhstan, Russia, Uzbekistan, Turkmenistan and Mongolia,[72] a fraction of their former range across the Eurasian steppe; 98 per cent live in Kazakhstan.[73]

These sandy-coloured animals are unusual-looking beasts, with a long, domed nose with a blunt end. If a Saiga was in a children's fable, its story would involve wedging its nose into a crevice to find a tasty blade of grass, becoming stuck and being levered out by a rhino that accidentally bent the Saiga's head and snapped off the end of its nose during the rescue.

This mass die-off of Saiga turned out to be due to *Pasteurella multocida*. Although these bacteria are naturally present in the animals, high temperatures, humidity and stress from giving birth caused vast numbers of the animals to die from haemorrhagic septicaemia in three weeks.[74]

While investigating the die-off, Kock and his team realised that PPR had reached Kazakhstan the year before and was circulating in sheep and goats. Kock and colleagues from Kazakhstan and the UK published about the presence of PPR and the government managed to quell the outbreak.[75] 'I'm very pleased about that,' he says. 'If the PPR virus had got into Saiga in Kazakhstan, you see how easy it could have been to have virtual extinction of the whole species. It was another demonstration that as a scientist, you've got to make these things public. You have to stick your neck out; occasionally you'll lose your head but it's really important.' Kock believes that if outbreaks of PPR and septicaemia had coincided, it is likely the species would be extinct today.

In Mongolia, Saiga weren't so lucky. Although Kock had warned that PPR might arrive, nobody took enough action. In late 2016, PPR devastated a population that had never experienced it before, killing as many as 80 per cent of the nation's 26,000 animals.[76] 'This is a goat-adapted virus, really,' says Kock. 'It was certainly impacting wild goats up in high mountain areas occasionally but we'd never seen it in antelope.' The Mongolian subspecies came close to extinction.[77] 'Hopefully the ones that survived are immune,' says Kock. 'But all biodiversity now is heavily threatened and its populations are not resilient. The final cause of extinction, I'm afraid, will be disease-related in many species because it has that capacity to affect a large number of animals quickly. I think it's a high priority that we are on top of it and produce a community to at least help to slow the process down.'

In 2021, following the *Pasteurella*-PPR double whammy, the world's total Saiga population, already under threat from habitat loss, the recent increase in fires on the steppes, and illegal hunting for their meat and elaborate twisted horns, was estimated at 124,000 mature animals.[78] The morbillivirus family had returned to its roots in the Central Eurasian steppe with a vengeance. Since then the population has recovered well and is back to 950,000.[79] Today, PPR is present in a broad band that stretches from West Africa across Arabia, on through Central Asia and the Indian subcontinent, and as far east as the Chinese coast. Ironically, what caused the disease to spread out of Africa was perhaps our success in vanquishing rinderpest. Goats and sheep caught rinderpest but tended not to suffer severe symptoms or die. The antibodies they gained probably protected them from PPR, a disease much more likely to kill them. When rinderpest became less common, these small animals no longer benefited from cross-protection and PPR had a chance to take off.

In 2014, the UN FAO and WOAH targeted the disease for global control and possible eradication because of its impacts on small livestock.[80] Scientists are working on improved vaccines for this livestock and wildlife plague; the target date for its eradication from the roughly 75 countries where it's active is 2030.[81] The first five years of the programme (up to 2021) were slated to cost nearly $1 billion, a figure that shrinks when you consider that each year the estimated global impact of PPR is between $1.4 billion and $2.1 billion.[82] Will we one day be able to erect a bronze statue of a Saiga antelope somewhere in the Eurasian steppe to mark the elimination of PPR? Or will we need that statue as a memorial for a species that once roamed the mountains in its hundreds of thousands before it succumbed to a disease helped on its way by humans starting to farm, getting closer to livestock, taking colonial conflict – and rinderpest – to Africa then eradicating it and accidentally helping PPR boom? The choice is up to us.

5

Tasmanian Troubles: Sheep Farming, Extinct Tigers and 'Distemper'

'Tyger Tyger, burning bright,
In the forests of the night;
What immortal hand or eye,
Could frame thy fearful symmetry?'
William Blake, 'The Tyger', 1794

In the flickering black-and-white footage, a dog-like animal paces around an enclosure, stands and scratches his shoulder with a hind leg, raises his head to sniff the air, and lies down on his side, eyes half-closed in the sunshine.[1] The dark, vertical stripes that run down his back and rump, his kangaroo-like stiff tail, and the giant yawn he gives, face-on to the camera, revealing vast numbers of wolf-like teeth in a long narrow jaw, show this is no ordinary dog. He's not a dog at all but a thylacine, a marsupial with 4 more teeth than a dog, 46 in total.[2] Also known as a Tasmanian tiger, the thylacine's expression as he looks into the camera is haunting. I want to reach into the picture and pat his warm fur, feel the muscles beneath his skin, experience the brush of his whiskers, hear the click of claws on the hard floor. Though that might not be wise – at some point during the filming he applied his impressive teeth to biologist David Fleay's buttock.[3] Besides, I can't. No one will ever experience one of these animals alive again.

This Tasmanian tiger filmed in a zoo in Tasmanian capital Hobart in 1933 or thereabouts was the second-last ever known of his species. The last known thylacine died on 7 September 1936 – this extinction is almost within living memory. And the thylacine's official extinction date,

50 years later in 1986 is within the memory of many more. The tiger fell victim to greed, ignorance, fear, scapegoating and quite possibly disease as well, a classic example of a longer-lived, larger species being more vulnerable to extinction. But why did an animal we valued enough to display as a visitor attraction in zoos go extinct?

It's hard to connect with just a few seconds of jerky, flickering footage of an animal that's so obviously restless. Only when I read about the thylacine and learn how this animal lived in the wild – its sleeping sites in caves and hollow logs and shrubs and rocky outcrops, the four young in each litter that inhabited the female's marsupial pouch, the way it hung around campfires tempted by the smell of cooking bacon – only then do I get a full picture.[4] And that's when I feel truly sad. Especially when I find accounts of what happened to that last individual in the zoo in Hobart. These animals didn't just flicker on screens, they ate and ran, their rigid tails perhaps providing balance while they turned. They cared for their young, were curious and played, fled from humans when they got too close or stood their ground and growled, their tails standing straight out behind them. Thylacines didn't only growl;[5] they had a husky coughing bark, yapped a double yap like a terrier when hunting, and made several other noises besides.[6]

In the early twentieth century, however, there was 'a swell of beliefs' that the animal was mute, according to Tasmanian tiger expert Bob Paddle from the Australian Catholic University. 'This came from the bushmen and the farmers who said, "There are a hell of a lot of them out there, just scientists can't find them because they can't hear them in the bush as they don't have a voice,"' Paddle explains. Through the 1950s and even into the 1970s, scientists wrote that the animals had no vocal ability.

Paddle, a zoologist who became a comparative psychologist investigating human and other animal learning, entered the world of thylacine research because in 1988 the Australian Psychological Society held its annual conference not at a university, but at a casino. Unable to browse journals in a university library during his spare time, Paddle headed to Hobart's Tasmanian museum and state library to 'find out everything I can about the extinct Tasmanian tiger as the largest marsupial carnivore that just might be relevant to my desire to run small marsupial carnivores in learning mazes'. Paddle had a ready supply of such animals in the form of fat-tailed dunnart (*Sminthopsis crassicaudata)* and *Antechinus* species at the family farm. But what he read about the Tasmanian tiger extended far beyond its ability to learn – Paddle not only discovered more about how disease helped push this animal to extinction, but also something that shattered one of his core beliefs.

Before this trip, Paddle had 'a great faith and belief in scientific objectivity, the Rolling Stones and the Geelong Football Club, to name my favourite three'. The week of reading in Hobart showed him that his belief in scientific objectivity was false. 'I discovered that scientific perspectives have changed about the animal since its extinction,' he says. 'That scientists are just as inclined to change their mind about the animal through the status of people who have written about them.' The idea that thylacines drank and sucked blood, for example, originated in a publication by an academic from the UK who was sent to Australia and became lost in the bush. The miner and trapper who found him told him 'all these bullshit stories about the thylacine as a vampire'. Tasmanians completely ignored the academic's book containing these details, but a professor rediscovered it in the late 1940s. 'We then have for the second time the idea by a serious scientist that the animal was a blood-drinking vampire,' Paddle says. 'And it just went from there. Once the professor said that, everybody follows. It was an amazing moment in time for me to discover that science is just as much a cultural phenomenon as any other artistic or creative endeavour that humans undertake.' Paddle was so fascinated that he wrote a PhD on the history and changing scientific pereception of the thylacine; he has yet to run small marsupial carnivores through mazes and probably never will.

The disease was the last in a long line of problems facing the thylacine. The animals had thrived in mainland Australia until two or three thousand years ago, when climate drying reduced the amount of vegetation for their herbivore prey to feed on and competition with the dingoes that arrived a few thousand years earlier forced them out. We know this, among other evidence, from the rock art at Ubirr in Kakadu National Park in the Northern Territory,[7] and 'Old Hairy', a thylacine from 4,000 years ago found mummified at the bottom of a cave shaft in 1966 on the Nullarbor Plain in southern Australia.[8] After their relatives vanished from the mainland and New Guinea, some 5,000 or so thylacines lived in relative peace in Tasmania, a wetter and dingo-free island the size of Ireland or West Virginia off the mainland's south-west coast. In relative peace, that is, until 1803, when British settlers and convicts arrived at Risdon Cove in the far south of the island.

FEARFUL SYMMETRY

While aboriginal people had called the thlyacine the lagunta, corinna, laoonana, ka-nunnah, and more, the settlers dubbed it the Tasmanian hyaena, dog-headed opossum, zebra opossum, zebra wolf, marsupial wolf, tiger wolf, striped wolf, Tasmanian wolf, Tasmanian zebra, dog-faced dasyurus, Tasmanian dingo, Tasmanian panther, Van Diemen's land tiger, greyhound tiger, bulldog tiger, or hyaena tiger. In 1824 scientists finally settled on the name *Thylacinus cynocephalus*, the first part of the name coming from the Greek for pouch, and the second part meaning dog-headed.

Having given the thylacine many a vicious name, the settlers blamed the animal for killing their sheep. Was it a case of give a dog a bad name and then hang him? Literally in the case of the famous photo 'Mr Weaver Bags a Tiger', which shows a hunter sitting next to a thylacine carcass that's been strung up from the ceiling. There was very little evidence of either blood-sucking or sheep attacks by the thylacine. Thylacines generally ate small wallabies like the pademelon, and sometimes kangaroos and the Tasmanian emu despite their relatively large size. The true sheep-killers were probably packs of domesticated dogs that had gone feral. (Feral dogs passed the buck in Europe too, where the wolf was often blamed for the dogs' sheep misdemeanours.)

Nevertheless, in 1830 the Van Diemen's Land Company that ran sheep farms in the island's North-west put a bounty on the head of this 'noxious animal'. Even though in the mid-1800s some people grew afraid that the thylacine might go extinct, the Tasmanian parliament followed suit in 1888, after lobbying by farmers, and set up an official government bounty scheme. Afterwards, a 'FARMER' wrote to a newspaper that 'The farmers could more reasonably ask for help to destroy the parrots, which are on the whole much more destructive'.

Between 1888 and 1909, some 2,209 of the island's estimated 5,000 thylacines were killed to claim the government bounty, worth several weeks' wages.[9, 10] Although this figure may be an overestimate – trappers would often hawk the head of a thylacine from private landowner to private landowner, claiming a bounty from each until their trophy smelled too bad.[11] Some reckon that roughly half of carcasses met this end.[12] Tiger habitat too was suffering, as farmers took more land for their sheep or businesses felled ancient forests for their timber. Not only was the tiger being deliberately exterminated, but its homelands were being destroyed too.

BOUNTEOUS EVIDENCE

Late in the nineteenth century, the beleaguered tiger faced another threat. As Paddle learned, a couple of scientists in the island's North-east noticed that it was harder to source specimens for Launceston's City Park Zoo and the Queen Victoria Museum.[13] For the eight years between July 1885 and June 1893, William McGowan, the curator of Launceston City Park Zoo, obtained 35 thylacines from northern Tasmania for display as an attraction, even though in the countryside the animals were hunted down as vermin. But for the next eight years, from July 1893 to June 1901, McGowan could only source 16 thylacines; for almost five of these years, from 21 July 1893 to 12 June 1898, he was unable to buy any live-caught thylacines from northern Tasmania at all.[14] Something besides bounty-hunters was harming Tasmanian tigers.

Police station records for north-eastern Tasmania back up McGowan's findings at the zoo. In the five years from 1891 to 1895, constables at St Helens police station paid out on eight adult thylacines, an average of 1.6 a year. In the first nine months of 1896, no one brought any thylacines to the police station at all. Why had Tasmanian tigers suddenly become so scarce? It's possible that they were hit by a disease. In October 1896 something strange happened. For the next six months, no less than five dead adults were taken to the station. Perhaps the disease had made it easier to catch or kill thylacines, or trappers had simply found dead bodies in the bush. These were the very last thylacines known to be presented for bounty at St Helens, even though the scheme ran for another twelve years.[15]

It Seems that the disease started in the North-east. Then records from other parts of the country began to show anomalies too. Over in the north-western tip of the island, hunting records kept by the Van Diemen's Land Company showed a similar pattern – a drop in the number of thylacines captured followed by a sudden increase in 1900 and 1901. It looks like the disease had travelled a couple of hundred kilometres from the North-east to the North-west and was temporarily making it easier to bag thylacines.[16] But again, numbers plummeted. Just seven thylacines were captured in the decade to 1913, and in the ten years to 1923 none were taken at all. The next two years saw four thylacines killed or captured, with the very last known thylacine on the estate trapped in October 1925.

Recalling this period in 1972, Harry Wainwright, one of the last 'tiger men' employed to kill thylacines at the Company's HQ in Woolnorth, said that the disease appeared in north-western Tasmania around 1899. To start with it affected thylacines living between the Arthur River and the Marrawah-Preminghana area; thylacines living to the north of Preminghana, a region that included most of the Woolnorth estate, initially remained healthy.[17] The disease appears to have spread across the whole of Tasmania. The state-wide government bounty records, which you'd expect to exhibit a decrease in thylacine captures over time, instead show an increase in thylacine killings and captures from 1899 to 1902.[18] The highest-ever government payout was in 1901, for 168 thylacines. These records indicate that juveniles suffered from the disease more severely. The scheme recorded bounties for 159 juveniles between 1888 and 1909, a period of 22 years. But just over two-thirds of these juveniles were trapped in the ten years that included the initial spread of the disease, compared to just over half of adults.[19]

UNDER THE SKIN

So what was this disease that was picking off thylacines or at the very least making them easier to catch? Launceston City Park Zoo superintendent McGowan noted in a letter that 'complete skeletons' of thylacines with 'damaged or rotten skins' were available, but fine, entire specimens were now almost impossible to find.[20] The thylacines were suffering areas of significant hair loss, skin lesions that sometimes bled, and mange, although there was little evidence of the skin hardening and folding you would expect with sarcoptic mange. Tasmanian devils were suffering too. Farmer Lewis Stevenson recalled in 1972 that 'devils ... got the mange like the tigers ... their hair fell out and left the black skins bare in the bad ones'. One Mr Fred Burbury of Parattah, a town in the east of Tasmania roughly 90 kilometres north of Hobart, told biologist Eric Guiler that a disease 'like distemper' swept through the Dasyures marsupial family about 1910, after which Tasmanian devils, native cats (also known as eastern quolls) and tiger cats (spotted-tailed quolls) as well as thylacines became scarce.[21] Paddle also suspects that the disease reached the mainland where it helped, along with the introduced red fox and feral cat, exterminate the native cat.[22]

Like Stevenson and Burbury, people tended to give the affliction one of two names – either 'mange', based on the skin symptoms, or 'distemper', due to the weakness, coughing, diarrhoea and weight loss produced by fever. At Hobart's Beaumaris Zoo, where the last film footage was shot, curator Arthur Reid, who was appointed in 1922, considered the thylacine disease was 'perhaps distemper that they had caught from the trappers' dogs', but his daughter Alison Reid was less convinced: 'there wasn't any sign of discharge from the eyes and nose like you see with a [distempered] dog ... they just died,' she recalled in 1980.[23] Besides, a letter to the *Tasmanian Mail* at the time had suggested that thylacines became more common only after wild dogs in the bush had died of distemper. 'The most convincing explanation that I've had from an academic veterinarian is that from its appearance, it was probably a lupus-like disease with illness, diarrhoea, bleeding, lesions on the body and the skin and the feet,' Paddle says. Such an auto-immune condition may have been linked to the stress for the thylacine of losing its habitat and being hunted.[24]

Unfortunately, museum curators later destroyed nearly all thylacine specimens with damaged skin, believing them to be infested with moths or beetles, so we have no chance of identifying any pathogen using modern techniques. Was it a distemper-like virus first transmitted to the thylacines by dogs shipped in by Europeans, or a mange caused by mites from those same dogs, or a toxoplasmosis induced by a parasite, or a lupus-like disease or a viral pneumonia? We'll probably never know. When a species goes extinct, it's often had such a tough time that it's hard to tell what exactly pushed it over the edge. Perhaps an infectious disease killed much of the population and then one or more other stressors killed off the survivors. Or perhaps other stressors caused a decline in numbers and a disease proved the last straw for the remnants of the population.[25]

Some biologists think that the evidence for a disease of thylacines is slim and that persecution, habitat loss and the decline of the thylacine's native prey due to hunting for fur and food, and competition with sheep, were enough to wipe out the animal.[26] Certainly, we're increasingly aware of the impacts of diseases on endangered species, and if there was an emerging pathogen, a depleted population that lacked genetic diversity and was stressed by being hunted and by losing much of its habitat would not have been best placed to withstand an onslaught of infection. So even if humans didn't import a disease or cause a disease by altering the landscape, we certainly didn't help the thylacine resist pathogens.

As the Reids' comments show, thylacines in captivity suffered too, although it took a while to diagnose the problem. 'In the early 1900s, there was suddenly a wealth of concern that the thylacines are turning on one another and killing one another,' says Paddle, who discovered many records of the thylacine and its behaviour from Melbourne Zoo while working for his PhD. Zookeepers were finding animals dead and with bloody lesions all over their body. 'For a while they thought, OK, we've got some strange social behaviour here because up till then there was no problem with thylacines – you got new thylacines, you just chucked them in the cage with the old specimen.' Thylacines didn't generally attack each other – we only have records of two incidents in zoos, despite 305 thylacines having been kept captive over the years. And one of these incidents was when a female attacked, but didn't kill, an adolescent male who made unwelcome advances.[27] 'But it suddenly appears at Melbourne Zoo that thylacines are turning up dead and they've got bites all over their body,' Paddle says. 'It took them about a year or so to discover that they're actually not fighting one another, that this is a disease that produces bloody lesions on the body and feet.' Then, as Paddle's trawl through the records showed, people realised the same thing was happening at other zoos too.

Across all age groups, tigers in zoos began to live less than half as long as their captive predecessors. Between 1896 and 1936, adults in zoos lived lives 43 per cent shorter than captives taken before 1896, while juvenile lifespan reduced by 69 per cent.[28] After 1896, animals arriving as juveniles had a captive longevity on average less than that of adults, the opposite to the situation before the disease emerged and to what you would expect. At Melbourne Zoo, where a pair of Tasmanian devils were prime suspects for importing the disease in June 1900, thylacines went down like nine-pins. In August 1900 the zoo held a pair of adults and their four cubs – the only cubs bred in captivity. Keepers added an extra thirteen animals over the next twenty months. But by December 1902, all but one of these nineteen animals had died. When one of the original thylacine cubs died in August 1900, it was recorded as 'eaten by others'. When one of the parents died bloody and damaged in May 1901, it was 'apparently killed by its mate'.[29] Soon keepers realised the tigers died with bloody, damaged skins even when their carers hadn't seen them fighting. The bleeding was most likely due to an attack by disease rather than a physical assault. Entries in the zoo's death book include 'Died from Cold', indicating an infection, and 'sore feet diarrhoea & weakness'. After a while, so many thylacines were dying that some didn't even make it into the death book. 'The specimens which died from disease very significantly weren't worthy of saving in museums,' Paddle says, explaining that the skins of animals that had enormous quantities of scabs or bloody lesions were of no interest. 'Sometimes zoos didn't even bother sending them off, they just threw them out themselves.' Luckily, the National Museum of Victoria did receive some corpses and was more fastidious with its records. Dying Tasmanian devils at Melbourne Zoo were also first thought to have killed each other; later their deaths were ascribed to 'mange'.

Over the years the severity of the disease waxed and waned. In New York Zoo, a newly arrived male thylacine recovered from his first bout of disease in December 1902. He fell ill again in February 1903 and once more in mid-May of that year but then survived disease-free for more than five years.[30] This animal seems to have been ahead of the trend. Until about 1903, both in the wild and in captivity, the disease appeared particularly virulent (damage-causing). Then, during the 1910s, captive animals more often acted like that animal in New York, showing mild symptoms for a while then recovering. In the late-1920s, in the records of Hobart and Melbourne zoos, the disease appeared to become more virulent once again. Between December 1927 and April 1928, Hobart Zoo was able to add nine more thylacines to the two it already held, probably because they were ill and easy to trap. One of the first to arrive, in January 1928, was a diseased male that was kept in quarantine until it died a few days later of 'a virus'. Despite the quarantine, the disease spread to the zoo's main thylacine cage and by the end of the year, only two of the zoo's eleven Tasmanian tigers had survived. Both continued, at times, to show signs of the marsupi-carnivore disease, and were visited by the zoo's veterinarian; one lasted a further ten months and died on 1st November 1929, the other in early January 1930.[31] In an interview in 1996, Alison Reid 'was persistent in her opinion that all thylacines present in 1928 eventually died of the epidemic disease'.[32] Some of the carcasses had skin in fair condition, others bad.

PROTECTION AT (THE) LAST

By this stage, at long last, more people were calling for protection for the thylacine. In 1914, Professor Theodore Thomson Flynn of the University of Tasmania, also known as Errol Flynn's dad, called for an island sanctuary for the animals.[33] As well as his concern for these large marsupials, Flynn senior was 'full of pranks' and produced a small marsupial from his pocket at a staff dance, to much consternation.[34]

Even though bounty records indicated tigers were dwindling, rural interests lobbied hard and the tiger didn't receive any protection until

1930. At that point the government banned hunting Tasmanian tigers only during the month of December, which was thought, wrongly, to be when the animals bred. Full protection didn't arrive until mid-1936. 'The species was totally protected for the last fifty-nine days of its existence,' wrote Paddle.[35] The anniversary of the death of the last known thylacine – 7 September – is now Australia's National Threatened Species Day. This 'endling' – the last surviving individual of a species – was female. However, thanks to false claims made by one Frank Darby that he had cared for this individual and that it was male, she became known as Benjamin.[36] Her end was unpleasant. Artist Vita Brown remembers seeing the animal and thinking it was very old, emaciated, lonely and wearing 'a look of despair' as it paced up and down. Brown was so upset that she left the zoo and never went back.[37] The animal lasted just four months in the harsh zoo conditions. Takings had fallen during the Great Depression and staff were laid off. When zoo curator Arthur Reid died, his daughter Alison Reid wasn't allowed to take over, it apparently being impossible for a female to do the job.[38] She effectively did the role unpaid until the powers that be prevented even that – they took away her key and she couldn't care for the animals after hours. Paddle recalls how Alison later spoke with genuine distress in her voice about the last weeks of her life at the zoo in 1936: 'Powerless, keyless and shortly to be dismissed from the zoo and turned out of her home, she listened at night to the distress calls of the zoo's remaining carnivores: the last thylacine, a Bengal tiger and a pair of lions, all too frequently locked outside in the open to face the cold, rain and snow of the Hobart winter.'[39] It's upsetting to hear about such mistreatment of individual animals, but we are unintentionally treating millions of wild animals this way by running the world without due care for nature. This 'last thylacine' was in fact the second-last thylacine, the male who appears in the film footage and many photos. When he died in May 1936, he was replaced by an elderly female who didn't benefit at any point from Reid's care; many of the remaining people working as keepers hadn't chosen the role – they were sustenance workers, members of the unemployed forced to toil at the zoo to get their benefits. This last-known thylacine's skin and skeleton were claimed by the education department at the Tasmanian Museum.[40]

In the wild, it's hard to know exactly when the Tasmanian tiger went extinct. By the late 1970s, we had camera traps but they didn't snap any thylacines. In 2023, a scientific paper estimated that the thylacine limped on until the 1980s.[41] But in 1980, a report that included the use of camera traps supported the 1979 classification of the thylacine as 'probably extinct',[42] and by 1982 the IUCN had classified the thylacine as extinct.[43] Thanks, in part, to guilt about the tiger's fate, some 50 per cent of Tasmania's land is now protected.[44] Some people believe that tigers remain to this day, hiding in some of Tasmania's deepest, darkest forests in the south-western wilderness. Hiding from humans, their farming, their persecution and the diseases they bring. From the big, bad wolf that we've become. Paddle is less convinced – in the 1920s, he says, thylacines were hit by cars and trucks and the drivers tied them to the front of their vehicle to show what they'd killed. But nobody hits the animal on the roads these days. 'There's a chance I'm wrong,' he says, about the extinction. 'I would love to be wrong. It's important that all scientists keep an open mind about sightings and that people record this information because there are a number of species which have come back from the dead. There's a faint chance they're still around.'

I'm in no position to cast doubt on people's sightings. Some 25 years ago I spent New Year with a group of friends in Scotland, pretty much as far north as you can go. A few miles down the road in Altnaharra, at about the same time, the temperature reached a record-breaking -27.2°c.[45] On the drive up we'd joked that we hoped the cottage we'd rented had heating. This didn't seem nearly so funny when we arrived and found it didn't. Ice formed on the inside of the windows and my shampoo turned too viscous to leave the bottle. Not that it was warm enough to wash. Or sleep well – you can't wear clothes on your nose. On one of the crisp, clear, short winter days, not long before the sun set, we went for a walk in a wood. There, trotting purposefully about 50 yards ahead of us down the snowy path in a gap between the lines of conifers, was a big, shaggy dog, its fur coloured like that of a dark Alsatian. It was totally alone, no sign of an owner. 'Is that a wolf?' I and my friend Su turned and said to each other. 'I didn't think they lived here.' The last wolf in Britain is thought to have been shot in Scotland in 1680.[46] And I'm still not sure, to this day, what animal I saw. Part of me still believes it was a wolf.

Tasmanians too are seeing animals that may not be there. Thousands of thylacine sightings have been reported across Tasmania, mainland Australia and New Guinea.[47] 'Either many people are wrong almost all of the time, or the thylacine is alive and well but in sufficiently low numbers, and sufficiently remote locations, to have avoided detection (a possibility, or even a probability, that might continue indefinitely),' writes David Owen, author of Thylacine: The Tragic Tale of the Tasmanian Tiger.[48] Certainly, when red foxes arrived on Tasmania in 2001, having hitched a lift from Melbourne on the car ferry or been introduced maliciously, the task force responsible for their destruction had a lot of trouble finding them. As part of that campaign, the team put out trail cameras and analysed the DNA of carnivore scats (droppings). There was not a single sign of the thylacine.[49]

Chances are that the thylacine is lost. In Tasmania, the main place it's prominent today is on the state coat of arms, where despite its government-assisted extinction based on fears that it was killing sheep, a pair of thylacines stand rampant supporting a shield display-ing a picture of a sheep. Even without a disease, habitat destruction and deliberate killing would have put the thylacine at risk. But the disease reduced the options. Without it, the tiger's journey to extinc-tion would have been slower. Attitudes may have had time to change and a captive breeding programme using animals from zoos could have maintained numbers. The only zoo that managed to breed captive thylacines was in Melbourne, where the team kept a young female in a cage alone for a year before introducing an adolescent male that came from a completely different area to her and so was definitely unrelated.[50] In the only other case of two unrelated, unmated ado-lescents being introduced, at Mary Roberts's Hobart Zoo, the young male was replaced after only two months by an adult male who'd already mated, Roberts having decided that this older male was more desirable. 'Every other place where biological garden curators tried to establish breeding, they said, "Oh, the male's the most important",' Paddle says. These curators added a female to a male's cage, a tactic that didn't ever work. In the wild, it seems that females established a territory and chose an adolescent male from the selection that turned up – records from indigenous people in the 1830s and 1840s back this up.[51] 'Let's face it, the knowledge was there to save the spe-cies,' Paddle says. 'The knowledge of how to breed them was present in Melbourne Zoo in 1899. It just wasn't put into practice.' People assumed that marsupials were inferior to placental mammals and were bound to die out anyway. This 'placental chauvinism' combined with the 'patriarchal assumption that if we're going to breed them, we need to get a good male to which we add females' proved fatal. By 1908, the successful breeding of thylacines at Melbourne Zoo completely disappears from the literature, Paddle says, even though before this it had been mentioned in books, magazines, newspapers and the zoo's annual reports. Too late and apparently full of unbounded optimism, in 1966 the government bought Maria Island off Tasmania's east coast as a sanctuary for captured thylacines.[52]

TRULY LOST?

If one project has its way, the island could still become a haven for a modern version of the thylacine. Researchers at the University of Melbourne, together with Colossal Biosciences, plan to de-extinct the thylacine by placing DNA from museum specimens inside the eggs of its closest living relative, the Tasmanian devil.[53] It will cost hundreds of thousands, if not millions of dollars to partially recreate an animal people earned five shillings from killing. Nick Mooney of Tasmania's Parks and Wildlife Service wrote in a letter to Tasmanian newspaper *The Mercury* in June 2002: 'A snappy, technological quick-fix such as cloning extinct animals tempts many from the slog of fundamental environmental conservation'.[54] Better our sheep farming, our misplaced scapegoating, our weapons technology and our introduction of a disease or stressful circumstances that helped a disease thrive hadn't destroyed the thylacine in the first place. At the time, some simple legislation would have been able to save the animal. And today we could channel the money that it would need to de-extinct the thylacine into saving Tasmanian species that are still with us, such as the Tasmanian azure kingfisher (*Ceyx azureus diemenensis*)[55] or the eastern-bandicoot (*Perameles gunnii gunnii*), a long-nosed, rabbit-sized marsupial that disappeared from the Australian mainland thanks to predation by introduced red foxes and loss of its habitat to sheep farming.[56]

In a quicker fix still, in 2021 an expert team digitised and colourised the 1933 footage of the thylacine.[57] He seems more real, somehow, in colour. But he's just a collection of pixels, not cells. That's nowhere near as good as the real thing.

6

Black Feet and Black Death: Long-distance Trade, Plague, Prairie Dog Days and Ferret Futures

'I am hurt.
A plague o' both your houses! I am sped.
Is he gone, and hath nothing?'
William Shakespeare, *Romeo and Juliet*, 1597

In a mid-nineteenth-century oil painting by naturalist John James Audubon, a long-bodied mammal peers down from a grassy mound into a nest of speckled grey eggs. In the instant it's caught on canvas, this ferret's yellowish beige fur tones with the landscape and its dark Zorro-like bandit mask shades out its eyes as it looms over the nest, one front foot forward, one spread wide, poised ready to snatch an egg with its black paws.[1] This picture is not all that it seems. Although he collected 23 species of mammal as he travelled America in 1843, Audubon didn't find a black-footed ferret (*Mustela nigripes*). Knowing what we know today, that's no surprise – the species spends 90 per cent of its time underground.[2] Instead, Audubon based his painting on a pelt from Wyoming sent to him by a fur trader.[3]

Needing to portray a living, breathing, 3-D animal with only a skin to base it on, Audubon looked to the behaviour of one of the two other ferret species, the Steppe or Siberian polecat from eastern Europe and central and eastern Asia, and the European polecat from western Eurasia and North Africa, which humans domesticated as the ferret. What he, nor anybody else at the time, didn't know was that black-footed ferrets rarely eat eggs. Audubon got away with this blunder but his portrait still got him into trouble – some people thought he'd made this elusive animal up entirely in order to sell books.[4] Of course, the tribes of the Great Plains, an area of flatlands to the east of the Rocky Mountains spanning roughly the middle of the continent, had been aware of the black-footed ferret for thousands of years, using its skin for medicine pouches, headdresses and sacred objects.[5] But it seems Audubon wasn't able to prove that he hadn't invented the species – perhaps the pelt had gone missing. It would take another few decades before an official publication confirmed the ferret's existence.[6] On this front, Audubon was in the clear, although today he is a controversial figure for numerous other offences, including slavery, racism and scientific fraud.

A century later, this member of the mustelid family, which includes stoats, weasels, otters, martens, badgers and wolverines (the long-clawed version that dwells in cold climates, not the ruggedly charismatic retractable-clawed version that dwells in Marvel universes) had become even more elusive. As well as spending a lot of time underground, the ferret is now one of the most endangered, and rarest, mammals in North America. At one stage the species almost disappeared, victim of a number of woes including habitat loss to farming and no fewer than two infectious diseases.

At one time, the black-footed ferret could be found – if you timed it right – throughout the Great Plains and the valleys between the Rockies, in an area that stretched from Canada to Mexico. When Audubon visited, it's likely between half a million and a million ferrets lived on millions of acres of grassland.[7] At night, the ferrets would have scampered slinky-like across the grasslands, unseen by Audubon as they darted from burrow to burrow in low leaps and springy bounds. Their long thin bodies and short black legs are perfect for living and travelling underground. 'It's a very harsh environment, the short grass prairie – high winds, rain, snow, heat,' says Kimberly Fraser, who is outreach specialist at the National Black-footed Ferret Conservation Center in Colorado, US. 'There aren't trees out there, there aren't bushes. It's short grass, three, four inches.' So there's no shade in an area where summer temperatures can top 30°c. Black-footed ferrets can't handle temperatures above 24°c, but underground stays cooler, at about 13°c. The ferrets don't do their own dirty work; they use burrows made by prairie dogs – small-eared ground squirrels. Then they eat them. 'It's evolution in its finest hour,' says Fraser. 'The ferret does not excavate, it just moves in. It seeks shelter from the weather and from predators and then their grocery store is right there, they don't have to go too far to get food.' Above ground, ferrets are at risk from owls, coyotes and badgers, among others.

Today we know that prairie dogs, not eggs, make up the bulk of the black-footed ferret's diet. At the time of Audubon's visit there may have been as many as five billion prairie dogs.[8] Of the five species, three live in black-footed ferret territory: the black-tailed prairie dog in the east of the ferret's range, the white-tailed prairie dog in the north-west and the Gunnison's prairie dog in the south-west. The bulk of black-footed ferrets, some 85 per cent, lived in black-tailed prairie dog habitat, 8 per cent in Gunnison's home range, and 7 per cent in white-tailed prairie dog territory. (I appreciate that reading about the black-tailed prairie dog and black-footed ferret in the same chapter is confusing – just thank your lucky stars we're not using the Lakota name for the ferret, which translates as black-faced prairie dog, perhaps in recognition of the close links between the two species.)

Apart from their tails, all the prairie dog species look similar. Largely beige in colour, they resemble a squat, pot-bellied meerkat or the chubby lovechild of a guinea pig and a squirrel. When they're not sitting on their hindquarters resting their front legs on their bellies, they feed blades of grass into their mouths as they grasp them in hefty-clawed paws – you can see why black-footed ferrets are wary of prairie dogs that are fully awake. As highly social creatures, prairie dogs live in 'coteries' that consist of one or two breeding males, several breeding females and their new pups. If conditions are good, females nurse the pups communally. This sounds idyllic, though if the rains fail and food becomes short, they flip to cannibalism.[9] This barbaric diet could save the adults so they can live to produce pups the next year, when there may be more rain, and more food. Life on the prairie can be tough. Perhaps that's why, to a human, prairie dogs look slightly sad. Or maybe it's the way their round dark eyes sit near the top of their heads, giving them long, hangdog cheeks.

During the day, prairie dogs call out when they see predators, warning others in the collection of burrows known as a prairie dog town. These squeaks are incredibly sophisticated, can convey that the approaching danger is due to a human who is tall and wearing blue,[10] and may even have their own grammar.[11] At night this system fails and black-footed ferrets are able to scramble into prairie dog burrows while their inhabitants sleep. If a ferret finds a sleepy resident, it generally dispatches it inside the burrow by biting the prairie dog's neck until it suffocates, the same way a lion kills a wildebeest. These ferrets are well equipped for this job; relative to the size of their skulls, the species has the longest canine teeth of any mammal species in the world. 'People are drawn in by this little tiny animal that is so cute,' says Fraser. 'And yet when they show their long teeth and they have a great chatter and a scream, people are quite taken aback with the ferocity of them. They're so charismatic but they will take your face off. If they were the size of German shepherds, we'd never leave our homes.' Their long teeth enable ferrets to take on prey as big as they are; black-footed ferrets and prairie dogs both weigh about the same as a bag of sugar. That said, if the prairie dog a ferret finds down the burrow turns out to be particularly large, discretion becomes the better part of valour and the ferret nips to another burrow in search of a smaller adult or some kits. Though that move is in itself risky – more time above ground means more risk from predators.

Some 40 years after Audubon visited the Midwest, European settlers moved in. Some wrote of their travels past prairie dog colonies that extended for miles, but the region would soon change.[12] First, settlers ploughed up the grasslands in the east, where there was more rainfall, to grow crops such as grain. On much of the rest of the land, people began ranching cattle. Although that didn't destroy habitat for black-footed ferrets and prairie dogs like ploughing did, many ranchers saw the prairie dog as taking grass from under their cattle's noses, especially during the area's frequent droughts, and prairie dog burrows as a danger to the legs of ambling cattle who weren't paying attention. In the early twentieth century, the director of the US Biological Survey, C.H. Merriam, didn't help matters by estimating, without apparent recourse to scientific data, that prairie dogs reduced land productivity by a massive 50–75 per cent.[13] By 1996, prairie dogs would be pardoned of this crime when a study showed that the animals keep grass short but improve its nutrients, digestibility and productivity.[14] Since the ranchers saw the prairie dog as an economic threat, they poisoned and shot it, piling the bodies into huge mounds. The attacks continue today, mainly as shootings. Back then, the US Biological Survey joined the poisoning spree. As a result of these deliberate killings, habitat destruction and disease, the black-tailed prairie dog's homelands have almost disappeared – they've shrunk some 98 per cent, from 40 million hectares to about 0.75 million hectares in 2004.[15]

This loss harms not only the prairie dog but the animals that rely on this keystone species – a species that plays a crucial role in the ecosystem – including the burrowing owls that use prairie dog burrows for shelter, mountain plover that nest on areas grazed by prairie dogs, bison that enjoy eating the nutritious young grasses that spring up in grazed areas, and prairie dog predators such as ferruginous hawks, swift foxes, rattlesnakes and black-footed ferrets.[16] Although farmers didn't persecute ferrets directly, killing their prey did the ferret no favours. Given that prairie dogs made up 90 per cent of the black-footed ferret's diet, bad news for prairie dogs was bad news for ferrets. Ferrets had to work harder to find food, and to spend more time with the risks above ground as they ran fruitlessly from burrow to burrow. That wasn't all. From the 1930s, prairie dogs – and black-footed ferrets – faced a new challenge.

BLACK-FOOTED DEATH

Back in 1901 a ship had sailed into San Francisco from Hong Kong. It turned out to be harbouring a flea-infested rat suffering from plague, just one victim of the world's third pandemic of this disease. This 'Modern Pandemic' began in 1855 in China's south-central province of Yunnan; by 1894 it had spread, with the assistance of troops and refugees from a rebellion, to Canton and Hong Kong. That was also the year that scientists determined which bacterium causes plague when Alexandre Yersin and Kitasato Shibasaburō each separately identified the pathogen at once. Only Yersin's name entered the history books, as his description was more accurate. Since the 1970s these oval-shaped bacteria have been known as *Yersinia pestis*. Before this pestilence arrived from Hong Kong, North America had been lucky – it had escaped the previous two pandemics.

Plague emerged in Central Asia at least 12,000 years ago,[17] probably as a result of people starting to farm and store large quantities of grain, which attracted rodents whose fleas spread *Yersinia pestis* bacteria to people. In the past, 'rat-fall' of dead rodents from the ceiling would herald a plague outbreak; it still does in villages with thatched roofs. It seems likely, though we don't know for certain, that the outbreak of disease the Bible describes in a Philistine city in the eleventh century BCE was plague.[18] Mice died alongside residents who suffered swellings in their groin and thighs – typical symptoms of the bubonic form of plague, which sees plague bacteria multiply in the lymph nodes in the groin, armpits or neck, creating swellings known as buboes after the Greek word for groin.

In all three of the world's major pandemics of plague, fleas spread the disease between humans and the black or brown rats that lived alongside them in what's known as the domestic cycle of transmission. Using fleas to transmit itself from animal to animal is a cunning move by the bacterium, but it requires the bacterium's own biology to be cunning too. Insects and mammals are very different but *Yersinia pestis* must be able to thrive in both. To achieve this, the bacterium expresses different genes depending on the temperature – mammals have a body temperature of 37°c while flea bodies are 28°c or less. In this way the organism optimises its survival and reproduction inside each type of host, even evading some of the host's immune defences. In some species of flea, after around five days *Yersinia pestis* bacteria have multiplied so much that they block the flea's foregut. That means the flea can't digest blood properly and vomits bacteria into the animal it's biting. Other fleas transmit plague without a blocked foregut.

The first plague pandemic, in the sixth century CE, probably started in Ethiopia or Egypt, spread to Constantinople via rats on ships loaded with grain, then on into Europe, Central and South Asia and Arabia. This Plague of Justinian, named after the emperor at the time, contributed to a 50–60 per cent drop in the region's population between 541 and 700 CE.[19] The next pandemic, known as the Great Plague, Great Pestilence or Black Death because of the associated gangrene and black extremities, began in the fourteenth century and killed one-quarter to one-third of Europe's population. This second pandemic likely began in central Asia then spread west along trade routes to Crimea. In 1346, when Tartar troops besieged the port of Caffa on the Black Sea in present-day Ukraine, many of them died of plague. The survivors catapulted their comrades' disease-ridden corpses inside the city walls, infecting inhabitants either by direct contact or through flea bites.[20] From ports like Caffa, plague spread into the Mediterranean on rat-infested ships then moved by land across Europe, where it broke out periodically for the next 300 years; one of its last major appearances was the Great Plague of London in 1665. In later outbreaks the death toll probably included a high proportion of the plague doctors who by then believed that a mask shaped like a bird's beak would draw germs away from the patient – birds being thought, at that stage, to spread plague. Though maybe the red-glassed spectacles designed to protect the doctor from evil or the black overcoat tucked around the bird mask to hide all skin successfully fulfilled their role.[21] As long as the air wasn't awash with bacteria from people dying of the pneumonic, lung-based, form of the disease.

Caffa wasn't the only time plague would be used as a weapon; during the Second World War, Japan dropped bombs packed with more than

30,000 plague-infested human fleas, along with grain and rice to attract rodents, onto two Chinese cities. This led to a couple of small outbreaks of bubonic plague – the first killed 120 people and the second 24.[22] In the 1950s and 60s, both the US and Soviet Union investigated plague in aerosol form as a bioweapon.[23] But even naturally distributed plague still kills, despite us having modern antibiotics. Plague is present on all continents except Antarctica and Australia, where it never established itself as a permanent fixture, not for lack of trying.[24] In the 1970s, Asia, particularly Vietnam, led the world in human plague deaths, but more recently this title-that-no-region-wants has gone to Africa, with a special mention for Madagascar, where some strains of plague have become resistant to multiple drugs, and Mozambique.[25] Between 2010 and 2015, some 3,248 people worldwide caught plague and 584 people died.[26] In the US, home of the black-footed ferret, 7 people, on average, die of plague each year.[27]

AMERICAN NIGHTMARE

The US is also home to coyotes. I remember being bitterly cold one night on a camping trip to Death Valley. The sleeping bag I'd borrowed was thin and I was too cold to drop off – the kind of cold where you grasp at the warmth inside your core by shrinking yourself, every muscle tense. A pack of coyotes yipped in the distance and I dazedly considered unzipping the tent and inviting them in so I could steal their body heat. Turns out that would have been a bad idea, not just because they might have bitten me or knocked over the tent, but because, although coyotes don't tend to suffer plague themselves, they may transport plague-carrying fleas.

Other wild animals in North America aren't so lucky. Plague infects over 200 species of mammal, though some are more susceptible than others. Mountain lions, bobcats and domestic cats tend to suffer badly and may transmit the disease to humans. Other carnivores are generally unaffected. Like coyotes, wolves, swift foxes and domestic dogs get off scot-free apart from having to produce antibodies. Handily, these antibodies mean a coyote, badger, raccoon, striped skunk or black bear can act as a means of testing hundreds of rodents for plague at once – if the predator has antibodies against the disease, it means at least one of the rodents it ate had plague. Rodents such as gerbils, marmots, deer mice and California voles show a range of responses – some individuals die but some become only mildly ill.[28] These 'maintenance hosts' may keep plague in circulation between outbreaks among 'amplifying hosts', rodents such as prairie dogs, ground squirrels, chipmunks and wood rats that are highly susceptible and die in huge numbers.[29] Plague bacteria were first discovered in wild rodents, as opposed to the rodents that live alongside humans such as black and brown rats, in 1895 in marmots in Mongolia and Russia.[30] In South America hosts include wild cavies and domestic guinea pigs, while in Africa three species of gerbil host plague. When plague arrived in North America in 1901, it spread north and east from San Francisco via rodents and their various species of flea, including ground squirrels, chipmunks, wood rats, deer mice, fox squirrels and voles, in what's known as a sylvatic transmission cycle (sylvatic means occurring in or affecting wild animals).[31] In the decades after it arrived, plague travelled 2,200 kilometres (1,400 miles) east in just over 40 years, moving at a speed of 45–87 kilometres (28–54 miles) per year.[32]

By the 1930s, plague had reached the western side of prairie dog and black-footed ferret territory. This was devastating as both prairie dogs and black-footed ferrets are highly susceptible; more than 90 per cent of individuals that become infected with plague will die. These days, plague has entered Kansas,[33] and seems still to be heading east.[34] Thanks to plague, habitat loss and deliberate slaughter, in the early 2000s numbers of black-tailed prairie dogs were estimated at 18 million,[35] a near 300-fold drop from the 5 billion of Audubon's times. White-tailed and Gunnison's prairie dogs experienced similar declines. In the 1960s one prairie dog town had stretched 35 kilometres; by 1998 the biggest prairie dog town in Oklahoma spanned 2.1 kilometres, some 17 times less. For the black-footed ferret, plague was a double blow. Not only did the disease kill the ferret but also its main source of food, which was already greatly depleted by human actions. 'What affected the ferret populations were the three Ps,' says Fraser, 'ploughing of the Great Plains, poisoning of the prairie dog, and plague the disease.' The ferrets couldn't win. In rainy years, fleas boomed and so did plague, but in dry years, prairie dog pups were scarce. It was die of plague or starve. This creature evolved its specialist prairie dog diet long before plague reached North America. The ferret's closest living relative, the Siberian polecat, resides in an area ridden with plague for millennia and has a much broader diet.[36] A polecat plague survivor can feed on whichever of its prey haven't just undergone an outbreak, whether that's large-toothed suslik (a ground squirrel), great gerbil or some other animal native to central Asia.[37] When Siberian polecats entered a plague-free

North America some hundred thousand years ago via the Bering land bridge, there was no need for them to stick to this broad diet.[38] They evolved into the black-footed ferret and specialised in feeding chiefly on prairie dogs. Later this turned out to be a mistake.

THEY SEEK HIM HERE

Female ferrets have territories of about a hundred acres to ensure they find all the prairie dogs they need to feed themselves and the three or four kits they produce each spring. These ferrets only group together to mate and to care for their young; a group of ferrets is known as a business. '[The rest of the time] they're solitary, nocturnal, they live on large expanses of landscape,' says Fraser. 'They have this kind of magical mysticism about them.' The black-footed ferret has always been elusive, as Audubon discovered at some cost to his reputation. Once plague arrived, the ferret became even harder to find, simply because so many of them had died. By the time US biologists looked for them in the 1960s, black-footed ferrets were thin on the ground. Some still lived at the US Air Force's Badlands Gunnery Range in south-western South Dakota, perhaps an odd choice for an animal wanting to live in peace. Once the Air Force learned about the ferrets, it agreed not to use prairie dog towns 'as impact areas', and only to control prairie dog populations if they threatened to spill beyond the edges of the range.[39] Between 1971 and 1973,[40] in an attempt to increase ferret numbers, the US Fish and Wildlife Service took nine animals into captivity. Though none of the resulting kits survived, this is how we learned that black-footed ferrets spend most of their time underground and mainly eat prairie dogs, consuming one animal – bones, teeth, fur, the lot – every two to three days.[41] We also learned to be extremely careful with modified live canine distemper virus vaccines in this species; four ferrets died after the weakened virus in the vaccine still proved too strong for them.[42]

Unfortunately, like the kits, the captive adults weren't in great health; they suffered cancerous tumours and diabetes,[43] perhaps because too few ferrets were left to maintain a healthy gene pool. The captive breeding and the reduced shelling and bombing were all to no avail. The wild ferrets disappeared in 1974 and the last captive ferret died in 1978.[44] With no more ferrets seen in the wild, it looked like the species had gone extinct. 'I'm old enough, I'm grey enough that I remember those headlines and feeling sad because black-footed ferrets, they have this kind of reputation and they were just gone,' says Fraser. By 1980 a US national tabloid was running an exposé about how taxpayers' dollars were being 'wasted' searching for an animal that 'may not exist'.[45]

LAZARUS WALKS

On 26 September 1981, ranchers John and Lucille Hogg woke in the middle of the night to the sound of their dog Shep growling and barking in the yard. They assumed Shep was fighting a porcupine, but when John went outside the next morning he didn't find Shep impaled by porcupine quills. Instead, a carcass lay on the porch floor, right next to Shep's food dish.[46] Fraser says she was lucky enough to meet John Hogg before he died. Hogg told her the fur on the carcass was soft but he didn't have any use for it so he tossed it away down by the creek. When he went back inside, John told Lucille it wasn't a porcupine but some kind of weasel. Curiosity aroused, Lucille recovered the carcass and decided it was so beautiful that she wanted it mounted in the living room. But the taxidermist in nearby Meeteetse in north-western Wyoming told John that he was going to have to confiscate the animal as 'you've got an endangered species on your hands'. The taxidermist called in the officials. 'We always loved John to tell that story,' Fraser says. 'And when he got to that part, he said, "And I never got the damn thing back."'

A series of coincidences – a young male black-footed ferret dispersed from the breeding population sixteen kilometres (ten miles) away and happened to head towards the Hoggs' ranch, Shep caught him and left him where John Hogg would find him, Lucille wanted him stuffed and the taxidermist recognised him – led to the rediscovery of a species. 'It would probably make a good Hollywood movie,' says Fraser. 'It's like links in a chain and if someone had just done something a little different, that chain would've been broken, and we wouldn't have black-footed ferret.'

About a month later, biologists found a live male ferret they estimated at less than two years old near Meeteetse.[47] They followed this male to a hole and live-trapped him when he popped out again a few hours later, fitting a radio transmitter on a collar around his neck to track his movements. Further investigation revealed that a small colony of black-footed ferrets had survived; a few years later, in August 1984, it numbered 129 black-footed ferrets.[48] The black-footed ferret was back, earning itself the nickname 'the Lazarus ferret' after the miraculous story in the Bible of a man restored to life four days after he died. 'That kind of turned the conservation world upside down that we had another opportunity,' says Fraser. 'How often are we given that?'

In 1985, just four years after the species' rediscovery, this last known colony faced another threat. Along with plague, the colony was hit by canine distemper virus (CDV), a morbillivirus like rinderpest and measles. Most of the ferrets in the colony died. That autumn, conservation workers took six of the surviving ferrets into captivity, but they all died of distemper. Later in the year, the team took six more ferrets, which survived but failed to breed in the spring of 1986. Estimates at this point indicated that not enough ferrets survived in the wild to form a viable population. So officials took all the survivors they could find – fewer than twenty, in total – into captivity in the hope of setting up a breeding programme.[49] By February 1987 the black-footed ferret was once again extinct in the wild. Not everyone was in favour of the breeding programme; some people, according to Fraser, said when biologists gathered up those last ferrets: 'Let them go. Do you see a dinosaur? Extinctions have happened, why does it matter?'

Fortunately for everyone apart from conservation sceptics, the captive ferrets bred. 'Scarface', a male who'd clearly battled with prairie dogs or other black-footed ferrets, did more than his fair share to keep the species going. These days, Fraser takes live black-footed ferrets on the road to visit some of the Black-footed Ferret Recovery Programme's 50 conservation partners in Canada, Mexico and 12 western US states. The partners gain a chance to meet an animal they might otherwise never see. 'It's the best job on the planet,' Fraser says. 'You haven't lived till you've stayed in a hotel room with a live black-footed ferret being nocturnal. They rock and roll about 2am.'

Slowly the captive breeding programme built ferret numbers back up. After four years or so, the first captives were released into the wild near Shirley Basin in Wyoming. These newly wild ferrets had effectively risen from the dead once more, in a double Lazarus. By 1994 nearly 230 ferrets had been set free at this site.[50] But like the ferret, plague is elusive. We don't fully understand where it goes between outbreaks. Does it hide in the soil? In fleas? In maintenance hosts where it circulates at barely detectable levels? Certainly, in areas where the disease was not active, when scientists treated prairie dog burrows for fleas and vaccinated ferrets against plague, more prairie dogs and more ferrets survived.[51] In 1995 plague reappeared in this area and the breeding programme stopped releasing ferrets.[52]

Even in captivity, the ferrets weren't safe from disease. In a mix-up in 1993, a captive ferret in Wyoming encountered captive white-tailed prairie dogs and died a few days later, presumably having eaten a plague-infected prairie dog.[53] A couple of years later, a new member of staff accidentally fed 30 captive ferrets pieces of prairie dog meat that were infected or contaminated with plague,[54] and had been taken to the centre for testing.[55] All but three of the ferrets died or went missing having expired in underground burrows.[56] This was the first time researchers truly appreciated how susceptible black-footed ferrets are to plague – domestic ferrets and other mustelids such as badgers, long-tailed weasels, pine martens, and striped and spotted skunks don't succumb.[57] But it wasn't the first time captive breeding has introduced disease. The domestic pigeons used to foster-rear chicks of the endangered Mauritius pink pigeon, for example, ended up killing the youngsters by passing on pigeon herpesvirus, which the domestic birds carried without showing symptoms.[58]

Today, all black-footed ferrets scheduled for release are fitted with an identity tag and vaccinated against rabies, canine distemper virus (using a vaccine that's safe for black-footed ferrets) and plague. A human plague vaccine turned out to be effective in black-footed ferrets and would even prove useful when the SARS-CoV-2 pandemic struck. 'When Covid hit, that scared us to death because we, even as humans, didn't know about Covid and all those mink were dying in Europe,' Fraser says. 'We really had strict protocols – we could not be in the same room with each other, we had to bring a pencil every day, we had to bring eating utensils, we had only N95 masks, we got tested every week. Here at the ferret centre, we did not want to be responsible for the extinction of a species.' According to Fraser, Toni Rocke of the National Wildlife Health Center in Madison, Wisconsin, who'd adapted the plague vaccine for ferrets, said, 'I'm just going to take out the protein for plague and plug in the spike protein for Covid.' The ferrets had a SARS-CoV-2 vaccine before humans did; the centre vaccinated between a half and two-thirds of its population just in case.[59,60] Ferrets born in the wild receive their routine shots thanks to Recovery Programme partners who capture them at night, take them to a medical trailer then return them to the burrow they came from. 'I'm sure they're thinking, *Well, that was an alien experience, what was that?*' Fraser says. Two weeks later, if the team can find them, the young ferrets receive a booster jab to gain near 100 per cent protection; plans are afoot for a slow-release vaccine that only needs one dose.

It's not feasible, however, to round-up and vaccinate prairie dogs against plague; they're simply too numerous. And without prairie dogs, ferrets have little to eat. Initially, conservation workers dusted prairie dog dens with an insecticide known as Delta Dust to kill plague-ridden fleas. But there were signs the fleas were becoming resistant and the insecticide killed other insects. Recently, scientists developed an oral vaccine that prairie dogs eat in bait laced with peanut butter and scattered across the landscape from vehicles or drones, saving time and money. The move could protect humans as well as ferrets. 'People take their dogs out and think it's funny that they chase prairie dogs,' says Fraser. 'That may be entertaining but your dog may come home with a plague-infested flea. Why take that chance? I don't think wildlife harassment is ever good – don't let your dog do that.'

FUTURE FERRET

After successful ferret reintroductions in eight western states, Canada and Mexico, in 2008 the IUCN upgraded the black-footed ferret's classification from extinct in the wild to endangered. The Meeteetse colony in Wyoming didn't have enough prairie dogs for the US Fish and Wildlife Service to return black-footed ferrets for 35 years. In 2016, amidst great fanfare, 35 young captive ferrets, one for each year of the hiatus, entered the wild near the site of their rediscovery. The Hoggs' son and his wife still live on the ranch where Shep caught the young roaming male and their home finally gained a mounted black-footed ferret, decades after the taxidermist confiscated the original target, when the Fish and Wildlife Service presented the couple with a thank you for their parents' help.

The release programme is ongoing. Over the years scientists have honed their techniques; young captive ferrets now practise chasing empty paper bags, live in outside pens within sight of wild prairie dogs before their release, and before they're let loose must prove they can kill and eat a prairie dog introduced to their pen. Thousands of ferrets, which live up to three years, have been released at 34 reintroduction sites. Today about 300 black-footed ferrets live in the wild and about the same number are in captivity. But there's a snag.

Of the original twenty animals taken into captivity, only seven bred. All black-footed ferrets alive today are effectively siblings or first cousins, and the population's genetic variation is 55 per cent less, on average, than in the 1980s.[61] This small gene pool could cause the species problems down the line – low genetic diversity has been linked to susceptibility to infectious disease.[62] Thanks to some extraordinary foresight, however, a solution is in view. Two of the original captives perished almost immediately – Willa, a female who died of unknown causes, and a male known only as Stud Book 2, or SB2 for short, who died from CDV. These days SB2 would have a better moniker. Staff at the recovery centre name ferrets as a perk of the job, and in addition to their stud book number and the ID number from their tag, ferrets receive a name based on chocolate, beer (Colorado is a big brewing state), friends, loved ones and pets, or something silly like 'booboo head'.

Neither Willa's nor SB2's genes entered the captive gene pool. Yet these are still available to us thanks to a chance meeting between Tom Thorne, the vet in charge of the ferrets, and Ollie Ryder of the San Diego Frozen Zoo at a banquet.[63] Ryder was collecting samples from many species, even though in the 1980s the technology to reproduce them didn't exist. Decades later scientists were able to clone the frozen sample of Willa's DNA and a domestic ferret surrogate carried two black-footed ferret clones to term. In December 2020 the surrogate underwent a C-section but the first kit to come out of the womb was dead. The second came out alive then appeared to die, despite efforts to revive her. She too was bagged up for disposal. 'One of our employees, Robin, just happened to walk into the room,' Fraser recalls, 'walked past the bag and said, "That kit just gasped." They pulled her out and revived her, and we had the first clone of an endangered species ever.'

This kit, Elizabeth Ann, had to have a hysterectomy for medical reasons and never got to breed. In 2023 two more Willa clones were born.[64] If Noreen and Antonia can produce young, they will increase the black-footed ferret's genetic diversity and help the species survive. Cloning SB2 has been more of a challenge. The canine distemper virus that killed him inserted itself into his DNA, but scientists have now worked out how to remove it. His first clones didn't survive but hopefully the next ones will – including SB2's genes would expand the gene pool of today's black-footed ferrets even further. 'Every species has the right to live,' says Fraser. In the longer term, the cloning team would like to increase black-footed ferrets' resistance to plague, perhaps by making their bodies express more antibodies against the disease. Such a clone would be able to pass on its genes, removing the challenges of vaccinating ferrets born in the wild.

Ultimately the recovery programme's goal is to have 3,000 adult black-footed ferrets living wild across 9 states, with at least 30 adults in all populations and 10 populations that have at least 100 adults.[65] The efforts to save the black-footed ferret have been Herculean. But how much more cost-effective would it have been not to destroy the bulk of this ferret's habitat for farming, poison its prey and ship in a lethal disease from the other side of the world? And what broader conservation measures could we have spent the money allocated to cloning on? 'I'm probably going to retire in five years,' says Fraser. 'The species will not be recovered but I hope there's many more new techniques in the toolbox because that's what it's going to take to really make it happen.' The captive breeding programme may have begun just in the nick of time. 'A thousand years from now, I don't think humans will be on the planet,' Fraser says. 'But I think black-footed ferrets will because they're that successful if we give them the right tools.' Plague is currently their biggest challenge for recovery.

7

Seals Go to the Dogs: Unusual Contact and Morbilliviruses out at Sea

'You could hear their clamour miles out to sea above the loudest gales.
At the lowest counting there were over a million seals on the beach,
—old seals, mother seals, tiny babies, and holluschickie [bachelors],
fighting, scuffling, bleating, crawling, and playing together,
going down to the sea and coming up from it in gangs and regiments,
lying over every foot of ground as far as the eye could reach'
Rudyard Kipling, *The White Seal*, 1893

At Mpala Research Centre, where I stayed in search of Grevy's zebra, the rains failed in 2017. Herders from the north moved into the park and other private land in desperate search of grass for their cattle. People were killed in the disturbances and the herders' dogs brought rabies and canine distemper virus (CDV). Mpala's pack of African wild dogs (*Lycaon pictus*) disappeared, presumed victims of CDV. Only a few thousand of these dogs remain, spread between isolated pockets scattered across western, central and eastern Africa and down into the South. The whole of Kenya, an area of just over 569,000 square kilometres,[1] more than double the size of the UK or four-fifths the size of Texas, may be home only to about 500.[2] When the rains failed, other African wild dogs in the Laikipia region, some 300 in total, vanished too,[3] as did members of the species elsewhere in Africa – in 2016–17, CDV killed three packs in South Africa and one pack in Tanzania.[4] During my visit to Kenya in 2020, I was extremely lucky to see a pack of African wild dogs by the side of the track in Tsavo West National Park. Their dark-rimmed beige ears rose like hoops above lean and lithe bodies blotched with dark chocolate and fawn. When one turned to look at something, ears pricked, amber eyes shining against dark fur that shaded into a long, slightly pointed muzzle, its profile was like a collie's (though African wild dogs are from a different genus to domestic dogs). After several minutes of information-gathering, all eight pack members loped silently back down the road with long, skinny limbs more like a gazelle's than a dog's. From time to time, they paused and stood; one near the back turned to give us a final stare. It's clear who was in charge. It wasn't us, and thankfully this time it wasn't a virus.

CDV is another morbillivirus, like the rinderpest that devastated the lives of humans, cattle and wildlife in Africa in the nineteenth century. This canine version may have originated in a bat[5, 6] or may have developed when a domestic dog in South America caught measles from a human sometime in the sixteenth to eighteenth centuries.[7] Then our domestic dogs spread CDV to Spain and on to the rest of the world.

These days it's not just canines who get the disease. CDV has all the attributes that often make a virus successful – as an RNA virus it mutates fast, unlike its generally slower DNA virus relatives with their extra error checks as they copy themselves. It's also multi-host and thrives inside many species. It's been found in more than twenty different families of animals, including individual species such as leopards, red pandas, rhesus macaques and giant anteaters.[8]

At some point, a morbillivirus even entered the oceans. In April and May 1988, almost all the female harbour seals (also known as common seals or *Phoca vitulina*) near the tiny island of Anholt, roughly halfway between Denmark and Sweden, aborted or lost their pups.[9] Soon many of these seals became sick and died, often showing signs of pneumonia.[10] By July, the outbreak had spread north and west through the Wadden Sea, an area that fringes the south-western coast of Denmark, the north-western coast of Germany and the northern coast of the Netherlands, across to the coast of Norway, and south and east into the southern Baltic. Harbour seals live around the globe, on the west and eastern coasts of North America, the north-western coasts of Europe, and the coasts of the North-west Pacific; they number around 315,000.[11] And the European seals were dying in their thousands. A photo published in *New Scientist* magazine in 1989 shows a couple of people in water-proofs and wellies dragging a seal carcass along a beach. Each holds a rear flipper in their hand and their heads bow down as they take up the seal's weight. The seal is upside down and its head scrapes along the shore.[12] By August, seals off East Anglia in eastern England were dying too. Sick seals appeared sluggish, didn't want to return to the sea, couldn't dive, coughed and had laboured breathing and nasal discharge. Some exhibited muscle tremors, uncoordinated movements, swellings around the upper parts of their neck and chest, diarrhoea, conjunctivitis or high fever.[13] By June the following year, 1989, the disease had killed 17,000 harbour seals in the North Sea and the Baltic Sea,[14] which is loosely enclosed by eastern Sweden, western Finland, Estonia, Lithuania, Latvia, Russia, Poland, Germany and Denmark. In total, some 18,000 seals died,[15] roughly half Europe's seal population.[16] Most of the victims were harbour seals, although 1 per cent were grey seals (*Halichoerus grypus* – a name that pleasingly translates as 'hook-nosed sea-pig').[17] Grey seals live in north-western Europe and the eastern coast of North America; like harbour seals, they number around 315,000. Grey seal males weigh as much as 300 kilograms (660 lbs) and females can be 200 kilograms (440 lbs) and up to 2.5 metres long; they're bigger than harbour seals, which weigh up to 100 kilograms (220 lbs) and are up to 1.6 metres long. And grey seals are, for reasons that aren't clear but may be connected to differences in their cell receptors (the places where the virus gains entry to the cell) or immune responses, less susceptible to the virus.[18]

This was the first mass seal die-off in Europe in recent times and caused much public concern about pollution.[19] It was also the first time that scientists were equipped to investigate. To start with, nobody knew what the disease was. Albert Osterhaus, currently at the Tierärztliche Hochschule Hannover (University of Veterinary Medicine Hannover), Germany, had discovered a new herpes virus in seals in the Wadden Sea a few years earlier and set out to investigate. Originally a vet, Osterhaus grew bored of veterinary practice and studied for a PhD in virology in the mid-1970s. 'All my peers said, "You're so stupid to go into a lab and study viruses that are almost all under control,"' he says. 'At that time, there was a perception among politicians and scientists alike that virus infections would be brought under control in the near future – perhaps with the exception of viruses in the animal world and in low-income countries. Some highly acknowledged scientists would say, "It's the era without viruses, 10, 20 years from now."' But Osterhaus has never had a dull moment since. Soon after he switched to virology, the world encountered HIV followed by Ebola, Hendra, H5N1 influenza, and Nipah virus, just to name a few.

In seals, at first, the culprit for the massive die-off in the North Sea looked like a herpes virus or a virus 'tentatively identified' as a picornavirus, a family that includes the viruses responsible for polio and foot-and-mouth disease – both were found in the organs of dead seals.[20] But when Osterhaus and colleagues immunised young harbour seals in the seal orphanage in Pieterburen, the Netherlands, with inactivated versions, the seals, even though they'd produced antibodies to both those viruses, still died from the mystery disease.[21] The race was on to discover the cause. Osterhaus had long debates with pathologists, and with Greenpeace, who suspected that pollution was to blame. Osterhaus thought the seals were dying from an infection. Greenpeace organised an emergency conference in London,[22] and Osterhaus worked to track down what could be causing a disease.

Aware that a couple of Swedish vets had noticed that seal carcasses showed signs similar to dogs infected with CDV, Osterhaus and his team tested blood from baby seals orphaned by the outbreak that had become ill at the Seal Centre Pieterburen. Sure enough, these seals all possessed CDV-neutralising antibodies, as did harbour seals in Denmark, Germany, Sweden and the UK.[23] By September 1988 the team knew that the disease outbreak was caused by CDV or a morbillivirus similar enough that antibodies against it also counteracted CDV. Greenpeace had been wrong, the researchers thought. Though not totally wrong, as later experiments showed. Meanwhile, further investigation showed the virus was related but not identical to CDV and it gained the moniker phocine distemper virus (PDV).[24] According to molecular analysis, PDV grew apart from canine distemper virus in roughly the seventeenth century.[25] It's not clear how marine mammals first met CDV and helped it evolve into PDV, it could have been via encounters between seals and dogs on beaches and coastlines, or perhaps by animal waste entering the seas.[26] However PDV arrived in the oceans, in 1988 the outbreak suppressed the seal's immune systems, allowing other infections to run rife and causing the initial confusion on the cause of death. Animals infected with canine distemper also often suffer secondary infections with bacteria and previous viruses.[27] As did children who caught measles before we developed routine vaccination, something that every physician at the time was aware of.[28] More recently, by testing unvaccinated children in the Netherlands, Osterhaus's team discovered that measles wipes out the memory cells that dictate our immune response.[29] By destroying these cells, measles – and other morbilliviruses like canine distemper virus – obliterates your immunity to other infections.

Phocine distemper virus probably reached the seals of Anholt via infected harp seals (*Phoca groenlandica*), a gregarious and numerous species (in 2011, the population was eight million) that lives off Norway, Iceland, Greenland, north-eastern North America and the northern coast of Russia.[30] These seals, in which the adults are black and silver grey and the pups are white, are not nearly as susceptible to the virus as harbour seals.[31] More than 80 per cent of breeding age female harp seals measured in Canada's Gulf of St Lawrence in the late 1980s and early 90s showed markers in their blood indicating they'd survived phocine distemper virus.[32] As with 'yearling disease' in wildebeest, harp seals generally only die from the disease this virus causes when they're juveniles, once the immunity their mothers pass to them has worn off.[33] But the more susceptible harbour seal had already suffered small outbreaks and deaths on the US east coast before the 1988 mass outbreak in Europe. And when people went back and looked, they found that phocine distemper virus had been circulating in seals, Atlantic walrus and polar bears in the western North Atlantic since at least the 1970s.[34]

Harp seals invaded the North Sea from 1986 to 88, perhaps migrating when overfishing around Greenland in the late 1980s meant the animals had to search elsewhere for food.[35] Harp seals were seen near harbour seal haul-out sites – areas where the seals left the sea to spend time on land and rest, moult or breed.[36] Here the virus probably spread in aerosol form,[37] when infected animals breathed it out and others breathed it in, or via infected discharges from the lungs or nose. Morbilliviruses mainly transmit in the air but can also infect through direct contact and from mother to offspring.[38] Then, having entered a naive population, perhaps already stressed by pollution and crowded together at haul-out sites, the virus was able to kill thousands.[39] 'One of the threats is that due to the changes in climate and also the Gulf stream etc., the migratory patterns of these animals change,' Osterhaus says. 'And this may kick off outbreaks.'

Seals, dolphins and other cetaceans had suffered mass die-offs every now and then in the past too – in several incidents in the British Isles in the eighteenth, nineteenth and early twentieth centuries, for example, large numbers of carcasses washed ashore.[40] Osterhaus says this is similar to the waves of measles seen in humans roughly every fifteen years before vaccines came into play. The researcher would also like to find out if mass strandings of other marine mammals are related to disorientation because of the virus. 'We have not been able to prove that yet,' he says.

'I'm not sure if it'll ever work.' In part, that's because it's hard to get fresh samples, particularly from whales, which have so much blubber that they overheat when they decompose, sometimes to the extent that their stomachs explode.

DIRTY WATERS

At the time Osterhaus was sceptical about pollution's harms, but he wasn't finished with the pollution/phocine distemper virus debate. To find out more, he and his PhD students fed one group of baby harbour seals from the seal rehab centre in Pieterburen herring from the relatively clean waters of the Atlantic, and another group with herring from the heavily polluted Baltic Sea,[41] which contained a tenfold higher concentration of polychlorinated biphenyls (PCBs) and other potentially harmful substances. Toxic chemicals like PCBs build up in seals' fat layers and can be passed on to the next generation. (If you're worried about seal cruelty during the experiment, you may or may not find it reassuring to learn that the team bought the polluted herring from Scandinavia, where it was originally intended for human consumption.) 'I remember that vividly,' Osterhaus says. 'This is about 40 years ago but the seals that they wanted to feed on the Baltic herring, they refused because somehow they smelled it wasn't good enough.' The team had to get the seals used to this new food. Then, after adapting techniques developed to assess humans infected with HIV, which had recently been discovered, Osterhaus and his PhD students tested the seals' immune system. Osterhaus didn't expect the two groups to show any difference. Much to his surprise, the immune systems of the seals fed polluted herring worked less well than those of the group fed relatively clean Atlantic herring; the effect of pollution on the seals was considerable. Both Osterhaus and Greenpeace may have been right – a virus caused the disease but pollution may have made seals less able to fight the infection.[42] That pollution is ongoing today and we still don't know the full extent of the role it plays in seal disease.

MIX AND MATCH

From the autumn of 1987, shortly before the outbreak of PDV in Europe, until October 1988, several thousand of the 80–100,000 seals in

Lake Baikal in Siberia died.[43] These Baikal seals (*Phoca sibirica*), the only seal species that lives solely in freshwater, suffered similar symptoms to the European seals, exhibiting paralysed hind extremities, diarrhoea, eye inflammation, and convulsion of the hind flippers.[44] When a team from Lake Baikal contacted Osterhaus, he and a couple of colleagues from the Seal Centre Pieterburen went to help investigate. Originally, they thought the disease was the same as the one killing seals in North-west Europe and dubbed the virus PDV-2. But that name turned out to be inaccurate.

For Osterhaus, visiting Lake Baikal was a fantastic experience – not only did the area have a Limnological Institute, seals and a totally different fauna, but an interesting culture too. 'We went there in the beginning of the wintertime,' he says. 'It was cold. Oh my God.' The trip was fascinating scientifically; the pathogen behind this outbreak turned out to be CDV.[45] 'The virus that is normal in terrestrial animals – carnivores – spilled over,' Osterhaus says. 'These were two outbreaks, contemporary outbreaks, but different (closely-related) causes.' Although CDV may first have entered the oceans sometime in the seventeenth century and evolved into PDV, the CDV behind this outbreak appeared to have spilled over into seals much more recently; the virus was genetically identical to a CDV strain circulating in Siberia's domestic dogs,[46] and local dogs had had 'close contact with seals' around the time of the outbreak.[47] This strain of CDV may have made its way south to Siberia inside the bodies of Arctic foxes or wolves;[48] sled dogs in northern Canada and Greenland had suffered a severe outbreak of CDV from October 1987 until February 1988, perhaps after migrating Arctic foxes (*Alopex lagopus*) spread the virus in the high Arctic.[49] As well as this mass outbreak, each spring at Lake Baikal, a few more seals die than you might expect. Scientists think these excess deaths may be due to CDV that's over-wintered. Sick seals emit copious snot and mucous, which falls to the ground and freezes when temperatures are low. In spring, the virus thaws and is ready to strike again.

This may not be the first time CDV entered a seal population directly. In 1955 in the Antarctic, more than 3,000 crabeater seals (*Lobodon carcinophagus*) died. At the time no virus was detected,[50] but some suspect that the bodies of unvaccinated sled dogs used during an American expedition to the South Pole were dumped in the water when they died of CDV, spreading the virus to the seals.[51] And it wasn't the last time either. In 1989, antibodies against CDV were found in Antarctic crabeaters, with a sled dog coming under suspicion for transmitting the disease.[52]

Just under ten years later, in spring 1997, around 7,000 endangered Caspian seals died from CDV along the western shore of the Caspian Sea in Azerbaijan.[53] In the year 2000, CDV again broke out in Caspian seals and 10,000 of the animals died.[54] This strain was different to the CDV from Lake Baikal and to any other known strain; researchers think it's the earliest strain of CDV found to date and split from other strains around 1832.[55] In recent times, morbilliviruses may have gained an additional route for infecting wildlife – domestic dogs vaccinated with live, weakened CDV virus may have spread the pathogen. 'We've a similar experience with vaccination of humans against smallpox using vaccinia virus,' Osterhaus says. 'This has eradicated smallpox but vaccinia virus poses a problem now in wildlife as it has slipped into rodents. So we have to be careful with live vaccines.' But Osterhaus doesn't have an opinion on whether this happened with CDV vaccines used in dogs – there's not enough proof.

CDV is not the only virus to come in many versions – multiple lineages of PDV may lurk in Arctic and Atlantic seals too.[56] In May 2002 a different version of PDV appeared in Europe,[57,58] again near Anholt, and spread to kill more than 30,000 seals.[59] Since this strain of PDV and the strain that circulated in 1988 were slightly different, it looks like the virus disappeared from Europe's oceans before being reintroduced.[60] Unusually for a morbillivirus, infection with one strain of this seal virus doesn't seem to offer protection to infection with another strain.[61] Other types of morbillivirus, such as rinderpest and PPR, provide cross-protection – if a small hooved animal has had rinderpest, for example, it's much less likely to suffer from PPR.

OTTERLY DIFFERENT

Until 2004, as far as we know, PDV had not struck the Pacific Ocean. That year, sea otters in Alaska (*Enhydra lutris kenyoni*) were found to be infected with the virus.[62] Otters washed up dead between 2005 and 2007 were also infected with PDV identical to the strain that had broken out in northern European seals in 2002.[63] The virus' travel to the Pacific probably involved environmental change and a seal relay race. Harp seals in northern European seas could have passed PDV to nomadic Arctic species such as ringed seals and bearded seals, which generally circle the pole. When climate change reduced sea-ice coverage, creating bands of open water, the ringed and bearded seals could have ranged further south than normal, into areas where they passed PDV to other species of ice seal like spotted seals and ribbon seals that roam even further south into the Pacific and that may, in turn, have passed the virus on to sub-Arctic species like northern fur seals, Steller sea lions and northern sea otters.[64] Sea otters already had enough to contend with, having been hunted by humans in the past and still suffering from oil spills, entanglement in fishing nets, shark attack, boat strikes, algal blooms, bacterial infections and infection with toxoplasmosis.[65]

Members of the morbillivirus genus have an astonishing capacity to jump to different species – seal to sea otter, dog to seal, cow to human, and so on. One day that could hamper our efforts to eradicate measles – CDV or another morbillivirus might reinfect humans, especially given that monkeys can catch CDV.[66] 'We have eradicated smallpox, and now monkeypox, or mpox as it is called today, comes back,' Osterhaus says. 'So once the immunity has waned, is there a risk that a related virus will fill the niche?' Back in the oceans, seals can catch at least three different morbilliviruses – the phocine distemper virus, canine distemper virus and, as scientists discovered in the late 1990s, one from yet another species.

OCEAN HABIT

Mediterranean monk seals (*Monachus monachus*) are some of the largest 'true', or earless, seals – at up to 2.8 metres they're about a metre longer than harbour seals, and can weigh as much as 300 kilograms (660 lbs), some three times more than a large specimen. Mediterranean monk seals look, to the seal-uninitiated, like the harbour seals you see around the coast of the UK. With round dark eyes, a pug-like short snout bearing a bristle of stiff whiskers, and huge nostrils the shape of upside-down tears, they have an expression that's both sad and haughty. Distinguished might be a kinder way to put it – these seals need gentle treatment. The animals gained their name when naturalist Johann Hermann remembered that people in Marseille on the French Mediterranean coast called the local seal '*moine*', French for monk, and saw a captive seal of what he suspected was the same species arch up on the edge of a pool in Strasbourg. 'In this posture, it looked from the rear not dissimilar to a black monk, in the way that its smooth round head resembled a human head covered by a hood, and its shoulders, with the short, outstretched feet, imagined like two elbows protruding from a scapular, from which a long, unfolded, black robe flows down,' he wrote – in German – in 1779.[67] A scapular, for those not in frequent contact with monks, is a garment that hangs from the shoulders and is about chest-width wide but open at the sides. Male Mediterranean monk seals have a black back with a white patch on their belly.

Like monks, these seals are rarely seen these days. The species is one of only two left from an original group of three – the Caribbean monk seal went extinct in the 1950s. And numbers of the Hawaiian monk seal (*Neomonachus schauinslandi*), also known locally as 'īlio holo i ka uaua, or 'dog that runs in rough water', are decreasing. In May 1997, the Mediterranean monk seals living off the coast of Mauritania on the western Saharan coast of Africa experienced a mass die-off,[68,69] of about half the 270 animals in this sub-population.[70] Other Mediterranean monk seals live in the Madeiran archipelago in the North-east Atlantic, and off the Mediterranean coast and islands around Turkey, Cyprus and Greece, where at the time the population was around 500.[71] Along with the Hawaiian monk seal and California sealion, this species is unusual for being at home in temperate and warm climates – most seals and sealions, and indeed the walrus, like it cold. Off Mauritania, the Mediterranean monk seals were in a war zone and had taken to hiding in two caves. 'The whole area was filled with bombs,' Osterhaus says. 'So this was just a small population that was thriving because they had not been killed deliberately by fishermen.'

Again, after the die-off there was debate as to the cause – an algal bloom came under suspicion.[72] But again the culprit turned out to be a morbillivirus; clustering in caves put the seals in close contact and meant any infection could spread fast. In the past these seals would have lived on open beaches, though thanks to human persecution and the growth of tourism, they've moved away from beach life in the Mediterranean.[73] This time the culprit was neither CDV nor PDV, but most closely related to dolphin morbillivirus (DMV).[74] 'It was just after an outbreak in dolphins had taken place locally,' Osterhaus recalls. Dolphin morbillivirus is a strain of cetacean morbillivirus (CeMV) and is more similar to rinderpest virus and PPR virus (PPRV) than PDV and CDV.[75] Cetacean morbillivirus is probably older than PDV and CDV,[76] perhaps evolving when waste from livestock entered the oceans or a distantly related species such as an even-toed hoofed animal somehow passed on rinderpest virus or PPRV.[77] We first discovered a strain of cetacean morbillivirus at around the same time as the 1988 outbreak of PDV in seals, when harbour porpoises (*Phocoena phocoena*) washed ashore in Northern Ireland

and the Netherlands infected with porpoise morbillivirus.[78, 79] Two years later, from July 1990 onwards, a dolphin morbillivirus closely related to this porpoise morbillivirus killed at least 400 striped dolphins (*Stenella coeruleoalba*).[80] Today, as well as porpoise morbillivirus and dolphin morbillivirus, we know of numerous other strains of cetacean morbillivirus, including pilot whale morbillivirus, Guiana dolphin morbillivirus and more.[81] Unlike seals, which tend to haul out onto the same small area of land in large numbers and sniff each other, cetaceans remain in the water throughout their lives, so any snot or mucous they emit when ill disappears into the ocean. Perhaps dolphins, porpoises and whales catch morbilliviruses when they swim alongside each other;[82] researchers who used a drone to take samples from the blowholes of humpback whales off Brazil found a cetacean morbillivirus in their 'exhaled breath condensates'.[83]

Back in 1997, when biologists realised that four seal pups in caves on Mauritania's Ras Nouadhibou (Cap Blanc) Peninsula were no longer being fed, probably because their mothers had died in the outbreak, they took them to a seal rehabilitation centre run by the Centre National de Recherches Oceanographiques et des Peches (CNROP).[84] In one photo taken at the rehab centre, the pups lie next to each other in a row as if swaddled in their own blubber – no need for blankets – almost ranked in size like a series of Russian dolls.[85] The female Mediterranean monk seal has a more lackadaisical parenting style than the Hawaiian monk seal and may leave her single pup alone in the cave while she looks for food. In her defence, she nurses her pup for up to five months, the longest of any true seal.[86] Hawaiian monk seals, on the other hand, remain constantly at the side of their pups but nurse them for only six weeks, perhaps understandable given that they can't eat during this time. Rehab centre staff vaccinated the pups against CDV to protect them from morbilliviruses and fed them milk and later fish before releasing them back into the wild.[87]

To date, as far as we know, only the Mauritanian population of Mediterranean monk seals has experienced mass mortality from a morbillivirus. Conservation workers are aiming to protect these seals by dealing with some of the other stressors they face. Sometimes that's all you can do to help a species at risk of disease – in the oceans vaccination is even more tricky logistically than it is for wild animals on land, and has almost never been attempted. That said, despite the cost and practical difficulties, in 2016 the US National Oceanic and Atmospheric Administration vaccinated Hawaiian monk seals, creeping up on the animals as they dozed in the sunshine on rocky beaches in the remote Northwestern Hawaiian Islands and injecting them with a needle attached to the end of a very long pole.[88] (Around 1,100 of the seals live in the largely uninhabited Northwestern Hawaiian Islands and about 300 live in the 'main' islands, alongside humans,[89] where they're more at risk of exposure to CDV from domestic dogs). Apparently, the seals barely noticed the vaccine enter their blubber. The team vaccinated 700 seals, roughly half the population, in the first ever vaccination of a wild marine mammal; today the focus is on vaccinating the pups born each year. For Mediterranean monk seals, researchers have recommended considering vaccination,[90] but currently only animals in rehab, like those pups in Mauritania, routinely receive jabs.[91]

When Osterhaus became involved with conserving Mediterranean monk seals in Greece with the Hellenic Society for the Study and Protection of the Monk Seal (MOm), the species was on the brink of extinction. 'I remember when we first went there, when you went to the Pláka in Athens, the place where all the tourists go, and you talked to people, nobody would know that in Greece there were seals,' he says. As well as tourists, the conservation team talked to local schools and local people, and the Greek government set up nature reserves that exclude big fishing trawlers, whose nets can entangle the seals. 'Originally, local fishermen would kill the seals because they saw them as competitors,' Osterhaus says. But through the project, the fishers realised that monk seals usually eat different species of fish to their quarry and 'if they keep the seals alive, they can continue their own artisanal [traditional] fishing'. These days, pupping caves are protected and seals are monitored.[92] 'Today when you go to Greece, everyone knows there is the monk seal,' Osterhaus says. 'It's a little bit like the panda for China. The monk seal becomes kind of a means to engage public interest a little more, get the media there, and educate young people.'

Today, thanks to their greater protections, these seals are thriving. And the Monk Seal Alliance is working to make sure it continues that way.[93] In 2021 the total population, including youngsters, was around 800, up from just 400–500 at the turn of the millennium.[94] Having been ranked critically endangered in 1996 and downgraded to endangered in 2015, as of 2023 this monk seal is now merely vulnerable, according to the IUCN.[95] In the Mediterranean, the monk seals are more dispersed than in the caves on Mauritania's Ras Nouadhibou peninsula and they don't tend to group together to the same extent. With luck, an outbreak of a morbillivirus would be less devastating and spread to fewer individuals.

BACK ON LAND

Elsewhere, on both land and sea, morbilliviruses continue to cause wild-life trouble. And trouble for humans, who experienced a resurgence of measles in the face of falling vaccination rates during the Covid-19 pandemic.[96] These viruses that emerged on an unprecedented scale when we started farming livestock, later spread to our domestic dogs thanks to close contact, then reached the oceans perhaps via encounters on beaches with our canine companions or livestock waste; even the oceans weren't safe from our disease-inducing ways. But there is a small piece of good news, for African wild dogs, at least. In 2018, nearly a year since any wild dogs had been seen at Mpala, a single female wild dog reappeared, was joined by two wandering males and produced a litter of five pups.[97] The 'Phoenix pack' soon became some twenty dogs strong and today there are at least three packs of wild dogs in the Laikipia region.[98] Though the animal is at risk from CDV (a vaccine for African wild dogs is still a work in progress),[99] rabies, habitat loss (these dogs need large territories to catch enough food), conflicts with farmers, road accidents and climate change (which is likely to harm pup survival),[100] this time the African wild dog hadn't gone locally extinct. And the world's two surviving species of monk seal, some with the aid of vaccinations, are still with us for now in the face of a morbillivirus that, probably with at least some human help, made it into the oceans.

8

A Devil of a Problem: Farming and Tasmanians Devilled by Cancer

'Beauty is mysterious as well as terrible. God and the devil are fighting there and the battlefield is the heart of man.'

Fyodor Dostoevsky, *The Brothers Karamazov*, 1879
Translated by Constance Garnett

Once the Tasmanian tiger had gone extinct, the Tasmanian devil, at the size of a small dog, became the largest surviving marsupial carnivore. The animal is carnivorous by name as well as nature – its formal moniker is *Sarcophilus harrisii*, 'Harris's meat-lover', after the scientist who described the species in 1807. To indigenous people on this large island to the south-east of the Australian mainland, the devil was known as the poorininah.[1] The first colonial settlers, meanwhile, heard its loud screeching at night and dubbed it Beelzebub's pup.

Dark-haired, with a stocky body and impressive canines, the devil truly does sound ferocious – comparative oncology and genetics professor Elizabeth Murchison of the University of Cambridge, UK, describes the loudest of its calls as 'a scary screaming sound, quite unique and bone-chilling'. I've only heard recordings but it seemed to me like a cross between the roar of a lion and the moo of an enraged cow. It's perhaps not surprising that the devil is rumoured to eat humans if they venture too far into the bush. This is actually true, but only if they're already dead – Tasmanian devils are highly effective scavengers with powerful jaws, possessing one of the strongest bites for their size of any land mammal. As we'll see later, those early European settlers would change life for the devil besides changing its name. And you can see why they called it a devil, the visual similarity can be as striking as the auditory. In a picture on the Save the Tasmanian Devil site, a programme set up by the Australian and Tasmanian governments in 2003 and accessible via an email address that starts 'DevilEnquiries', one imp – also known as a pup or joey – resembles the kind of devil you might see carved on a stone gargoyle or on a wooden misericord for a monk to perch his bottom on. The imp's head is almost the same size as its body and it's coiled up with its hind leg bent, as if it has a knee, and its hind foot resting on its chin, clasped beneath the 'wrist' of a front leg that looks astonishingly human apart from its coating of fine black hairs – it's holding its front paw in a fist and you can't see the claws. The only thing that doesn't look devilish is its expression – the youngster is sleeping peacefully, eyes shut rather than bulging in a malevolent grimace.[2] Despite its small size, this devil and its brethren would largely take over Murchison's career.

Even though she hails from Tasmania, Murchison didn't become knowledgeable about Tasmanian devils until she moved to the US to study for a PhD in genetics in the early 2000s. 'Everybody always asked me about them because in the US it's the only thing they know about Tasmania,' she says. Even that knowledge comes mainly from the cartoon character Taz, who Murchison thinks does not represent Tasmanian devils well. His spinning, she believes, probably originated when a Warner Bros. artist saw a devil conduct the stereotypical pacing of an animal trapped in a zoo. Though devils do have a reputation for eating huge amounts in one sitting – around 40 per cent of their bodyweight.[3] 'They kind of gobble whenever they get a chance because they're scavengers,' Murchison says. 'That's why they would eat an entire carcass.' There is, however, no evidence that Tasmanian devils – as the cartoon Taz does – love all music apart from bagpipes. Taz is ferocious, and while devils can be aggressive with each other, with humans real-life Tasmanian devils are much calmer than the cartoon character. If they encounter people close-up, these animals are docile and 'very, very scared'. When trapped by researchers, a devil may hunker down in the bottom of the trap with an air of resignation, all four feet clustered together, their sharp claws pale pink as if the blood vessels are showing through, and just a glimpse of the white band across its chest – like the coloured stripe on the V-neck of a school pullover – that some devils possess. With long whiskers fanning out from its long nose, such a devil looks like a cute oversized rat or sleepy hamster. Fawn circles ring its eyes while two small patches above its eyes and parts of its muzzle are speckled with fawn too.[4] Infectious disease ecologist

Hamish McCallum of Griffith University, Australia, has a theory as to why devils take being trapped in a tube so calmly. 'This is being a bit anthropomorphic,' he says, 'but they're the biggest, baddest predator in Tasmania. I think when something bigger than them grabs them, they just go catatonic. It's so beyond their evolutionary experience that they don't know what to do.' If you trap brushtail possums, on the other hand, they will 'fight like hell because they're expecting something bigger than them to be trying to eat them'.

DEVILISH PROBLEM

In 1996 a photographer working in North-east Tasmania had spotted that devils were suffering from a mysterious disease that caused tumours on their faces.[5] The condition killed the animals within six months, sometimes snapping off their jaws. The new disease spread rapidly south and across the east of the island before slowly heading west. Nobody knew what caused the affliction, which was dubbed Devil Facial Tumour Disease, or DFTD for short. Murchison wanted to search for a virus but was working in a different field and didn't know how to get started. In 2006 she went back to Tasmania to visit family and spent several days in the Tasmanian wilderness on a hiking trip. She'd often hiked before, hearing Tasmanian devils screech at night or seeing their droppings during the day. The devils themselves had largely been elusive – they're nocturnal and tend to avoid humans, so Murchison didn't see them often. On the way back from this particular hiking trip, Murchison and her family came across a roadkill Tasmanian devil. And it had a small tumour, almost like a pimple, on its muzzle. 'It was pretty extraordinary,' Murchison says. 'I was interested in studying this disease, I didn't know how to get into it and then this devil just was there.' Murchison's road to studying DFTD turned out to be roadkill.

Having had a devil with DFTD fall almost into her lap, Murchison took samples from the tumour and co-ordinated with field researchers to obtain additional tumour samples in order to study their genetics.[6] The DNA from the tumour and the rest of the devil's body didn't match, and she was able to confirm that the disease is due not to a virus, but a transmissible cancer, as others had recently discovered.[7] Some cancers, like cervical cancer in humans, spread between individuals via transmission of a virus – in this case the human papillomavirus (HPV) – that makes cells more susceptible to cancer. On a few rare occasions in humans, cancers have transmitted through the direct transfer of the cancer cells themselves. In 1957, for example, Chester Southam injected human 'volunteers' (many were unknowing and/or incarcerated) with live cancer cells.[8] A few people, mainly those who were already ill and had compromised immune systems, developed cancer.[9] This transfer of cancer cells was induced artificially but a handful of species suffer naturally transmissible cancers. Cancers are usually self-destructive lifeforms – by killing their host they kill themselves. But becoming transmissible lets them live on as a kind of eternally dividing zombie.

For devils the transfer occurs when they bite each other as they fight over mates or food, or during mating itself. To look at the tumours that result, researchers entice the devil using a piece of meat on a string into a trap made of plastic pipe with a door that drops down; they then shake it out into a hessian sack and poke its head out of the top. 'It was actually quite hard to get them to open their mouths, so you'd often have to blow in their face,' McCallum says. 'You would often be able to actually put a twig in underneath their tongue to lift it up to see if there was something there. In all this time, they might be gently hissing, but they would certainly not be struggling in any way. It's quite surprising, frankly.'

Normally, if an animal encounters cells from another animal, its body rejects them as foreign. That's why people who receive an organ transplant must take drugs that suppress their immune system. But Tasmanian devils don't recognise these cancer cells as coming from another being and their bodies let the cells stay, with devastating outcomes. For Tasmanian devils, the result of DFTD was even more devastating than the spread of Covid was for humans. Numbers of the animal crashed across most of the island; once the infection strikes, it decimates the devils, with numbers falling to just 10 or 20 per cent of the original group size.[10] Since the 1980s, the Tasmanian devil population has dropped by some 70–80 per cent.[11] The structure of the population deteriorates too; most of the adults disappear, leaving mainly juveniles.[12] The species is now endangered; in 2008, the IUCN assessed the Tasmanian devil population to be as low as 10,000 animals and decreasing, from an estimate of 130–150,000 in the mid-1990s.[13]

SPEAK OF THE DEVIL

So did the cancer in Tasmanian devils arise spontaneously and spread without human assistance? No prizes for guessing that the answer is no.

The actions of European settlers, who not only renamed the poorininah but also cut down forests to make farmland and created tracks that later became roads, may have altered the fate of the devil, both at the time and generations later.

At the time, the colonisers effectively gave the devil a bad name and then hung him. In Hobart Town, settlers complained that devils raided their poultry yards and put a bounty on the devil's head, along with those of the thylacine and wild dogs. In 1830 the Van Diemen's Land Company offered 2/6 (25 Australian cents) for killing a male devil and 3/6 (35 cents) for a female in order to clear the animals from its north-west properties. The bounty was in place until June 1941, some five years after the thylacine went extinct, when the devil was protected by law.[14] To this day, some sheep farmers in Tasmania believe that devils kill lambs.[15]

In combination with other population declines in the early nineteenth century and the late nineteenth or early twentieth century, most likely thanks to disease epidemics, the bounty would have reduced Tasmanian devil genetic diversity.[16] As did the devil's restriction to Tasmania – devils died out on mainland Australia roughly 3,500 years ago, long before Europeans colonised, when changing and drying of the climate, increasing human numbers, the spread of the dingo, possibly a disease that was unable to cross the Bass Strait to Tasmania, or all of the above forced the mainland devils into extinction.[17, 18] In Tasmania, the dingo never arrived and the density of aboriginal humans remained low. This reduced devil diversity made it more likely that when the very first transmissible tumour arose spontaneously in a female in the east of Tasmania,[19] the next devil she bit wouldn't reject the cancerous cells in her saliva as foreign – they were too similar to its own genetic makeup to spot. Those cells spread from devil to devil, undetected. And so that first female's genes live on inside DFTD tumours, albeit significantly mutated.

SEE NO EVIL

A bitten devil's immune system should have two lines of defence against these cells. Almost all cells in vertebrates express a molecule called major histocompatibility complex (MHC) class 1 on their surface. That MHC is the most variable part of the genome. 'So you would have completely different MHC to me,' Murchison says. 'And I would have completely different MHC to somebody sitting next door.' The immune system looks out for foreign MHC as its first line of defence. If T-cells, a type of white blood cell, detect alien MHC on the surface of a cell, they attack that cell. That's why organ donors must be matched to the recipient – to ensure both parties' MHC molecules are as compatible as possible. Even then the recipient will need immunosuppressive drugs so that their body doesn't reject its new organ. It's possible that the cancer cells in the very first devil to suffer a transmissible tumour didn't express any MHC class I. Or perhaps they did and in the early days, the devil's low genetic diversity meant that individuals couldn't easily detect the cancer cells as foreign to their bodies then over the years, the DFTD cancer cells evolved to stop expressing MHC class I at all.[20]

In theory, that should cause the immune system's plan B to kick in and attack cells that don't express MHC on the grounds that they're abnormal. The devil's transmissible cancer cells, however, don't activate this second line of defence. And it's not completely clear why not. 'We think it's likely because tumour cells are making molecules that dampen down the immune response in some way,' Murchison says. As a result of the failures of both plan A and plan B, the bitten devil's immune system doesn't recognise the cancer cells as foreign and they are free to take up residence in its body.

INTO TEMPTATION

The other way that Europeans transformed life for the devil initially seemed beneficial. By farming and creating roads, the settlers unwittingly created prime conditions for devils to thrive. 'If you have animal carcasses lying around, both from agriculture and roadkill, there's a lot for devils to eat,' says Murchison. 'They used to be known as the garbage bins of the bush because if there was a carcass, it would just disappear.' Turning forest into farmland also created prime habitat for wild grazing herbivores. Their numbers rose and numbers of devils, who hunt herbivores such as pademelons (a small wallaby), Bennett's wallaby and wombats, boomed too. When I first see a picture of a pademelon my heart skips a beat – the animal has turned its head side-on to look straight at the camera, its long, thin triangular face and dark round eyes the spitting image of the face that peered at me from the bush when I woke early on a camping trip in northern Queensland. We stared at each other for a couple of seconds, this animal and I, trying to work out what the other would do next, before the pademelon decided I was a danger, hopped into the vegetation and disappeared. It was probably right, even if, unlike a Tasmanian devil, I wasn't a direct threat. Traditionally devils lived on coastal heath and in forests, but they took advantage of these new fields and their grazers. Thriving on their newly abundant prey, the devils reached a record high in terms of numbers per square kilometre in the 1980s.[21] In stark contrast with the situation for the Grevy's zebra, our changes to the environment had made the Tasmanian devil's life easier. For a while, at least. And no doubt to the detriment of other, less adaptable forest-dwellers who had lost their habitat.

For the devils, everything went well until DFTD emerged. While humans first saw evidence of the tumours in 1996, Murchison's studies indicate they may have begun in the mid-1980s. The record-high devil population density, and the roads that let the animals travel more easily across the landscape, could have helped the tumours to spread from devil to devil. In some ways the situation for the devils is similar to the ways we've transformed connectivity for people living in remote tropical forests. In the past, if you picked up a disease from one of the animals you hunted, chances are you'd transmit it to only a few people in your village, and perhaps the neighbouring village, before the outbreak fizzled out. Nowadays, if you travel by road to the nearest town to sell your haul of bushmeat, you can pass the disease to others who might then head to a city and give it to people who might travel to another continent by air. 'If you imagine Covid had arisen thousands of years ago, it might have just remained in one population,' Murchison says. 'It wouldn't have spread so rapidly.'

In 2014, the plot for the Tasmanian devil thickened. A new strain of the cancer emerged, in a devil from d'Entrecasteaux peninsula in south-eastern Tasmania, near where Murchison grew up and close to the outskirts of Hobart.[22] Murchison says the discovery of this second transmissible cancer was the most surprising moment in her entire career. 'Until then we thought that devils were just unlucky and there was an extremely unlikely circumstance that this happened – just bizarre,' she says. But the emergence of another transmissible devil cancer indicated that the disease is a natural phenomenon that has probably happened to devils before and will almost certainly happen again. 'Devils have some kind of susceptibility to this type of disease,' Murchison says. 'It really changed my perception.'

The two strains of devil-transmissible cancer arose from the same type of tissue – Schwann cells, which wrap around nerve axons (nerve fibres) in the peripheral nervous system to protect and insulate them.[23] If the nerve is damaged, Schwann cells leap into action to recruit the repair factors that will bring healing, dividing frequently. 'We think these devil cancers probably both arose from a nerve injury where these Schwann cells were busy repairing the damage and then failed to switch off and become normal again,' Murchison says. 'It looks like there's some inherent vulnerability in that cell type.' This vulnerability could be specific to devils; transmissible cancers are seldom seen in other species.

The second strain of devil tumour has not yet spread out of the isolated peninsula near Hobart. Murchison hasn't had much time to study it, but chances are that this strain will span the island – there's no practical way to stop it. 'The only options would be some kind of culling, which I don't think would be popular and would probably not work anyway, or building a giant fence, which just is not feasible,' she says.

The devil has already changed its behaviour to cope with the onslaught of this first strain. Before the cancer emerged, female Tasmanian devils began to breed when they were two years old and typically had three litters of imps – one each breeding season – before they died at around five or six. (Each litter has a maximum of four imps, as the mother's pouch has four teats; she may give birth to as many as thirty young but only the first four to reach a teat survive.) In diseased areas, nearly all devils older than two die from the disease and female devils have started to breed younger – in five sites where DFTD is present, sixteen times more females than before are able to breed in their first year.[24]

This precocial breeding gives Murchison hope that devils may be able to persist at low densities, with young animals reproducing and surviving the cancer for the nine months their offspring need to become independent. 'The question is whether that's a stable state or whether other factors could cause devils to die out,' Murchison says. The disease could potentially also disappear locally, enabling populations of devils in the area to thrive once more. 'Then disease probably would come in again,' Murchison speculates. 'So at the moment it's very unclear.'

SAVING THE DEVIL

Only one or two populations remain free of the first strain of the disease and, according to Murchison, 'it's probably just a matter of time before they get infected as well'. With this in mind, in 2006 the then three-year-old Save the Tasmanian Devil programme set up an insurance population of devils in captivity by taking 122 founder animals from the wild, in four intakes between 2005 and 2008.[25] Many of these devils were taken from North-west Tasmania, as the devils there were disease-free at the time.[26] By 2017 the programme had 700 devils on its books. Today, they're distributed among some 40-odd zoos in Tasmania and the Australian mainland, the 116 square kilometres (45 square miles) of Maria Island off eastern Tasmania, Tasman-Forestier Peninsula in South-east Tasmania, which has been fenced off to prevent diseased devils entering, and a number of fenced enclosures in parks.[27] In 2020, Tasmanian devils returned to live on the mainland for the first time in thousands of years when 26 adults were released into a 990-acre protected area in New South Wales.[28] Some people want to release a selection of these insurance animals back into the wild to increase genetic diversity. 'My view is that that's a pretty foolish thing to do,' says Hamish McCallum, who worries that releasing insurance animals, who have never encountered the cancer and so are fully susceptible to it, into the mix could slow down, and perhaps even reverse, any evolution of resistance to the disease by existing wild Tasmanian devils. As with anything conservation-related, the situation is complicated – on Maria Island the devils have destroyed the island's short-tailed shearwater colonies and wiped out its little penguins.[29, 30] The world of conservation brings devilish choices.

BETTER THE DEVIL

The only real way out from the bind the devil's in would be a vaccine. But developing a vaccine is tricky, as the immune system doesn't have much to latch on to. 'These DFT [devil facial tumour] cells don't really have anything on them that differentiates them from any of the devil's normal cells,' Murchison says. 'It's asking the immune system to see something which is basically itself, which is a big challenge.' Researchers are currently hunting for subtle differences between devil facial tumour cells and normal devil cells. As yet there's no vaccine that works, although some approaches have primed the devils' immune system to recognise DFTD cells as foreign and made it possible to treat tumours with immunotherapy until they disappear.[31]

Without a vaccine, the devil's in a fix. 'The bottom line is, I don't think the outlook looks very good,' Murchison says. Like the thylacine before it, the Tasmanian devil may vanish. For now, Murchison is seeking to understand whether different tumours evolve different sub-lineages that behave differently, 'like we've seen with Covid'. She's also keen to discover whether the tumour will evolve to become more transmissible or more able to escape the immune system. Either of these could push the devils even nearer to the brink.

HANGERS-ON

For now, the Tasmanian devil is teeming with life – it can be home to up to 28 species of parasite, including nematode worms, flatworms, protozoans and ectoparasites that live on its skin such as mites, ticks and fleas. Five of these twenty-eight parasite species live all or part of their life cycle only on Tasmanian devils, so if the devil goes extinct, these five could vanish too.[32] The *Dasyurotaenia robusta* tapeworm found in the devil's small intestine was already classified as rare under the Tasmanian Threatened Species Protection Act of 1995.[33] Unless it can adapt to live inside another species, its time might be up.

The Save the Tasmanian Devil programme aims to conserve not just the devil but also its parasite ecology.[34, 35] If you disrupt an animal's interactions with parasites you might unwittingly harm that very same animal, perhaps by altering the development of its immune system, disrupting the wider ecosystem or making other parasites – and the diseases they can bring – more prevalent. The programme is also taking care not to introduce devil parasites and novel pathogens to humans and domestic wildlife.

The decline of the devil is altering the web of life in other ways too. As devil numbers drop, the activity of feral cats, which tend to avoid devils, is increasing.[36] That's a problem for the small and medium-sized native mammals, such as the eastern quoll, swamp rats, bandicoots and potaroos, that these cats eat. And it's possible that reduced numbers of devils could let introduced red foxes thrive, again with ominous consequences for many mammals, including the devils themselves.[37] The disappearance of the devil may even trigger a cascade of other diseases, boosting the number of Tasmanian herbivores such as Tasmanian pademelons infected with the intracellular parasite *Toxoplasma gondii* thanks to the increase in cat numbers – cats are the only known host in which this parasite can reproduce.[38] And, without so many devils gobbling them up, carcasses tend to stay in place 2.6 times longer, potentially encouraging the spread of diseases to other wildlife, livestock and humans.[39] As just one example, carcasses provide food for maggots and increase the risk of blowfly strike to sheep.

So it's fortunate that transmissible cancers only affect a handful of species. 'Transmissible cancers are a bit of an oddity in nature,' Murchison says. 'As cancer itself is pretty common both in humans and other animals, that suggests it's really unlikely for a cancer to make that step to being a transmissible cancer.' Such a cancer needs not only a way to escape the immune system but also a way to transmit, and 'probably those transmission mechanisms are not present very often'.

CANINE CANCER

As a result of her work on the Tasmanian devil, Murchison grew interested in the only other transmissible cancer in a mammal. Known as canine transmissible venereal tumour (CTVT), or Sticker's sarcoma, this cancer spreads when dogs mate. Thanks to the anatomy of the dog penis, dogs have a 'coital mating tie' that physically locks them together for some time. 'That likely means they are particularly at risk of a sexually transmitted cancer because it gives time for cells to contact,' Murchison says.

Cells from the tumour of the very first dog to suffer from this cancer – the 'founder dog' – are still alive, in a sense, inside tumours in other dogs today, even though the rest of the founder dog's body died millennia ago; the tumour cells are the oldest mammalian lifeform that we know of. 'When we look at the tumour cells, we actually have the DNA from that dog,' Murchison says. 'It's pretty crazy.' From this 'ghost DNA', Murchison reconstructed what the long-lost founder dog may have looked like – pricked ears, long snout, barrel chest, lean belly, long legs, athletic stance a little like a coyote or a dingo. Studying dog genetics indicated that this dog lived in Siberia or thereabouts,[40] some 8,500 to 4,000 years ago.[41]

By analysing mutations in the DNA of dog tumours from around the world, the team traced not only where the tumour originated but also how it evolved as it spread around the world.[42] For the first few thousand years of its existence, the cancer seems to have remained isolated in the same region as the founder dog, most likely in north-eastern Siberia. Then CTVT spread slowly west, first to India then to other parts of Asia and Europe, potentially as dogs travelled with the humans who were trading goods along the Silk Route. From around 500 years ago, the cancer began journeying by sea, on board dogs on board ships. Early colonisers brought it to the Americas, although it may have already arrived inside dogs from Siberia that entered North America thousands of years ago and became the Native American breed of dog.[43] This breed has almost completely disappeared. It would make a neat story if it was killed off by the very CTVT that its ancestors gave rise to, but there's no evidence either way. Today, dogs in the Americas are descended from European dogs, and the DNA of the ancient Native American breed that was closely related to those early Siberian dogs remains 'alive' only inside CTVT cells.[44] In terms of today's breeds, the ancient Native American dog is most similar to northern breeds – Spitz dogs such as the Alaskan Malamute and the Siberian Husky. Basically, sled dogs. Some descendants of those Native American dogs may live on, Murchison believes – we just haven't taken genetic samples from enough dogs to find them yet.

A few hundred years after CTVT reached North America, in the nineteenth or perhaps even the eighteenth century, the disease spread around the world still faster, in just a couple of events. 'Probably there were just a few dogs on boats that went to lots of different places,' Murchison says. 'They mated with lots of local dogs and that was it.' These sailor dogs had at least one girl or boy in every port, and by roughly a hundred years ago the disease was everywhere. Today, in the developed world the removal of free-roaming dogs has largely eliminated CTVT, but there has recently been an increase in the UK due to the adoption of dogs without adequate health checks from areas such as Romania.[45] Fortunately, these days, we are able to treat the disease with chemotherapy and most dogs recover. We can help with the fallout from our enabling the disease to spread around the world, by land and sea.

ALL AT SEA

So far, the animals we know to suffer from transmissible cancers are Tasmanian devils, dogs and, slightly incongruously, shellfish. In 1969 scientists discovered that mussels from the North-east Pacific were suffering a leukaemia-like disease known as disseminated neoplasia. Cells in the shellfish equivalent of blood, the haemolymph, would multiply then infiltrate other tissues and stop them working. Today such diseases have been implicated in mass die-offs of wild and commercial populations of various different shellfish. And we know that some of these leukaemia-like cancers transmit to other individuals; sometimes they even jump between closely related species.[46] At least eight species of marine bivalve molluscs around the world, including mussels, clams and cockles,[47] suffer from bivalve transmissible neoplasia, as it's known, and the disease is a suspect for causing mass mortality events. It's likely that shellfish suck in cancerous cells from the seawater around them as they filter feed for particles of plankton. Here too man may be, at least in part, to blame for the disease's spread after it appeared spontaneously in one or more unlucky individuals.

To start with, our pollution of the oceans may make shellfish less healthy and more susceptible to disease, although there's no strong evidence for this as yet.[48] And in 2019 researchers reported that mussels from two species – *Mytilus chilensis*, which chiefly lives in South America, and *Mytilus edulis* from Europe – both suffered from a transmissible cancer that had probably arisen in a third species, *Mytilus trossulus*, that typically dwells in the northern Pacific.[49] That means the cancer cells had travelled from the Northern to the Southern hemisphere and across the Atlantic and Pacific oceans. That's quite a feat, particularly when you take into account that mussels don't live in the tropics and ocean currents don't flow across the equator. It's pretty unlikely that the mussels managed to transport these cancer cells without assistance. One feasible explanation is ships, carrying diseased mussels on their hulls beneath the waterline or in their cargo.[50] It's hard to prove this happened but ships are known to have mixed mussel species around the world, introducing the European *Mytilus edulis* to Canada, the Mediterranean mussel to Chile and Argentina, and *Mytilus trossulus* from the northern Pacific to Scotland, among others. It's highly likely that our shipping trade, as it did for dogs, spread transmissible cancers to shellfish worldwide.

Chances are that we haven't yet discovered all the species that suffer transmissible cancers. To become transmissible, a cancer needs not only a way to escape the immune system, but also a way to transmit – devils bite each other's faces, dogs lock together during mating and shellfish filter feed. Although these mechanisms are fairly rare, in susceptible species like devils and certain marine bivalves, transmissible cancers can arise repeatedly. One study considered the odds of transmissible cancers arising in animals using the Drake Equation – the same mathematics used in 1961 to determine the chances of finding intelligent and communicative extra-terrestrial life elsewhere in the Milky Way.[51] According to these calculations, more species than we know about do indeed have transmissible cancers, and more bivalve species have transmissible cancers than mammals; bivalves don't possess MHC molecules and have a less complex immune system. The ocean could be a fruitful place to look in general, Murchison believes; aquatic organisms might die from leukaemia-like diseases without exhibiting any obvious symptoms of cancer.

For now, Murchison would like to prevent the Tasmanian devil becoming the first known animal to go extinct from cancer. That might sound like a natural phenomenon, but chances are that colonial settlers helped the cancer spread when they built roads and cleared forests to farm, inadvertently enabling devil numbers to boom. As devil researchers Hamish McCallum and Menna Jones wrote, 'In the words Oscar Wilde put into Lady Bracknell's mouth, to lose one large marsupial carnivore may be regarded as a misfortune; to lose both would look like carelessness.'[52]

9

Fungus and Frogs: Lab Animals, Pets, Food, and Trading Amphibian Disease Internationally

'Eye of newt, and toe of frog,
Wool of bat, and tongue of dog,
Adder's fork, and blind-worm's sting,
Lizard's leg, and howlet's wing,
For a charm of powerful trouble,
Like a hell-broth boil and bubble.'

William Shakespeare, *Macbeth*, 1623

Amphibians are struggling. Roughly 200 species have gone extinct since the 1970s.[1] And 41 per cent of the survivors are threatened with extinction, a far higher proportion than for birds (12 per cent) and mammals (26 per cent).[2] The true number in trouble is probably even higher, as many amphibian species have yet to be assessed by the IUCN. Herpetologists (specialists in amphibians and reptiles) first noticed amphibian declines and extinctions in various parts of the world – most notably Australia, Central America and South America – in the 1970s. But not until 1989, at the very first World Herpetology Congress in the UK's cathedral city of Canterbury, did scientists realise that these disappearances might be linked. In many areas, frogs were suffering habitat loss, pollution and poor freshwater management. Many of the losses, however, were in completely pristine environments and so were dubbed 'enigmatic'.[3] It would take twenty years to work out why amphibians were having heart attacks and dropping dead even in areas almost untouched by humans. Herpetologist Karen Lips told wildlife health expert Andrew Cunningham that the frogs basically just turned to stone – they still looked alive, but when you picked them up they were dead. At first, experts debated whether amphibians were in decline at all. Then people put it down to acid rain or ultraviolet radiation let through by the thinning ozone layer or another environmental change or something humans were doing.[4] Very few people suspected a disease, particularly as the frogs showed no sign of illness. At the time, nobody thought infectious disease was important for wildlife conservation. It was a real puzzle.

As the frogs' plight worsened, that thinking would change. In 1988, Cunningham took a job as a pathologist at the Zoological Society of London, the organisation behind London Zoo. Soon people began phoning in about dead frogs in their gardens, worried they'd done something wrong such as poisoning them by creosoting the fence to stop its wood from rotting. 'Initially I dismissed it, but more and more reports were coming in and some people were very emotional and wanting something to be done about it,' Cunningham says. Together with professional herpetologist Tom Langton, in 1992 Cunningham set up the Frog Mortality Project so that people could report their findings and send in bodies for post-mortems. The deaths, the pair discovered, were due to a virus newly arrived in the UK, probably from North America. It was the first discovery of a ranavirus, a virus containing strands of DNA rather than RNA, killing amphibians in the wild in Europe.

In 1996, Cunningham travelled to Australia at the invitation of the World Organization for Animal Health ranavirus reference laboratory to research the newly discovered UK ranavirus and compare it to other ranaviruses. While he was working there, another researcher, Alex Hyatt, approached him about an investigation into unexplained frog deaths in Queensland. 'Amphibian disease, mortalities and pathology was a complete Cinderella area,' Cunningham explains. 'It wasn't even considered to be veterinary work – I was a vet and people thought it very strange that a vet would be working with amphibians. And so Alex said, "You know something about frog disease stuff, do you think you can help?"' As we saw with rinderpest, wildlife vets are often in short supply.

HOPPING SICK

Hyatt's masters' degree student Lee Berger, now a professor at the University of Melbourne, Australia, was investigating the deaths of frogs found in the Big Tableland in the Queensland rainforest: the sharp-snouted day frog *Taudactylus acutirostris*, a small (2–3 centimetres long) brown frog with a white-edged snout that looks like it should lift up, like the triangular cover on the spout of a metal teapot, to reveal the insides of its head; the slightly larger common mist frog *Litoria rheocola*, typically about 3 centimetres long for males and a little more for females, with a beige back, cream undersides and enormous round eyes; and the larger still, at 5 centimetres, waterfall or torrent tree frog *Litoria nannotis*, its back awash with swirls of mottled brown and fawn, and almost indistinguishable from the rocks it clings to. By 1996 at least 14 species of stream-dwelling frog in eastern Australia had disappeared or declined by more than 90 per cent during the previous 15 years.[5] Berger examined the dead frogs' internal organs and even their eyes and brain. But there seemed to be no reason why the animals had died. 'We can't see anything wrong with them at all,' Cunningham recalls Berger's supervisor telling him. 'Berger's not going to get her master's unless you can help her.'

At this point Cunningham's habit of visiting other zoos while on holiday came in handy. A few years earlier he'd dropped in on Melbourne Zoo. When the zoo vets learned Cunningham worked with amphibians, they told him about their problems keeping one of the frog species, the sharp-snouted day frog, in the zoo alive. This species was nearing extinction by 1993 when an expedition failed to find any adults but took the last remaining tadpoles they could locate into captivity at Melbourne Zoo and James Cook University. At both locations, the tadpoles successfully metamorphosed into adults but the froglets only survived for two or three weeks. In Melbourne, the zoo vets suspected they weren't feeding the creatures what they needed, and asked Cunningham to look at their health and pathology records. The team had sent dead frogs to a couple of pathology labs, but when they looked at thin sections of the frogs' tissues under the microscope, the labs had found nothing conclusive. The first lab said it could see nothing wrong apart from an overgrowth of fungus in the skin. Since the frog's skin had not reacted by becoming inflamed or damaged, the scientists in question assumed the fungus grew after the frog had died. The second lab, on the other hand, couldn't find anything wrong apart from a protozoan (single-celled animal) invasion of the skin, which they also assumed happened after death because the frog had not reacted. Aware that under an optical microscope some protozoa look like fungi and some fungi look like protozoa, Cunningham realised the labs were probably seeing the same thing. 'One's calling them a protozoan and one's calling them a fungus,' he recalls telling the Melbourne vets. 'And it's really strange to see such a consistent change. So maybe there's something going on here.' The mystery would soon be solved but too late for the sharp-snouted day frog; none of the animals in captivity survived and the species is thought to have gone extinct in 1997.[6]

When Cunningham heard about Berger's frogs, he suggested she look at the skin. Sure enough, Berger found the same tiny organisms that looked like protozoa or fungi. Examining them under an electron microscope revealed that each organism was a chytrid fungus, named after the Greek for 'little pot' because of the container for its spores. To determine if the presence of this fungus was a coincidence or a killer, researchers exposed healthy frogs to the skin of an infected dead frog. The healthy frogs picked up the fungus and died in the first known infection of a living vertebrate by a chytrid fungus.[7] Generally, members of the Chytridiomycota phylum live in soil or water and feed on decaying organic matter; up to this point chytrid fungi had only been found as parasites in fungi, algae, vascular plants, rotifers, nematodes and insects. Today the disease the fungus causes in amphibians is known as chytridiomycosis (kit-RID-ee-oo-mike-OH-sis), mycosis meaning a fungal infection.

The fungus, later dubbed the doomsday fungus or, almost as terrifyingly, *Batrachochytrium dendrobatidis* (*Bd* for short), affects all types of amphibian – frogs, toads, salamanders, newts, caecilians (legless amphibians that live underground and frankly look a lot like worms), the lot. But the fungus only affects parts of the body containing keratin – the protein also found in human skin, hair and nails. In tadpoles only the mouth parts contain keratin, which may be why tadpoles typically survive the fungus until they metamorphose.[8] In adult frogs, *Bd* attacks the skin but not the nasal cavity, mouth, tongue, conjunctiva or intestines. Ultimately chytridiomycosis would be named the worst infectious disease ever recorded among vertebrates 'in terms of the number of species impacted, and its propensity to drive them to extinction'.[9] In 2021, with the last individual having been seen in the wild in 1997,[10] the IUCN assessed the sharp-snouted day frog as extinct, although the two other species whose deaths Berger investigated are faring better – the mist frog has been downlisted from endangered to near threatened,[11] and the torrent tree frog is of least concern.[12]

WORLDWIDE WEB(FEET)

While Cunningham was working with Hyatt and Berger in Australia in 1996, he received a call about dead frogs that had appeared in western

Panama.[13] In the Fortuna Forest Reserve, the variable harlequin frog *Atelopus varius*, a poisonous yet striking combo of black, orange and yellow splodges, the white-spotted cochran frog *Cochranella albomaculata*, with a body in delicate green covered in tiny pale dots, the brown-coloured yellow-bellied Chiriqui robber frog *Eleutherodactylus cruentus/ Pristimantis cruentus* and many more all croaked. As Cunningham was busy in Australia, he arranged for the carcasses of these Central American frogs to be sent to a veterinary pathologist colleague in the US, D. Earl Green at the Maryland Animal Health Laboratory. When Cunningham told Green about the fungus in the skin of the Queensland frogs, Green found the same organism in his frogs. The pair realised it was likely that the very same pathogen was killing frogs in both Australia and Panama, despite the 14,500 kilometres (9,000 miles) between them. And when Cunningham, Green and other researchers met in the US and compared the organisms found in Australian and Panamanian frogs under an electron microscope (molecular genetics not yet being available), they were able to confirm that the same organism was killing frogs on different sides of the world. In 1998, the researchers published a paper that explained the mystery of the missing frogs,[14] with Berger as lead author. Berger was never awarded her master's degree – she earned a PhD instead.

'Disease is now considered to be a significant threat to species conservation,' says Cunningham. 'But at that time it wasn't, so it was quite a breakthrough, not just for amphibians, but for conservation biology in general.' Many people remained sceptical even after publication of the team's paper,[15] with one eminent herpetologist saying chytridiomycosis was a fungal infection of the skin and querying how many people have died of athlete's foot.[16] 'That just shows you what we were up against trying to get people to accept that this was a real problem for amphibian conservation,' Cunningham says. Skin is important to any animal and even more so to amphibians, who use theirs to take in oxygen and sometimes even liquids via a 'drink patch' on their undersides. Amphibian skin also acts like our kidneys, regulating the concentration in the blood of ions such as potassium, sodium and calcium. If a frog is infected with chytrid fungus, its skin can't do this job and these ions leak from its body. For potassium in particular, this is a big problem – blood potassium levels affect how well muscles work in any animal. 'The most important muscle in your body is your heart,' Cunningham says. 'Amphibians basically die of a heart attack when they've got this infection of their skin.' Some species suffer small sores at the end of their toes, but often frogs show no signs of illness whatsoever when you find them dead; that's why it was so hard to work out what was killing them.

These days, chytridiomycosis has won the dubious accolade of being the infectious disease that has wiped out the most species;[17] some 90 species of amphibians that were susceptible have gone extinct.[18] Part of the problem might be that spores of *Bd* can survive in the environment, at least for a while, meaning it's easier for the pathogen to make a species go extinct.[19] And fungi can reproduce fast and infect many individuals before that species' population density declines so much that the pathogen can no longer spread.[20] What's more, some species can survive with the infection and act as reservoir hosts, providing a constant source of infection even when the population density of a susceptible host species has declined.[21] *Bd* has contributed to the declines of at least 410 other amphibian species, 124 of which have seen reductions of at least 90 per cent.[22]

TRADING FUNGI

Berger, Cunningham and their colleagues had pinned down the culprit for the mystery deaths. But where did *Bd* come from and why did it start killing amphibians in the 1970s? The fungus has at least six different versions, or lineages. The hypervirulent form of *Bd* killing amphibians around the world is known as the Global Panzootic Lineage (GPL). It's a hybrid and probably didn't evolve naturally. We don't know where it originated but genetic tests indicate that the *Bd* it emerged from evolved in Asia,[23] most likely on the Korean peninsula. This area is a hotspot for *Bd* diversity and amphibians there don't die from *Bd* infections, indicating that they have lived alongside *Bd* for a long time. Whereas other *Batrachochytria* species and *Bd* lineages are typically hundreds of thousands of years old, *Bd*GPL emerged in the early twentieth century,[24] according to scientists' readings of its genes. That's about the time that amphibians were first traded internationally, by ship, plane, road and presumably rail for use as food, pets and in medical research. 'It is likely no coincidence that our estimated dates for the emergence of *Bd*GPL span the globalisation "big bang" – the rapid proliferation in inter-continental trade, capital, and technology that started in the 1820s,' writes Simon O'Hanlon of Imperial College London and his colleagues in their paper.[25] It looks like different lineages of chytrid fungus mixed and matched when traded amphibians were brought together, forming a hybrid lineage that, unlike its parents, was highly virulent to amphibians. 'In some ways, it's a human-constructed pathogen,' says Cunningham. 'I term it as Frankenstein's monster because it actually is a monster of human creation.' We didn't deliberately create this fungus in a lab, but our trading practices did, inadvertently, help it come into existence.

Another form of human creation may have contributed to making this monster. From the 1930s until the 1960s, the African clawed frog (*Xenopus laevis*) was the standard provider of pregnancy tests; if you inject a female of this species with urine from a human, she will respond to any pregnancy hormones in the urine by laying eggs. At the time, this was the most reliable way to deduce human pregnancy and these frogs were installed in labs around the world. As one of the earliest species to be traded internationally, the African clawed frog may have contributed to the spread of chytrid fungus, especially as these frogs become infected but generally don't become ill. South Africa was the original suspect for being the source of virulent *Bd*, both because the nation exported African clawed frogs for pregnancy tests and medical research, and because it was home to several lineages of chytrid fungus.[26] In 2004, an African clawed frog collected in the Western Cape lowlands in 1938 became the earliest known case of chytridiomycosis when researchers retrospectively analysed samples from museums.[27]

The amphibian trade didn't just supply pregnancy testing; the food, pet and medical research trades in amphibians continue to this day. Many cultures have a long tradition of eating amphibians, but this local consumption was, until chytrid fungus hit, largely sustainable. The international trade less so. From 2015 to 2020, the US imported 14.5 thousand tonnes of North American bullfrog *Lithobates catesbeianus* in the form of wild-caught or farmed live frogs and frozen meat.[28] Although the bullfrog is native to the US, it has been released and farmed in many countries and the imports came from Mexico, Ecuador and China. The species is tolerant to chytrid, becoming infected but rarely showing symptoms of disease; Cunningham suspects its introduction elsewhere, whether deliberate or accidental, has taken the infection to new areas. The US also imports three other species of frog for food, the East Asian bullfrog *Hoplobatrachus rugulosus* from Thailand and Vietnam, Forrer's leopard frog *Lithobates forreri* from Mexico, and the pig frog *Lithobates grylio*, a US native that's farmed in China. Meanwhile, the European Union, another massive area of frog consumption, particularly in France, imported 40.7 thousand tonnes of frogs' legs, mainly in frozen form, between 2010 and 2019.[29] Depending on their size, the legs must have once belonged to between 814 million and two billion frogs, nearly three-quarters of them from Indonesia, a fifth from Vietnam, and others from Turkey and Albania. Meanwhile, in a disturbing trend in the pet trade, collectors track down newly discovered species of frog using location details provided in the species description and sell them at high prices. Some 352 amphibian species were on sale as pets in Germany in 2017–18; 46 of them had only been described in the preceding ten years, one only three months earlier.[30]

MIDWIFERY MIX-UP

Even conservation workers have inadvertently spread *Bd*. At Jersey Zoo in 1991, during a captive breeding programme for the Mallorcan midwife toad *Alytes muletensis*, in which the male carries piles of eggs on his hind legs until they're ready to hatch, some of the Mallorcan midwife toads died, and to start with it wasn't clear why – this was seven years before *Bd* was identified as a killer of frogs. In 2004, when a young Mallorcan midwife toad died of the by-now-identified infection in the wild, researchers looked at samples from the captive breeding programme and found *Bd* in both the Mallorcan midwife toad and the endangered *Xenopus gilli* Cape Platanna frog that had been housed in the same room and can carry the *Bd*CAPE lineage without becoming ill.[31] It looks like midwife toads from the breeding programme in Jersey that were reintroduced to Majorca took *Bd*CAPE they'd picked up from the African frog with them.[32] 'If you're going to keep animals for reintroduction programmes, probably the worst place to keep them is in a zoo because they're likely to come into either direct or indirect contact with the parasites and pathogens of other species,' says Cunningham.

These days in European zoos, some species whose offspring are destined for release into the wild are held in biosecure areas that have no direct or indirect connection to the main part of the zoo. 'It's best if they're not in a zoo at all,' Cunningham adds. 'But it costs a lot of money and expertise to keep these animals, and if they weren't there at all, they'd probably be extinct.' In Mallorca, the conservation workers did at least attempt to clean up after themselves. By transporting midwife toad tadpoles to a lab in plastic bottles and bathing them in an antifungal drug every day for a week, as well as draining infected ponds, treating them with Virkon S disinfectant and waiting for them to refill naturally, a team managed to eliminate the infection.[33] This was possible purely because the Mallorcan midwife toad was the only amphibian host in an isolated ecosystem on an island.[34] And it looks from follow-up studies as if the pathogen managed to persist in the face of the clean-up after all.[35]

CARIBBEAN CHICKEN

Cunningham's amphibian studies have taken him far and wide. He's currently working in the field with Darwin's frogs in Chile and the mountain chicken frog (*Leptodactylus fallax*) in the Caribbean. This frog, which has eyes that are dark apart from an upper arc of bronze, and a black under-eye stripe that flicks backwards and down towards its reddish-brown back and powerful forelegs, eats insects, snakes, small mammals and even other frogs.[36] Officially called the Giant Ditch frog,[37] the mountain chicken may get its nickname from the way its young froglets stay near their mother, like chicks clustering around a hen. These froglets receive care in their earlier life stages too – both parents guard the foam nest and its eggs, and the female feeds the tadpoles infertile eggs each week.[38,39] Cunningham enjoys working in the field – 'it's fantastic, who doesn't love amphibians?' – but the mountain chicken can be hard to find. To track one down, you need to wait until the male is calling to attract a mate to his chamber, where he hopes they'll one day guard eggs together. 'If you find the male, you may be lucky and find one or two females that have also been attracted to his call,' Cunningham says. Though your difficulties aren't over yet – at about 20 centimetres long and a kilo in weight, some 40 times heavier than a European common frog,[40] mountain chickens are one of the largest frogs in the world and their chunky beige thighs banded with darker brown can leap the frog six or eight feet into the air. 'You have to be very quick,' says Cunningham. 'A couple of leaps and they've gone so you've got to creep up on them and pounce.'

One reason the mountain chicken is hard to find is that few survive in the wild. When chytrid fungus arrived on the Caribbean island of Dominica in 2002, the population experienced one of the fastest range-wide species declines ever recorded – more than 85 per cent of the island's mountain chickens disappeared in less than 18 months.[41] The authorities identified ways to prevent the fungus reaching the mountain chickens' other hold-out on Montserrat, including removing amphibians before shipping produce, euthanising any tree frogs or other amphibians found in shipments of goods from Dominica, and awareness-raising for exporters, importers and the public.[42] Dominica started to remove amphibian stowaways and contaminated soil from produce for export to Guadeloupe and Martinique in 2007, but Montserrat did not require these measures.[43]

The *Bd* fungus reached the island of Montserrat in 2009. 'You could see the wave of the disease going through,' Cunningham says. Chytrid hit a population that had already plummeted, possibly due to lava flow and ash fall from the eruption of the Soufriere hills volcano in 1995.[44] Cunningham and colleagues put themselves ahead of the wave and took 50 mountain chickens into a captive breeding programme. Other individuals they caught and bathed in a solution of the anti-fungal drug itraconazole up to five times a week over seventeen weeks.[45] This improved survival and reduced infection rates but did not provide long-term protection.[46] And it was labour-intensive. 'You can't go around spraying antifungals in the wild because fungi are such a hugely important component of ecosystems,' Cunningham says. 'Even if we had an antifungal that just killed chytrid fungi, we wouldn't be able to use it because there's a whole range of different chytrid fungi that are important for activities such as breakdown of vegetation and nutrient cycling.' Besides, frogs may live on land, up trees, in water and underground, so it would be tricky to access all species.[47] Today, the only mountain chickens on Montserrat are reintroduced and semi-captive, protected by heat treatments such as rocks to bask on or solar heating mats or corrugated steel sheets in their ponds that raise temperatures above 28°c, hotter than the fungus can survive,[48,49] as well as treatment with antifungals most years.[50]

Back in Dominica, hunting mountain chicken was banned in 2003, despite the frog being a source of income for hunters, restaurants and the tourism industry.[51] And in 2016 the crapaud, as it is known locally, was replaced as the official national dish of Dominica by callaloo, a stew containing vegetables and coconut, to discourage people from relying on the frog for food.[52] Despite its removal from the national menu, the mountain chicken is still a key feature of the Dominican coat of arms, featuring in the top-right quadrant.[53] At one point, researchers thought the frog had been wiped out. Then they discovered that around a hundred had survived and were breeding, despite being infected. Evolution had done its job – the frogs appeared to have developed genetic resistance to chytridiomycosis. In 2017, however, Hurricane Maria killed 90 per cent of these survivors. 'It's just awful what those frogs have had to put up with,' Cunningham says. 'We're just talking about a handful of animals left in the wild now. They're perilously close to extinction in the wild. In fact, they're perilously close to extinction, full stop.' A count by the Mountain Chicken Recovery Programme in 2023 found 21 live mountain chickens.[54]

JUMPING TO IT

Around the world, biologists have tried various techniques to battle *Bd*. To the uninitiated, a vaccine might sound like the answer, but it's much harder to make a vaccine against fungal diseases than bacterial or viral ones; most attempts at immunising amphibians to protect them from *Bd* have failed.[55] Bacteria themselves may have the answer; in some amphibians, certain types of these organisms inhibit *Bd* on the skin. After a dunking in a probiotic bath, mountain yellow-legged frogs (*Rana muscosa*) from the Sierra Nevada had lower loads of the fungus, and a year later nearly 40 per cent of them had recovered.[56] It remains to be seen whether probiotics pose any risks to ecosystems, and how practical they might be to develop and apply, but if they do work, they would be more cost-effective, more ethical and less controversial than treating the environment with *Bd*-destroying chemicals.[57]

To date, no reintroduction from a captive breeding programme has fully succeeded, as chytridiomycosis has always returned, after lurking in other amphibians or the environment. Such programmes will need measures to deal with chytrid in the environment before they release their captives. They might even need to cull reservoir and superspreader hosts.[58] Worldwide, we can probably only take 50 threatened amphibian species into captivity, even though estimates indicate 943 species are in need of captive breeding programmes.[59] As well as probiotic therapy, other measures could help, such as boosting the numbers of micro-predators like zooplankton that might eat the fungus' zoospores as they use their tiny tails to propel their single-celled bodies through the water in ponds and streams in search of new victims, altering salinity of water bodies to kill the fungus, or draining ponds and killing predators of amphibian larvae so that there are more tadpoles in the first place and more survive, even if many still get infected and die.[60] In 2024, Berger and colleagues in Australia built 'frog saunas', consisting of ten-hole masonry bricks inside greenhouses. Here infected green and golden bell frogs (*Ranoidea aurea*), an endangered species, reached a body temperature high enough to clear the chytrid fungus. The team is now installing saunas at Sydney Olympic Park, where one of the largest populations of the green and golden bell frog survives.[61]

Some amphibians, like Dominica's mountain chickens, have developed resistance naturally. Elsewhere, researchers are hoping to artificially insert genes from other frogs into highly susceptible species, creating genetically modified (GM) frogs that are resistant to chytrid fungus. These 'designer' frogs could, for example, secrete antimicrobial peptides (protein molecules) from their skin and live with protective bacteria that inhibit the fungus, as the southern leopard frogs (*Rana sphenocephala*) native to the eastern US do.[62, 63] Cunningham has his doubts about the GM strategy, however. 'It might be possible in one or two cases of species that are extinct in the wild and [we] have no other way of getting them back into the wild,' he says. 'We're talking about over 500 species of amphibian that are going to be wiped out by this fungus, so that's a massive undertaking if you're going down a GM route. I can't see it being feasible for all those species. And whether a resistance gene in one frog will be a resistance gene in another species, I think that's a big ask.' Cunningham thinks GM frogs are worth pursuing but novel ways of killing *Bd*, and only *Bd*, might be more useful for all the species of amphibians at risk. As would genetically modifying the fungus so that it's no longer virulent to the frogs, perhaps by discovering what made the original lineage of *Bd* become hypervirulent then ensuring that *Bd*GPL stops being virulent whenever it reproduces and eventually fizzles out.[64]

DARWINIAN DELIGHTS

For some species, no such molecular sophistication is required – a simple fence might do the trick. The southern Darwin's frog (*Rhinoderma darwinii*), a native of temperate rainforest in Argentina and Chile whose males keep tadpoles inside their vocal sacs until the resulting froglets are ready to leave their mouths,[65] is a homebody. 'You can find the same frog almost to the exact position year after year,' Cunningham says. 'They do wander around but then they go back to their little area that they know.' The frog, which ranges from emerald-green to brown in colour, with a black and white belly, is so susceptible to chytrid fungus that it dies within weeks. That means it probably can't sustain the infection without input of more *Bd* from other, more resistant species. Since the southern Darwin's frog doesn't move far, it's relatively easy to fence in an entire population and remove any other amphibians. And that's what Cunningham and colleagues are trying in Chile. Only time will tell whether chytridiomycosis fizzles out within the fenced area – 'it's one of these long slow-burn experiments', according to Cunningham. In the meantime, the frogs must battle fragmentation of their habitat by pine and eucalyptus plantations, for timber and paper, that dry out the ground. 'The remaining populations are not joined up,' Cunningham says. 'And they're just gradually getting smaller as chytridiomycosis eats away at them.'

As well as these scientific solutions, we could also stop international trade making the problem even worse. Regulation isn't glamorous or exciting, but it can be effective. Together, the food, pet and medical research trades spread *Bd* around the world; the fungus has reached virtually everywhere that amphibians live. By allowing the larvae of insects such as mosquitoes to remain uneaten, the trade in amphibians may also have helped other diseases spread.[66] Governments tend only to act, Cunningham says, if there's a public health or economic incentive – biodiversity conservation isn't enough of a motivation. Nations seem to pay more attention to the spread of fungal pathogens that affect plants than those that affect animals, valuing crops more than wildlife.[67] Many policy makers may feel the horse has already bolted now that *Bd*GPL is on the scene, so there's no point controlling imports. They'd be misguided – it's just as important as ever, if not more so. We're likely still creating hybrid lineages and if one of these is even more virulent and attacks species that are currently resistant, amphibians will be in yet more trouble. In 2008, after intense lobbying, the World Organisation for Animal Health listed *Bd* as a pathogen of international concern. That means governments must monitor for the fungus and try to keep it out if it hasn't already arrived, but few of WOAH's 180 members have imposed the resulting sanitary requirements on traded amphibians,[68] and some make comments like 'we should be reducing trade barriers, not increasing them'.[69] Perhaps a greater public profile for the harm fungal diseases cause nature will galvanise policy makers into improving – and acting – on the regulations.[70] Measures such as taking skin swabs at import centres like airports, for example, could show if an amphibian is infected with chytrid fungus so that further tests can reveal which lineage it has.[71]

Even for frogs that seem to be surviving in the face of the fungus, there may be a *Bd* time bomb. Individuals from some species, like the mountain chicken, dropped dead almost as soon as the fungus arrived.

But the Darwin's frogs in Chile took 15–20 years to show a population decline. 'For other species, it may be even longer,' Cunningham says. 'We shouldn't be complacent, just because we're not seeing dead animals, that the population isn't going to go extinct eventually. We could be looking at five years, we could be looking at fifty.' We don't even know how long the fungus can survive when it's not on amphibian skin or whether it can live, or reproduce, in the muck at the bottom of a pond or in damp soil. 'We've only known of this entire genus of pathogen since the mid-1990s,' says Cunningham. 'That's not very long at all in the history of investigating pathogens. There's a huge amount that we need to learn.'

FIRE-FIGHTING

Unfortunately, frogs are not the only amphibians under threat from a chytrid fungus. The fire salamander, which lives across swathes of central and southern Europe, has a body splotched with black and gold. At about fifteen centimetres (six inches) long, it's the shape of a massively obese newt. In 2010, fire salamanders in the Netherlands became locally extinct after suffering chytridiomycosis.[72] Not, this time, because of *Bd*, but another chytrid fungus – *Bsal*,[73] thought to have been introduced via the pet trade from Asia,[74] where it's co-existed with salamanders for millions of years and amphibians have evolved resistance or tolerance.[75] Infected animals tend to have tiny skin lesions that look like they've been blasted by a miniature shotgun.[76] Like *Bd*, *Bsal* (scientists choosing, for some reason, not to name it *Bs*) may kill by altering the way the skin handles ions, inducing cardiac arrest. Or it may kill by reducing the ability to breathe through the skin – salamanders rely on this much more than frogs, especially species from the lungless salamander family.[77] Or maybe by letting bacteria enter the bloodstream through damaged skin. Again, we don't yet know enough.

As well as in the Netherlands, *Bsal* killed huge numbers of salamanders in Belgium, Spain and Germany; this last country is now thought to be where the fungus arrived.[78] The fungus has also infected many of Europe's alpine newts and smooth newts,[79] but has not yet reached the wild in Austria or the UK, so British newts and the salamanders near Schallaburg Castle are currently safe. Nor has *Bsal*, as far as we know, reached North America. Which is something to be thankful for, as North America has more than half the world's salamander species, including the gloriously named snot otter (*Cryptobranchus alleganiensis*), after the mucous that coats its skin.[80] As for *Bd*, treatment options for *Bsal* are limited – it's possible that delivering a vaccine through the nose, adding probiotics to the skin, or introducing extra micropredators such as zooplankton to the environment to eat the fungus's zoospores could help, but none has yet been proven on a large-scale. In an attempt to keep *Bsal* away, it's no longer legal to import live salamanders and newts into the US,[81] but frogs can also carry *Bsal* and make up 94 per cent of amphibian imports to the country.[82] Perhaps the most effective way of protecting North America's salamanders is to develop testing and certify imported amphibians as being clear of *Bsal*.[83] Clearing infection from amphibians in the pet trade using heat or antifungals could also be an option, though not all amphibians tolerate heat or these medications, and it would certainly be wise to encourage owners not to release their pets into the wild.[84] A new surveillance system introduced alongside the import ban indicates that, so far, *Bsal* has not reached the continent. Should the fungus arrive, the North American *Bsal* Task Force has set up rapid response plans to prevent its spread.[85] Let's hope they don't need to use them.

FLYING FUNGUS

A different fungus has, however, invaded North America, where it's devastating the lives of bats. *Pseudogymnoascus destructans*, or *Pd* for short, has been in Europe and Asia for millennia, where it caused so little harm that nobody even knew it existed until it reached the US.[86] In 2007, in a cave near Albany, New York, biologists doing a routine winter survey of bat numbers noticed that bats were growing sick and dying.[87] The bats that were still alive were coated with patches of white fungus, most noticeably on their muzzles, ears and wings. On checking their records, the team found that photos taken nearby the year before showed bats with white powder on their noses.[88] It looked like the pathogen had been in North America since at least February 2006; the evidence now points to an arrival in the early to mid-2000s.[89] Unusually for a wildlife disease, this routine monitoring means scientists can tell fairly accurately when the fungus arrived and have been able to track its spread across the country. To start with, it wasn't clear whether the fungus killed bats or simply infected animals that were already ill. But soon it was all too apparent that the fungus was lethal. When scientists knew to look for it, they found *Pd* in bats in Europe, Russia, China, Mongolia and Japan too.[90] We don't know exactly where the *Pd* that reached the US came from, although it's genetically most similar to a sample found in Ukraine.[91] Spores could have been carried to the US by a person on their clothing or outdoor gear,[92] by accidental transport of an infected bat, or by the import of cave-aged food items from Europe.[93] The whole sorry problem may be due to cheese.

From the caves of New York state, the fungus moved slowly west, north and south. Bats probably spread *Pd* themselves, apart from a mysterious leap in 2016, when the fungus jumped to Washington state, some 2,100 kilometres from the nearest contaminated bat hibernation site in Nebraska, no doubt with assistance from humans.[94] By 2021 the fungus had reached 39 US states and 7 Canadian provinces.[95] Twelve species of North American bat suffer white-nose syndrome.[96] Some colonies of the little brown bat (*Myotis lucifugus* or MYLU for short) saw all individuals die and carcasses litter the floor of the cave; in others 90 per cent perished, and the little brown bat is now at risk of local extinction.[97] The northern long-eared bat *Myotis septentrionalis* suffered badly too, disappearing from much of its range. In March 2023, the US Fish and Wildlife Service reclassified this bat as endangered,[98] directly as a result of the population crashes caused by white-nose syndrome.[99] A third species, the tricoloured bat *Perimyotis subflavus*, which has hairs that are dark at the base and tip and lighter in the middle, has also suffered dramatic losses, with numbers at some winter colonies dropping 90–100 per cent.[100] Previously, caves in the eastern US held bats in the hundreds of thousands, much greater numbers than caves in Europe or Asia. Many millions of bats have been lost here, in 'one of the most devastating infectious disease outbreaks in wild mammals to emerge over the past Century'.[101]

The fungus grows best at temperatures in the low-to-mid teens Celsius; it thrives in the cool, moist caves where bats hibernate over-winter. It takes a couple of years for levels of the fungus in a cave to build. Then, when bats enter, they pick up spores that colonise the bare skin on their nose, ear membranes, wing and tail membranes when their immune systems become less active during hibernation. Bats with white-nose syndrome rouse from hibernation over and over again, perhaps because the infection irritates the skin, where it creates small pockets,[102] or because it makes the bats thirsty. The repeated waking burns through their fat stores and many bats don't have enough energy to survive until spring. Others emerge from hibernation then die because their immune systems went into inflammatory overdrive and damaged their tissues.

Remarkably, it looks like natural selection is fighting back. In some cases, remnant populations of little brown bats have survived and numbers are stable or even increasing slightly. To help more bats survive, researchers are trying to understand how these populations have persisted, closing infected caves to tourists and cavers, developing decontamination guidelines and working on treatments. Making caves colder so bats wake less often could improve survival, probiotic treatments may stop the fungus growing, a vaccine is at pilot stage, and ultraviolet light systems that bats turn on as they enter and leave the cave may kill the fungus.[103] The big question is how you deploy these solutions for large numbers of bats. And are they still helpful once populations have crashed and undergone intense natural selection? The northern long-eared bat is in such a dire situation that every little intervention could help it avoid extinction – protecting these bats from wind turbine blades could also make a big difference now that numbers are so low.[104] But for other species, perhaps it's better to increase the number of insects and improve environmental conditions rather than focus on white-nose syndrome. Ultraviolet lures that attracted insects, for example, made it easier for little brown bats to feed, building up their fat reserves.[105] Simple geography is arguably providing the best current hope. *Pd* is present across the US, apart from the South where the warmth means bats don't need to hibernate, and a few pockets in the West. Many bats in the Rocky Mountains overwinter not in caves but in small rock crevices on talus slopes – aggregations of rock debris – in a microclimate where the fungus may not fare so well. It's possible they'll survive better.

Crucially, the fungus has not yet reached Australia, though cavers, tourists or researchers could accidentally take it there at any time. The authorities in the US and Canada have issued guidance for cave visitors on how to decontaminate their clothes and gear,[106] and when the International Congress of Speleology (scientific study of caves) took place in Australia in July 2017, organisers insisted the hundreds of cavers and scientists visiting from abroad did not bring kit, boots or clothing that had been inside caves infected with *Pd*.[107] Scientists reckon it's very likely to almost certain that *Pd* will reach Australia, and that Australian bats will be exposed in the next ten years.[108] Up to eight species live in caves in South-east Australia that have the right conditions for *Pd*, and seven of them are already at risk of extinction.[109]

Pd, *Bd* and *Bsal* are just three modern examples in a long line of pathogens that humans inadvertently introduced by trading animals, whether for food, lab work, pets, medicine, to boost birdsong and more, not to mention the pathogens brought by stowaways on board those ships or planes. Many species that we thought were lost because of habitat loss or land-use change were actually lost to disease, Cunningham believes. 'Because it's not been on the radar of conservation biologists until very recently, and it's not been an easy thing to investigate – it's still not – disease often goes under the radar,' he says. That's why raising awareness is crucial. 'I'd love to have done that back in the nineteenth century when people started moving things around, and the Victorians were bringing all these exotic plants into the country,' Cunningham says, 'and to have that embedded in our culture by the time we got to the twentieth century.' Then *Bd*GPL may never have arisen.

10

City Catches: Foxes, Mange and High-Density Living

'From the desperate city you go into the desperate country, and have to console yourself with the bravery of minks and muskrats.'

Henry David Thoreau, *Walden*, 1854

In Bristol in the early 1990s, if you bought chips on the way back from the pub, you might glance behind you in the dark suburban street and find you were being followed – by a red fox some eight feet behind, tempted, like you were, by the smell of warm salt and vinegar. The boldest would persist until you threw them bits of your dinner. In 1994, however, chips became safer – sarcoptic mange broke out and killed 95 per cent of the city's foxes.[1] And chips were in part responsible for this horrifying death rate – thanks to the city's easily accessible food, foxes here lived in high numbers and at high densities. And they had never encountered this mange before. The disease spread like wildfire and 'burned' away the foxes' skin.

Sarcoptic mange is caused by the *Sarcoptes scabiei* mite, an arachnid less than half a millimetre in size. As you'd expect from a member of the same class as spiders, these *Sarcoptes scabiei* mites have four pairs of legs. More surprisingly, the first two pairs have suckers and the last two pairs have long bristles that trail behind. With short spines across their back and a broad chunky body, these mites look like a very tiny cross between a malevolent pincushion and a rhino. Part of this small mite's long name comes from the Latin scabere, meaning 'to scratch'. For those now feeling itchy it won't be much reassurance to learn that we know of more than 55,000 species of mite and at least 50 of them cause different types of mange;[2] *Notoedres* mites, for example, cause notoedric mange, which can be fatal to squirrels, particularly if they lose their fur in winter.[3]

When a female mite burrows into an animal's outer layer of skin to lay her eggs at the end of a tunnel, at first she suppresses the animal's immune response,[4] but later she triggers it. An animal that's not too badly affected might scratch, bite, lose hair and suffer some patches of thickened skin, wrinkling and inflammation. 'It's amazing how these tiny creatures can have such a huge impact,' says Emily Almberg, a disease ecologist at Montana Fish, Wildlife & Parks, US. 'It is tough to see animals in such discomfort, scratching, and in such poor condition,' Almberg says of the wolves she studied early in her career. 'It's a nasty, nasty infection.' In red foxes and several other species that are particularly susceptible to mange, including wombats, the parasite leads to much more than a distressing scratch. During the later stages of infection, these animals' immune systems go into overdrive and cause huge amounts of inflammation, resulting in 'crusted mange' – thickened skin full of cracks and fissures that become infected with bacteria and yeasts and smell foul. Eventually the animal may find it difficult to see, hear and eat, so that it grows emaciated and dies.[5] It's relatively unusual for a parasite like a mite, tick or worm to cause an immune response and threaten life; more often they cause chronic, life-long infections.[6] And, unlike with some viruses or bacteria, despite their immune response, individuals don't gain long-term immunity to mange mites and can be reinfected by other animals of the same or a different species, such as an untreated domestic animal, or by mites in the environment.[7] Mange is hard to get rid of.

The disease spreads by mites crawling from one animal to another when they're in close contact, or through mites deposited at shared dens, or perhaps at shared feeding sites; people feeding Bristol foxes in their gardens may inadvertently have helped the disease to spread.[8] When they're not on or in an animal's skin, the mites don't live more than a month,[9] depending on the temperature and humidity, but that may be long enough to encounter another animal, perhaps in a den or burrow, and get under that individual's skin. In Bristol, foxes kept dying from mange even once the population had plummeted, indicating that transmission was frequency-dependent,[10] and that burrows may have been to blame.[11] By 2004, ten years after the outbreak, Bristol's fox population had only recovered to 15 per cent of its 1994 peak.[12]

MITEY START

Before the 1994 outbreak in Bristol, the UK had seen at least five outbreaks of mange in red foxes – in Somerset, a county to the south of Bristol, in 1906, on the south coast of England in 1914, in South-west England in 1918, the West Midlands and Wales in 1931, and the Sussex/Surrey border and London continuously since the 1940s.[13] The earlier outbreaks were generally small and local. In some cases they were due to foxhunt associations buying in and releasing foxes so people on horses could chase packs of foxhounds running after the fox, two species against one.[14] In London, mange became endemic in the 1940s. It may not be a coincidence that scabies, which is caused by the same mite in humans, was rife during the Second World War, particularly among soldiers and those living in overcrowded conditions.[15] Itching due to the mite is even half-seriously said to have been the reason Napoleon always posed with his hand inside his coat.[16] Scabies is still rife today, at any one time around 100 million people have the disease.[17]

Foxes have lived in UK cities at least since Victorian times, when we have reports of them in London. The boom in urban fox living began between the wars, when cities grew larger and leafy suburbs of semi-detached houses blossomed around their edges, particularly in the South-east.[18] The UK has one of the highest urban fox populations in Europe; in the US, raccoons are the mammals that tend to take up this city scavenger niche. We don't know how mange spread from London and its environs to Bristol in 1994. There, as in London 50 years earlier, the disease became endemic. Later, it spread to other UK cities.[19]

According to a book by veterinarian Gottlieb Heinrich Walz published in the kingdom of Württemberg, in what's now Germany, in 1809, both Virgil and Cato mentioned mange in sheep, with Cato believing it to be caused by hunger and continuous rain.[20] The first known wild animal victim was a red fox diagnosed in 1861 by a German vet.[21] But mange had been around for thousands of years, perhaps even many thousands of years.[22] What's not clear is which species the mites started burrowing into the skin of. Did these mites evolve to live in humans or their ancestors, then crawl onto dogs when they became our companions some 14,000 years ago, then wander on into the wild?[23] Others believe the mites may have travelled the other way, from dogs to humans.[24] Or perhaps they came from another domestic animal or a wild species. It's too far back to know, though transport via human or domestic animal would certainly explain how the mite was able to spread around the world.[25]

We don't know a huge amount about the mange mite's travels. The disease it causes seems to have been endemic in Spain since the nineteenth century, and spread across Scandinavia in the 1970s and 1980s.[26] There this tiny mite affected not only its red fox victims but altered the whole ecosystem – once their predators were missing, numbers of fox prey such as hares, grouse, capercaillie and roe deer increased.[27] Similarly, in the Andes in Argentina, when numbers of vicuña, a relative of the llama, plummeted in 2015 due to a mange outbreak, fewer Andean condors, which had scavenged on the carcasses of vicuñas killed by puma, used the area. Meanwhile, grasses and other plants on the open plains where vicuña are relatively safe from predators flourished.[28] These days mange is expanding its geographic range in Europe, North America, Asia and South America.[29]

SKIN DEEP

In the last few decades mange also seems to be infecting more species – we've found mange in an average of nine new hosts each decade since 1970,[30] particularly in Australia and North America.[31] The disease is affecting animals more severely too.[32] In essence, mange is an emerging panzootic, although it's not clear why.[33] One theory is that more species are coming into contact with one another as a result of global change, making it easier for pathogens and parasites to spill over. 'The expansion of humans to every corner of the globe and the intensification of agriculture and livestock production could contribute in some cases to increased movement and contact between species,' says Almberg. 'The world has become more connected.' We've also restored many wildlife populations from the lows they hit during the early twentieth century; these higher population sizes and population densities may help mites spread and enable outbreaks to grow larger. 'It seems like this is becoming, if nothing else, a more well-documented or noticed phenomenon,' Almberg adds, 'where we're seeing some of these larger disease outbreaks leading to conservation concerns.'

Sarcoptes scabiei is one of the most generalist skin-living parasites, or 'ectoparasites' targeting mammals.[34] These mites infect at least 148 species of domestic and wild mammals, including herbivores, carnivores and humans.[35] The dog family, hooved mammals, marsupial mammals, some of the pig family, rabbits, squirrels, raccoons, black bears, porcupines and primates all get sarcoptic mange. The mites that cause the disease live on six continents and come in at least fifteen different varieties or strains.[36] Different strains tend to prefer different hosts – it's possible they're adapted to deal better with that host's immune response.[37] 'Early on, there was a lot of discussion about whether these were different subspecies of mites adapted to different host species,' says Almberg. 'But I think the consensus now is that it's one species of parasite with local variants that may have adaptations to or preferences for certain host species. But many of them are able to infect multiple species given the opportunity.'

In some cases, as revealed in a scientific paper with the horror-movie title 'The curse of the prey',[38] animals can pass mange mites on to the predators that kill them. In Masai Mara, Kenya, three different varieties of *Sarcoptes scabiei* mite infect cheetah – one that only lives on cheetah, one similar to mites found on Thomson's gazelles, and the third similar to mites found on both wildebeest and lion. Cheetah's favourite prey are Thomson's gazelle and wildebeest, while lion favour wildebeest. It's pleasing to think that the prey animals get a – very small – chance to take revenge. But for endangered species this mite-y ability to infect multiple hosts is a problem – the endangered animals may continue to be infected by other species even when their own numbers have dropped very low. Some 40 of the species that catch mange are 'in some suboptimal conservation status according to the IUCN'; two are critically endangered, eleven endangered, eighteen vulnerable and nine near-threatened.[39]

CALIFORNIA DENNING

Far from the chip shops and red foxes of Bristol is California's San Joaquin Valley. Part of the state's Central Valley, this region lies to the east of San Francisco and runs around 400 kilometres south broadly parallel to, and some 100 kilometres inland from, the coast. It's the only home of a threatened subspecies of kit fox. Listed as endangered in 1967,[40] the San Joaquin kit fox (*Vulpes macrotis mutica*) resembles a cross between a fox cub and a kitten. The animal is dainty, just 30 centimetres tall at the shoulder with a slim tan-coloured body, delicate legs, a dark tip on its slender fluffy tail, a long thin nose that divides round cheek-bones, and tiny button eyes. Even its foot pads are unusually petite for a fox.[41] The only thing large about this animal is its ears – if it could fold its ears down over its face, the fox could pretty much hide behind them. Though these ears are not for hiding but for keeping cool. In their natural habitat, kit foxes mainly eat kangaroo rats but will also munch on mice, ground squirrels, rabbits, ground-nesting birds and, at certain times of year, insects.[42] They hunt at night and shelter from the desert heat during the day in dens that they dig themselves, borrow from other animals, or convert from human structures such as culverts, old pipelines or banks for sumps or roadbeds.[43]

Around 5,000 San Joaquin kit foxes remain. 'This species can't be in a hilly area or it can't avoid predation,' says disease ecologist Janet Foley of the University of California Davis. 'It can't see predators and get down into a burrow fast enough. So it was already a little bit limited.' These days the hot, flat land in the Central Valley is valuable not just to the kit fox, but also to humans for agriculture, cities, oil exploration and infrastructure such as arrays of solar panels. 'As soon as you start ploughing all of it and putting in orchards and fields and crops, the animal can't be there anymore,' Foley says. 'It's just gotten pushed, pushed, pushed to where there's very little space left for it.' Since 1930 this fox has seen its range halve. Today it lives mainly in the southern and western parts of the valley, with some individuals nearer the coast.[44]

Some kit foxes headed for the streets of Bakersfield, a city near the southern end of the San Joaquin Valley with a human population of more than 400,000.[45] The sprawling city, according to Foley, is one of the most conservative in California. It's not necessarily somewhere you'd associate with conservation. Nevertheless, many people fed the roughly 400 kit foxes in Bakersfield, where some of the animals even slept on porch patio furniture.[46] 'We would see the little burrows that they dig near a street corner,' Foley says. 'And then you'd see chicken bones where somebody would go to a fast-food outlet and just leave food for them.' Like Bristol's red foxes in the early 1990s, the kit fox population in Bakersfield reached a high density. Since the kit foxes here had plenty of predators, including coyotes, it's likely that their high numbers were due to the amount of garbage and other food they could get their paws on. While Bakersfield has fox-proof bins, plenty of people leave chicken out for the kit foxes, while others throw their fast-food bags, complete with leftover French fries, out of their car doors. Others feed feral cats. 'I love feral cats,' Foley says, 'but feeding feral cats means all the wildlife that's going to come eat the cat food is going to be in that environment and trade diseases too.'

In 2013, to the consternation of both human and kit fox residents, mange broke out in Bakersfield. The source of the mites was probably a kit fox den that had previously belonged to an introduced European red fox or a native Sierra Nevada red fox. Genetic tests indicated that kit fox mange mites and those from other fox species were closely related, but Foley and the team couldn't get a huge number of samples so this result may not give a true picture. The mites from coyotes were also 'not that different' and coyotes carry multiple strains of sarcoptic mange mite, one of which could be slightly more infectious to foxes. Another option is dogs. Or mange may have spread from another species to a kit fox in the countryside, after which, once the disease reached the city, the high population density meant it transmitted between kit foxes fast.

TOXIC ENVIRONMENT

Besides providing a high density of foxes, cities can help mange thrive in other ways. Mange becomes more prevalent in animals that are under social stress, not getting enough or the right kind of food, or during difficult times such as drought or winter.[47] And cities can provide some or all of those conditions. In Japan, for example, raccoon dogs in the urban and suburban landscapes of the Tama Hills area of Tokyo are much more likely to suffer mange than raccoon dogs in the surrounding forest.[48] The same goes for other diseases; sometimes in urban areas of Florida, the American white ibis, normally a wetland bird that eats fish and aquatic invertebrates, feeds on bread that people give it in parks and on refuse at trash heaps, a diet that reduces its ability to fight bacteria.[49]

Then there's another problem. In 2002, in the coastal mountain ranges to the north and west of Los Angeles, notoedric mange broke out in bobcats and mountain lions. The disease, caused by the *Notoedres cati* mite, which also infects the ears of domestic cats, had previously only ever been seen as isolated cases in the wild. The bobcats with mange suffered severe encrustation on their head and shoulders and often mange on much of the rest of their body too, including their hind legs. The animals grew thin, became more active during the daytime, and eventually died. The prime suspect behind the outbreak? Anticoagulant rat killer that the animals encountered in the grounds of private homes, golf courses, office parks, schools, water utilities, apartment complexes and parks.[50] Bobcats mainly eat rabbits, woodrats, pocket gophers and ground squirrel, so it looks like non-target species ingested the poison too.[51] All nineteen bobcats that died of severe mange had been exposed to anticoagulants, many to more than one compound. And the total anticoagulant level in these animals was higher than in bobcats that died of other causes. Perhaps by causing chronic anaemia, the anticoagulants made the bobcats more susceptible to mange. As a result of the outbreak, bobcats disappeared from some patches of habitat and were slow to return.[52]

Only one of the bobcats studied near Los Angeles died of poisoning by the rat-killing chemicals.[53] But two out of four mountain lions died from internal haemorrhaging due to the poison; both animals had spent most of their last month in the most developed parts of their home range and eaten coyotes in addition to their usual mule deer prey, which tend not to consume mice, rats or rat poison. Members of the cat family are up to 100 times more resistant to certain compounds, including the anticoagulant brodifacoum, than dogs such as coyotes, in which anticoagulant poisoning is a leading cause of death.[54] Both poisoned mountain lions were also suffering from mange, although the condition was less advanced than in the bobcats that died of it. The other two mountain lions died in fights with members of their own species; their bodies contained two or more different anticoagulants. Wildlife may be suffering a hidden epidemic of direct poisoning from such rat-killers; many don't show outward signs of poisoning, and to prove it, you'd need to find their body before it decomposed and analyse its liver using high-performance liquid chromatography.[55] We need alternative pest control methods to protect non-target wildlife from both direct poisoning and exposure-related mange.

WOLF – AND MITE – COMEBACK

Stress and population density make a difference in the wild too. This was the case at Yellowstone National Park, where Almberg studied for a PhD on infectious diseases – including mange – in wolves. Almberg loved following packs of these animals. 'I cherished those days when I was out in the field every day,' she says. 'Wolves are fun to watch, they're so social.' Parents bring sticks and other items to the den for pups to play with; surveys after the pack had left would reveal 'shirts, CDs, bottles, caps of various things, shoes'. One time, Almberg saw a few yearling wolves interact with a skunk that had woken from its winter torpor. 'We were watching with scopes and you could see the wolves jump back from the skunk and then roll in the snow,' Almberg says. 'They were covered in scent. It must've been extremely stinky.' Even though it was surrounded, the skunk held its own against the yearlings for 30 minutes, flipping up and spraying, until an adult female turned up 'all business' and swiftly dispatched it.

Unusually, mange hadn't reached Yellowstone naturally. State vets deliberately introduced the disease into the Greater Yellowstone Ecosystem back in 1905 in an attempt to control wolf numbers.[56] It appeared to have worked, along with other predator control efforts, and by the early twentieth century wolves were gone. After wolves were reintroduced from Canada in 1995, this mite-borne disease returned more slowly to the wolves than viral diseases did; mange re-emerged in the species in 2002, appearing first outside the park and then inside the park in 2007.[57] Wolves and coyotes tend to get classical mange, which is not nearly as severe as crusted mange. But it still makes the animals uncomfortable, at the very least. 'We'd see wolves mid-hunt, stop and scratch,' Almberg says. 'The itching was just so overwhelming and so unbearable.' And pups were 'noticeably smaller, lethargic and constantly scratching'.

The disease hadn't been recorded inside the park at Yellowstone since wolves had disappeared. So where it had reappeared from was a mystery. Perhaps wolves inside the park came into contact with wolves from outside, or with infected coyotes or foxes that had gone under the radar and acted as a reservoir. With wolves gone, coyotes reached much higher densities and may have supported higher levels of mange mites, providing an intense source of spillover to wolves when they returned.[58] High densities likely affected the spread of mange in wolves too. In the northern third of the park, elk are numerous and the wolves that prey on them also grew in numbers. Once mange reached this region it spread much more quickly, and several wolf packs completely disappeared 'because they all lost hair and by the wintertime, they were in terrible shape', as Almberg puts it.

Over time, the symptoms of mange here have become less pronounced and biologists see fewer severe cases. 'Some of that could be attributed to lower densities of wolves,' Almberg says. 'Some of it could be attributable to intense selection pressure on individuals that may have been more susceptible.' In other words, only wolves less susceptible to mange may have survived to reproduce their more mange-resistant genes. At Yellowstone, the team decided not to treat wolves for mange, in part because of the limited options but also because sarcoptic mange had been present in North America before its deliberate introduction to the park.

SOLVING SAN JOAQUIN

For the endangered San Joaquin kit fox, however, the decision to treat was another matter. 'If the host is an endangered species and [the mange mite] is host-specific, then chances are mange could be treated,' Foley says. 'The mite could go extinct and then the host could be OK. But if you've got a situation where twenty different species have it and one of them is an endangered species, they're in really bad trouble.' While some individuals of some species can clear mange themselves,[59] kit foxes don't seem to mount a successful immune response.[60] So Foley, Brian Cypher, Jaime Rudd and others set out to treat the kit foxes with Seresto anti-mange collars, which are designed to treat mange, fleas and ticks in dogs over a period of eight months. The team attached the devices to solid leather collars so that the kit foxes couldn't chew them off; they managed to capture and collar about 10 per cent of the foxes. Treating kit foxes for mange is hard work – first you find a burrow and set a big metal trap. Not surprisingly, finding itself in a trap makes a kit fox cross. 'No carnivore wants to be handled by a person,' Foley says. Foley's colleague Jamie Rudd would put dog chew bones made of cloth along the wires of the cage so that the kit foxes didn't break their teeth. 'Chewing, chewing, chewing, just furious' Foley says. 'But they get this little chew toy and they're fine. When you take them out, as long as they are handled quietly – they put a muzzle and a face cover on so that they can't see – then they just sit there and wait until you're done.' For individuals in the early stages of mange, it's relatively easy to kill the mite. But if the individual has stopped eating and has a low body temperature, helping is a lot more difficult. The skin of animals with many mange lesions no longer works properly and protein leaks from the animal's blood into the environment. 'Those animals need really aggressive ICU [Intensive Care Unit] treatment and still may not make it,' Foley says.

If you had all the resources in the world, according to Foley, you'd start in one relatively isolated corner of the city, trap all the kit foxes and put anti-mange collars on them, that would kill any mites for at least three months. Once you had that pocket under control, bearing in mind that the foxes would have pups in February or March that would later change locations, you'd start work on the next area of the city. And then the next. Another way to do it would be to use a baited treatment station, tempting a kit fox in with chicken then dripping a small amount of selamectin anti-mite treatment drug onto its back. 'You would treat it without even touching it,' Foley says. 'But you'd have to figure out issues – you don't want to overtreat the same animal. That's probably what you would do, is get the same animal every night.'

Researchers working with wombats have developed a couple of innovative treatment techniques for mange that don't involve trapping the animal. European settlers brought sarcoptic mange mites to Australia,[61] although the very first wombat known to be infected was found at the Muséum National d'Histoire Naturelle in Paris in 1817. At the time its mites appeared identical to *Sarcoptes scabiei* mites found on a human; it's possible the animal picked up mange during its transportation to the museum.[62] These days, mange is one of the biggest killers of wombats, along with cars and habitat destruction.[63] In northern Tasmania between 2013 and 2016, an outbreak slashed numbers of bare-nosed wombat by 94 per cent.[64] And wombats living at high density, particularly in the remains of riverside or wetland forests next to agricultural grasslands, may be more at risk.[65] When I first read about the 'pole and scoop' treatment method,[66] I had all sorts of fanciful ideas regarding scooping up a wombat in some kind of net like an oversized butterfly. But the 'pole and scoop' is much more down to earth – a plastic cup attached to the end of a lengthy pole and filled with parasiticide chemical. The operator positions the scoop over the back of the mangy wombat and pours the chemical between its shoulders and down the length of its spine, at which point a wombat that's relatively healthy may well lumber away.[67] An alternative is to put up a flap,[68] perhaps made from an ice-cream tub lid, outside the wombat burrow and incorporate a small tin filled with parasiticide that drips onto the wombat as it pushes through the flap.[69] The burrow-flap was thought effective only for treating individuals.[70] However, recently, the technique seems to have been effective for treating the population as a whole, provided that enough burrows are fitted with flaps.[71] 'I think that's a really neat prospect,' Foley says.

Most treatments for mange to date have focused on killing mites, which will stop the host animal's allergic reaction. 'But there's a lot of interest right now in vaccines against ticks,' Foley says. 'And if we can make vaccines against ticks, we certainly can make vaccines against mites.' Such vaccines would encourage the host animal to make antibodies against proteins in the tick or mite's saliva in order to prevent the parasite attaching. They might even kill it. In some situations, animals become immune to arthropods such as ticks. In order to raise ticks for research, scientists may feed these arachnids on a laboratory rabbit and wait for them to reproduce. But this technique only works twice per rabbit – by the third time, the rabbit has developed antibodies and T cells that kill the ticks. In some ways, preventing mange is more straightforward than preventing a tick-borne disease such as Lyme disease, which is carried by bacteria inside the tick – with mange, you're dealing directly with the arthropod itself. Though in North American cities, the death of wild canids such as coyotes and foxes from mange may increase the risk of humans getting Lyme disease by boosting rodent numbers.[72]

In kit foxes, that first mange outbreak in 2013 fizzled out before mange returned the following year. The disease is still around today, though not at the peak levels of the first outbreak. Nowadays, not all kit foxes become infected with mange. Animals that don't interact with other family groups might be protected, or it's possible that some animals have developed slight immunity. The kit foxes are no longer at their peak, however; by 2020, by which time sarcoptic mange had become endemic and was barely detectable, the population of Bakersfield kit foxes had roughly halved.[73]

So what does the future look like for this fox? In 2019 the town of Taft, 60 kilometres (38 miles) away from Bakersfield, also suffered an outbreak of mange, this time spread by a kit fox itself. There are fears mange could spread into the rural population, though as the rural animals live at a lower density perhaps it would spread less easily. Foley is hopeful that in Bakersfield and Toft the cycles of mange will remain damped. 'Hopefully, it'll never get as bad as it got, but it's probably not going to go away,' she says. If an outbreak did escalate, we have protocols and knowledge about how to intervene, enabling faster action. Foley reckons we'd need to fit a high proportion of kit foxes, perhaps as much as 90 per cent, in affected areas of the city with long-acting treatment collars.

Since it would be very difficult to eradicate mange, a good approach, according to Foley, is to support the overall health of the kit fox population by letting animals behave as naturally as possible in the city – moving as far as they need to move, not being crowded into one den because nowhere else is available, not eating garbage or interacting with house cats. Basically ensuring the animals are in 'as healthy and as natural and as clean an environment as possible'. Meanwhile, mange continues to develop and spread. In 2012 a man in the US caught mange from his dog,[74] although zoonotic scabies tends to be self-limiting and not to require treatment.[75] And in 2018 farmers in Switzerland's Jura mountains saw a mangy fox in their cowshed. Their oxen, goats and dog caught the disease and then passed it on to four people.[76] Researchers fear such transmissions could become more common.[77] The more that mange mites, and indeed other pathogens, switch host to new species, the more likely the mites are to evolve and become better at switching.[78] It's in our interest to stop that happening. Thanks to our cities and the way they enable foxes and other species to live at high densities, our mite-y problem may be coming back to bite us.

11

Warming and Westerning: Climate Change, American Crows and West Nile Virus

'Crows crowd croaking overhead,
Hastening to the woods to bed.'
John Clare, 'Summer Evening', 1820

In New York City, the summer of 1999 was sweltering. So sweltering that people slept outside, collected containers of water for their plants and let their swimming pools go stagnant.[1] Then things turned horror-movie weird. In June, the city's American crows (*Corvus brachyrhynchos*) began to act drunk and drop dead.[2] As did fish crows in the streets,[3] and a bald eagle, a snowly owl, flamingos and cormorants at the Bronx Zoo.[4] Horses on Long Island became ill with encephalitis (inflammation of the brain).[5] And so did humans. In August and September 1999, 59 people in the New York City area, many from the borough of Queens, went to hospital with similar symptoms, more than 60 per cent of them with encephalitis and 27 per cent with muscle weakness. Seven people died.[6]

It was time to investigate. To begin with, the fact that the viruses that cause encephalitis tend to stick to known geographies seemed like an asset. In the US, for example, the St Louis encephalitis virus (SLEV) generally used to be the culprit, while Japanese encephalitis virus haunts South East Asia and the Indian subcontinent. The blood test kit designed for identifying insect-borne viruses in North America indicated that the St Louis encephalitis virus had indeed made these people sick. And the results from another blood test fit that diagnosis too.[7] It was the first time this flavivirus had been seen in New York. Its usual stamping grounds are the Mississippi Valley, along the Gulf Coast and, more recently, the South-west;[8] the last major epidemic broke out in the mid-to-late 1970s.[9] Doctors didn't have time to wait for confirmation by testing samples of the virus itself, rather than patients' blood samples. At the behest of the mayor, the borough of Queens implemented public health measures, everything from distributing mosquito repellent at the US Open tennis tournament to deploying mosquito-killing chemicals from trucks and the air. Soon cases had emerged in Brooklyn, the Bronx and Manhattan, and these boroughs kicked off mosquito control too.[10]

In the first four months of the outbreak, New York City lost almost 6,000 American crows.[11] And these deaths were a conundrum. Although birds host St Louis encephalitis virus, this pathogen had never, to our knowledge, killed birds before. Nor had any of the other flaviviruses – a genus named after the Latin for yellow, in honour of yellow fever – that use birds as amplification hosts,[12] a means for the virus to multiply to high levels. By coincidence, a meeting of the Unknown Encephalitis Project was taking place in Albany, New York state, at the same time as the thought-to-be-St-Louis-encephalitis outbreaks in birds, horses and humans.[13] Most cases of human encephalitis are never diagnosed, and this project was sending specimens for lab tests to remedy that. During the meeting, virologists from the New York State Department of Health gave samples from patients with encephalitis caught during the New York City outbreak to a scientist from the University of California, Irvine. The scientist and their colleagues analysed the genes of viral material they found in the samples, which indicated that the Kunjin virus from Australia, a subtype of West Nile virus, was the culprit; the hospital doctors' test kit hadn't included Kunjin virus as it shouldn't have been in North America.[14] This virus not only came from the other side of the world, but had never caused large outbreaks of disease in either humans or birds. The researchers dubbed the New York City virus Kunjin/West Nile-like virus.[15] It was all very confusing, even to the experts.

Eventually, a Chilean flamingo from the Bronx Zoo helped solve the mystery. Sequencing virus samples taken from the body of this bird showed that the virus was West Nile itself.[16] This RNA virus is adapted to live inside the bodies of mosquitoes, birds and mammals and to deal with their different temperatures and biochemistries – no mean feat. It's small, roughly 50 nanometres in diameter, and broadly spherical.[17]

As you might expect from the diagnostic confusion, West Nile virus is in the same group – the Japanese encephalitis virus complex – as St Louis encephalitis virus. West Nile virus shouldn't have been in the US and generally didn't kill large numbers of birds. Nevertheless, when the team revisited the human blood samples, their antibodies reacted far more strongly to West Nile virus than to St Louis encephalitis virus. The mystery virus behind the bird and human deaths was West Nile, far from home and acting strangely. Its previous identification as Kunjin was understandable as, at the time, the only gene sequence available for West Nile virus was from Lineage 2, whereas both Kunjin virus and the West Nile virus from New York City are from Lineage 1 and more closely related to each other than to Lineage 2. Its identification as St Louis encephalitis virus was also logical; humans produce the same antibodies to all species in the flavivirus genus and West Nile virus had never been seen in the western hemisphere before[18].

TRANSATLANTIC TRAVEL

So how did this virus reach the Americas? First discovered in Africa, in a sick human in Uganda's West Nile province in 1937, for the next couple of decades West Nile virus caused outbreaks of fever in humans in Africa and the Middle East. The 1950s, for example, saw epidemics in Egypt and Israel, among others.[19, 20] Then the virus expanded its range, perhaps introduced into Europe by a bird migrating from Africa,[21] and broke out in people in many African, Middle Eastern and some Mediterranean countries roughly every ten years. Then something changed. In the 1990s a new wave of human outbreaks with significant rates of illness and death ensued:[22] the virus caused an epidemic of encephalitis in 400 humans in Romania,[23] killing around 40 people,[24] as well as infecting 1,000 people in Russia of whom at least 40 died,[25] and bringing outbreaks to Algeria, Tunisia, Israel and Sudan. Horses in France's Camargue region and northern Italy also fell ill;[26] in Europe, horses are the only other mammals besides humans known to grow sick from West Nile virus, though they're generally less badly affected.[27] Around one in ten horses becomes ill, and 20–40 per cent of these animals die,[28] whereas around one-fifth of infected humans suffer a flu-like illness and one in a hundred develops encephalitis and dies.[29] Around the world, small mammals such as voles, mice, squirrels, chipmunks, cottontail rabbits and bats, domestic sheep and cattle, larger mammals such as grey wolves and skunks, and even amphibians such as the lake frog have been found with the infection, as have reptiles such as alligators and crocodiles.

Until 1999 the West Nile virus' range encompassed Africa, southern and eastern Europe, the Middle East, India, Central Asia and Australia. Outside the US, the pathogen hadn't caused mass deaths or outbreaks of disease in wild birds, perhaps because the birds had evolved to live with the infection, though people may simply not have noticed occasional losses or small outbreaks.[30] Significantly, in 1998, West Nile virus broke out in Israel in storks and domestic geese and for the first time that we know of killed birds.[31] It looked like the virus in this region had become more deadly to birds. Then, somehow, it travelled to the US.

ARRIVAL

The virus in New York City, geneticists discovered, was almost identical to the virus that infected these domestic geese.[32] So how did the Israeli version make it across not only the Pond, but also the Mediterranean Sea? Maybe infected mosquitoes, eggs or larvae nabbed a ride on a ship alongside other cargo.[33] The Port of New York and New Jersey is the second busiest port in the country, after Los Angeles,[34] and processes the most containers by weight on the US east coast. The *Aedes albopictus* mosquito invaded the US from Asia by stowing away inside wet used tires, as well as reaching the US and Netherlands in shipments of 'lucky bamboo plant'.[35] Or maybe West Nile arrived inside an infected bird, more likely one that was imported rather than flying itself in, although birds, including Eurasian wigeon, garganey, green-winged teal, tufted duck, ruff, and little or black-headed gulls,[36] do migrate from the eastern hemisphere to the west.[37] Infected humans and other mammals are probably off the hook – they don't generally have high enough levels of the virus in their blood to infect mosquitoes; they're not competent hosts and are a dead-end, sometimes literally. This virus needs birds. We may never know exactly how it arrived in the US or even precisely when. As West Nile spread across the US, it became clear that disease didn't break out in humans until the year after the virus reached an area's mosquitoes and birds. So maybe West Nile reached the New York area in 1998 or even earlier, rather than the spring of 1999.

AMERICAN NIGHTMARE

Whenever it arrived, among birds the virus was most devastating for American crows. Before 1999, numbers of American crows had increased steadily for two decades. Thanks to the virus, by 2005 the population in some regions had declined by up to 45 per cent.[38] To some, crows are vermin. And people in cities may have encountered large, noisy and messy crow roosts.[39] But these birds are highly intelligent, tool-using, social creatures, some of whom have double lives – a crow may spend part of the day in its home territory in a city with its family and the rest with a flock in the countryside eating waste grain.[40] 'West Nile virus, because of its local impacts, wiped out entire communities, entire families of crows,' says disease ecologist Shannon LaDeau of the Cary Institute of Ecosystem Studies, US. It's heart-breaking to think of the disease devastating the family life of these intelligent birds. With black feathers, black legs and black eyes, American crows might look monochrome from afar. But their beauty is in the detail – the bright glint in an eye, a velvet patch of feathers next to the gleam of a strong beak, the pattern of sunshine on glossy blue-black contour feathers, an almost hidden scruff of hairy feathers under a belly, and the delicate grey scales on sturdy legs and claws. West Nile virus is almost always fatal for these birds,[41] whereas some 75 per cent of infected blue jays die and just over half of fish crows.[42] Of the more than 300 bird species in the US susceptible to West Nile virus,[43] corvids, including crows, jays, magpies and ravens, and birds of prey such as eagles, hawks and owls are most susceptible.[44] The bird species that escape most lightly show no symptoms at all; in cities the American robin and house sparrow are likely the chief reservoir hosts – animals that carry the virus and can transmit it to other species but are often symptom-free. At the other end of the scale lie neurologic disease symptoms such as impaired vision, dull thinking, head tilt, paralysis, seizures, loss of control of bodily movements, and death within as little as one or two days.[45, 46]

A bird is typically infectious for a week before it dies or recovers from West Nile virus. Only some species of mosquito that bite that bird will become infected with the virus themselves. In New York City the main vectors are two species that are resilient to human-induced changes in the environment: the common or northern house mosquito *Culex pipiens* and the tiger mosquito *Aedes albopictus*. In summer it takes the virus about a week to pass through the mosquito's gut and become available to infect other animals. As well as spreading via mosquito bite, West Nile virus may transmit when birds come into close contact, perhaps when feeding or preening each other, or when birds eat infected mosquitoes or prey.[47] The few mosquitoes species that thrive in urban areas are more likely to bite humans and other animals as well as birds. 'If they're doing their business in the city, humans are one of the most available vertebrates,' as LaDeau puts it.

Because crows are so big – up to about half a metre long – people tend to notice when they die.[48] So these birds act as a sentinel that reveals that smaller birds are suffering too and people may be at risk.[49] Crows are also useful to humans in other ways, for example by predating on other species and stopping them becoming overpopulated and suffering disease epidemics. This might be the case with squirrels, which 'get these weird mange diseases', according to LaDeau. It's possible, although LaDeau isn't aware of data on it, that crows taking young squirrels from the nest keeps the population in check and reduces disease prevalence. Similarly, crows also remove dead biomass. Near Lake Michigan, crow numbers dropped due to West Nile virus at about the same time as a fish die-off. Normally, researchers believe, crows would have gobbled up all the dying or dead fish. This time, however, there were more fish than the few remaining birds could eat and piles of stinking rotting fish littered the landscape.[50]

After birds died of West Nile virus in North America, closer examination revealed that birds of prey in Europe, including goshawks in Hungary and eagles in Spain, regularly succumbed to West Nile virus.[51] And dead birds from a flock of juvenile white storks that reached Israel in 1999, likely blown off-course by unusual hot, strong winds during their autumn migration, had West Nile virus infections that they'd probably picked up in Europe.[52]

LASTING WINTER

When West Nile virus arrived, people thought 'we just have to survive until the winter and then the environment will take care of things', LaDeau explains. Most people, even those running pest control agencies, didn't think much about mosquitoes in US cities – the focus had been on controlling mosquitoes in areas such as beaches where people wanted to relax outside. Complaints about mosquitoes in urban areas were 'channelled through these control agencies that took care of bedbugs and other "pests", often with a very limited budget', according to

LaDeau. Nobody compiled data on what species of mosquitoes were present or when they were active, 'it was just a pest nuisance response'. And few people believed mosquitoes survived the winter. But the hope that West Nile virus would peter out proved wrong. In the face of all expectations the virus returned in the year 2000.

We're still not entirely sure how the virus manages to survive. Adult mosquitoes may go dormant during the winter.[53] 'They're kind of bears in that they don't hibernate, but they shut everything down for periods of time,' LaDeau says. If temperature and moisture levels rise, the mosquitoes snap back into action. Then there are sewers, relatively warm and cosy underground havens (as long as you can take the smell). Here mosquitoes spend the whole winter with their bodies on a go-slow – the virus may take a month rather than a week to be ready to infect other animals; in the temperatures outside, the virus can't replicate at all.[54] Or the virus itself may go dormant, hiding for the winter inside the warm organs of a bird that looks like it's recovered. If that bird's immune system weakens, perhaps through stress, the infection could reactivate.

In the past, people didn't pay New York's American crows much attention. We didn't know where these unglamorous birds went in winter, or how they lived. Once we realised their role in transmitting West Nile virus, however, that changed; over the next decade or so researchers discovered that this species flocks together in winter, in large communities that include other species, and sometimes lives alongside active mosquitoes, although it's possible the birds transmit West Nile to each other directly via their droppings during winter.[55] Sometimes the birds form their winter roosts in cities, sometimes in the countryside, in gatherings of a few hundred up to two million birds.[56] Often they head a little south. Once we knew more about the wintertime exploits of these part-time city-dwellers, it no longer seemed so impossible for West Nile virus to persist through a New York winter, then spread. Given enough introduced pathogens and enough time, the 'unlikely happened', according to LaDeau. The fact that the virus arrived, got a foothold and spread across the US in just four years spurred great interest in disease ecology and 'actually changed the discipline itself'.

This new-found interest in crow ecology also revealed that American crows in and near cities in the north-eastern US fared worse in the face of West Nile virus than their rural relations.[57] Even crows in suburbs did better than crows in city centres. In the countryside, mosquitoes have very definite seasons, with each species peaking at a slightly different time. The insects lay their eggs in pools of water in the spring and each population peters out as the environment dries and the adults desiccate or can't find water to lay eggs. In the city, on the other hand, humans regulate how much water is available – if it's very wet, we remove water, and if it's very dry, we add it. That means mosquitoes, although there are fewer species, can be active for a lot longer and their populations tend to peak multiple times. Cities also tend to be warmer than the country-side. To amplify, West Nile virus needs many mosquitoes to bite many infected birds that are competent hosts so that the virus reaches high levels inside their bodies. Each female mosquito takes about a week to turn the proteins in the blood she's sucked into eggs and only then is she ready for another meal. So the longer that mosquitoes are around, the better West Nile virus can spread. What's more, mosquitoes and crows in urban places may more often rub shoulders than their country cousins. City mosquitoes are 'very well adapted to being right up close to where humans are living and are breeding in trash containers that hold water where you may also find lots of crows', LaDeau says. Not all birds cope with city life but those that do tend to be competent hosts for West Nile virus, which means that other, poorer hosts don't 'dilute out' the West Nile virus amplification cycle. It's also possible that city crows are more stressed and their immune systems work less well.[58]

In an attempt to quell West Nile virus, for the first several years of the outbreak teams sprayed Central Park with insecticide. 'Rats and mosquitoes are two things that humans have tried forever to eradicate in various places,' says LaDeau. 'And the success rate is very low.' Sure enough, the mosquitoes survived and human diagnoses revealed that the virus headed out from New York, spreading further each year to the west, north and south. 'You could actually see the wave,' says LaDeau, 'which is such a unique data set for an invasive species. We rarely have that kind of response.' Within 4 years, West Nile virus was in all 48 contiguous US states, likely spread by migrating birds and dispersal of resident birds.[59] In the northern half of the US, *Culex pipiens* (the northern house mosquito) was a major driver of transmission, while its close relation *Culex quinquefasciatus* (the southern house mosquito) drove transmission in the southern half. In many areas of the plains and west, *Culex tarsalis* helped too.[60] In 2003, when West Nile virus reached the traditional homelands of *Culex tarsalis* and St Louis encephalitis virus, cases spiked to nearly 10,000;[61] *Culex tarsalis* can fly further than the *Culex pipiens* that spread most of the disease in New York.[62] Within ten years, the virus had ventured into the very tip of South America, though for reasons unknown, large human outbreaks only seem to occur in the

US and Canada.[63] In 2012 the virus peaked again, making more than 5,000 humans ill. By 2013, West Nile virus was responsible for more than 37,000 human cases in the US and was the leading cause of mosquito-borne encephalitis in the US and Canada.[64] By 2019, the virus had killed more than 1,500 people in the US.[65]

CHANGING CLIMATE

Involving, as it does, insects, birds and potentially also mammals, the transmission of West Nile virus is complicated. As climate warms, West Nile virus could well replicate faster, so that mosquitoes become infectious more quickly after exposure.[66] As an ectotherm or 'cold-blooded' creature, the mosquito's life cycle depends strongly on the ambient temperature. Heat a mosquito, and it will probably develop faster, breed faster and bite more often.[67] Currently, the populations of invasive mosquitoes introduced to cities on the US East coast rise much faster in the south than in the north, where many don't survive the winter.[68] So hotter winters may offer less dampening of West Nile virus transmission; after the virus first arrived, American crows fared worse after warmer winters.[69] Climate change may lengthen the transmission season for West Nile virus too; in northern North America the transmission season lasts three months, whereas in Florida and other southern states it's eight months long.[70] As temperatures rise, the transmission season will likely lengthen in the north too. West Nile virus transmits most efficiently at 24–25°c.[71] Based on the average summer data for 2001–16, around 70 per cent of the US population lives where it's cooler than this; these people are likely to see more transmission of the virus as temperatures rise.[72] And birds in those regions will too.

What's more, mosquitoes may expand their range to higher latitudes and move higher up mountains, entering new ecosystems. And birds' migration routes may shift, potentially introducing the virus into yet more new areas. As West Nile virus reaches new areas, it may cause die-offs in new species of bird.[73] Yet if temperatures rise too far, mosquitoes may find it too hot. And how rainfall will change, and the effects of that, are hard to predict. Drought depends on both temperature and rainfall, and might hamper mosquito breeding or make people more likely to store water, creating mosquito habitat; it might kill mosquito predators or concentrate nutrients in water.[74] On the bird front, drought could force birds to move to mosquito habitat, or kill juvenile birds so that there are more mosquitoes relative to each surviving bird.[75] Based on predictions of drought, climate change could double the number of human West Nile virus cases each year in the US by the mid-twenty-first century, and triple case numbers in areas where people have low immunity.[76] Or sudden downpours could wash away mosquito eggs, larvae and pupae.[77] We probably won't know until it's too late.

In Europe, 15 per cent of land is currently at risk of West Nile virus; in 2023, 709 people in Europe caught West Nile virus locally and 67 died.[78] The area at risk could double by mid-century as the climate changes, putting an extra 161–244 million people and countless birds in danger, particularly in Western Europe.[79] The disease is already endemic in Italy and Greece.[80] It has yet to reach the UK, where the South Kent marshes may be first in line unless we get our climate-change act together.[81] The related virus usutu, which infects birds and mammals including humans,[82] spread from Africa to Europe in 2001, where it's now found in nations such as Germany, the Czech Republic, Hungary, Italy, Portugal, Spain, Austria and the UK, in blackbirds and a sparrow.[83]

At the same time as climate is changing, our world is increasingly connected and pathogens and vectors such as mosquitoes are more often reaching new locations. 'We certainly have seen that with Zika and chikungunya, viral pathogens that cause enough illness that we recognise them,' LaDeau says. 'That's going to be a tiny proportion of what's actually introduced every year.' Zika generally brings adult humans mild symptoms but may cause a high temperature, headache, joint pain and rash, and harm developing babies, causing their heads to be unusually small. Chikungunya may also be mild or cause symptoms such as fever, joint pain, headache, nausea and rash. Currently, most travellers infected with mosquito-borne diseases like these return home to New York City outside the mosquito season as people don't tend to travel to areas where such infections are endemic in the heat of summer. So the pathogen arrives but can't hitch a ride to its next victim with a mosquito; at the moment, new diseases probably aren't taking hold because of all the factors that need to align before they can transmit. But climate change could make this simpler by, for example, helping mosquitoes survive longer. 'The unknown with climate change is how long will [the mosquito season] get?' LaDeau says. 'When will that period of time that a mosquito can be active for become very similar to low latitudes, where they're more likely to be active year-round? That's when I think some of these diseases will cross the threshold of increasing concern.' The same would apply for wildlife diseases.

VIRUS VANQUISH

Humans have the option of protecting themselves against West Nile and other mosquito-borne diseases by removing small receptacles of stagnant water to limit breeding sites, wearing mosquito repellent when outside, making sure their homes are mosquito-free, using bed nets, and encouraging their urban authorities to control mosquitos. City-dwelling birds also benefit from municipal mosquito control but otherwise have few options. Horses, the second most badly affected mammal after humans, can receive a vaccine to protect them from West Nile virus. No vaccine has yet been approved for humans, though some have advocated the development of an oral vaccine for American robins that could stop the virus transmitting by reducing its levels in the robins' blood.[84] None of the attempts to create oral vaccines against West Nile virus for birds have been successful so far.[85] Without an oral vaccine, it's not practical to vaccinate birds in the wild on a large scale, with the except of endangered species. The California condor was already faring badly before West Nile virus arrived thanks to human harms such as hunting during the mid-nineteenth-century California Gold Rush, and poisoning by lead ammunition and the pesticide DDT. By 1982 just 22 individuals remained; conservation workers took them into captivity for a breeding programme that thankfully managed to boost numbers. But raptors are particularly susceptible to West Nile virus. And as well as being bitten by infected mosquitoes, as scavengers these condors could ingest the virus when eating dead infected birds and small mammals. Before West Nile virus reached the west, a team vaccinated all remaining California condors against West Nile virus in a move that may have saved the species from extinction.[86] Captive susceptible birds such as birds of prey in zoos have received vaccinations too. The most important thing we can do for other birds is 'protect natural habitat and help maintain low stress for wildlife so that whatever comes next in the form of a pathogen is not the thing that pushes it over the edge', LaDeau says. Pathogens are rarely the single cause of an extinction.

To stop West Nile virus spreading into new areas and the virus transmitting faster in areas it's already reached, we could mitigate climate change. For more than twelve years I wrote about climate change science, watching with growing frustration at our lack of action as scientific papers poured in describing the trouble the planet is in due to rising carbon dioxide levels. After a five-year break from preparing daily climate science news, in the spring of 2024 I listened to a talk by Earth system scientist Johan Rockström of the Potsdam Institute for Climate Impact Research (PIK), Germany, for the Oxford Smith School of Enterprise and the Environment. Things haven't got any better and at times I sat circling my ankles, too distressed to focus. Climate denial has never seemed more appealing – I absolutely do not want what I heard to be true. Though it clearly is. What alarmed me most was hearing that in 2023 the ocean warmed more than it ever had before and that towards the end of April 2024, its temperature had risen even higher. And we don't know why. Is the ocean losing its ability to mop up heat after us? Until now, the more we've punched the planet, the more she helps, Rockström said. So far, though I'm paraphrasing massively, the biosphere has done a colossal amount of work to counteract the harm we've thrown at it by burning fossil fuels. The ocean and the ecosystems on land have absorbed much of the excess carbon, and oceans have taken up a large amount of the excess heat. Both these actions have protected us from a huge amount of temperature rise. But we don't know how much more the Earth can take. Could this new ocean warming be a sign we're moving closer to tipping points beyond which this help stops and there's no way back to normal?

On the plus side, we know what we need to do. To fight climate change, Rockström says, we must halve carbon emissions by 2030. To put ourselves on track we can put a global price on carbon, stop 'losing' intact natural carbon sinks like tropical forests and keep the remaining rainforest biomes healthy, make sure the largest emitters, the US, EU, China and India, join forces and get serious on cutting emissions and oil and gas production, and impose moratoria on deep seabed mining and Arctic oil exploration. It's time for delivery, not negotiation, as Rockström put it – the agreement from COP28 in Dubai already has a blueprint for pretty much everything we must do to protect the climate.

CROW ABOUT

These days the American crow population has stabilised, indicating that some birds have developed some immunity.[87] The drop in crow numbers may have helped. Many of the areas that saw the most crow loss had seen numbers rise steeply before the virus arrived. 'My guess is that roosts were just overpopulated,' LaDeau says. In these crowded conditions, crows had both greater exposure to the virus and weakened immunity.

In 2017 and 2018, a team at Cornell University's Janet L. Swanson Wildlife Hospital nursed five American crows infected with West Nile virus back to health in captivity,[88] through treatments such as fluid therapy, extra B-vitamins and antiparasitic medication.[89] This was the first time this species had survived the virus, even with treatment.[90] The released female crows may be able to pass on some level of maternal immunity to their offspring, so we may be able to add wildlife rehabilitation to our crow protection kit.

American crows are still numerous, despite their decline after we shipped in West Nile virus and the risks of wider and easier spread posed by climate change. But in Hawaii, where West Nile virus has yet to arrive, the Hawaiian crow is already extinct in the wild. Traditionally, the bird leads human souls to the Big Island's Ka Lae cliffs for the afterlife. If West Nile virus reaches these islands, as it could easily do in an infected mosquito or bird arriving by plane or ship (between seven and seventy West Nile virus-infected mosquitoes may reach Hawaii by plane each year)[91] or in a migrating Pacific golden plover, the last of the Hawaiian crows could disappear, along with some thirty of the hundred-odd bird species listed as threatened or endangered in the US. And who will look after human souls then?

Right: A hornbill at Mpala Research Station, Laikipia, Kenya. (Liz Kalaugher)

Below: Grevy's zebra at Mpala Research Station, Laikipia, Kenya. (Liz Kalaugher)

Plains zebras elsewhere in Kenya. Note the wider stripes, horse-shaped ears, and absence of white belly and white rump. (Liz Kalaugher)

African elephants in Kenya – modern megafauna. (Liz Kalaugher)

Woolly mammoths. (By Charles Robert Knight, Public Domain, via Wikimedia Commons)

Maclear's Rat, formerly of Christmas Island until its disease-related extinction. (By Joseph Smit – Proceedings of the Zoological Society of London 1887, Public Domain)

Above: A Hawaii 'amakihi. This bird was banded and released back to the forest. (Kahn, Noah/USFWS, Public Domain)

Right: The now extinct honeycreeper the Hawaii 'akialoa (*Hemignathus obscurus*), male & female. (By John Gerrard Keulemans, 1842–1912. Walter Rothschild, The Avifauna of Laysan and the neighbouring islands with a complete history to date of the birds of the Hawaiian possession. London: R.H. Porter, 1893–1900, Public Domain)

Above: An 'i'iwi in Hakalau Forest National Wildlife Refuge, Hawaii. (Schmierer, Alan/USFWS, Public Domain)

Above: Rinderpest outbreak in South Africa, 1896. (Public Domain, via Wikimedia Commons)

Left: Lesser kudu in Laikipia County, Kenya, January 2020. (Liz Kalaugher)

Buffalo, Kenya, early 2020. Buffalo were also susceptible to rinderpest. (Liz Kalaugher)

Above left: Wildebeest, Kenya, early 2020. Wildebeest were susceptible to rinderpest before its eradication. (Liz Kalaugher)

Above right: Saiga antelope. (Navinder Singh, CC BY-SA 4.0 <https://creativecommons.org/licenses/by-sa/4.0>, via Wikimedia Commons)

Tasmanian tiger at Beaumaris Zoo in Hobart, 1933. (Public Domain, via Wikimedia Commons)

Above left: The picture from 1869 known as 'Mr Weaver Bags a Tiger'. (Public Domain, via Wikimedia Commons)

Above right: Thylacine with three cubs, Hobart Zoo, 1909. (Public Domain, via Wikimedia Commons)

Tasmanian wolf and cubs. (The New York Public Library, 1903. *Tasmanian wolf and cubs*)

Black-footed ferret at the Black-Footed Ferret Recovery Program in Colorado, US. (Hagerty, Ryan/USFWS, Public Domain)

Black-footed ferret chasing a prairie dog. (USFWS, Public Domain)

Above left: Black-footed ferret by a burrow entrance. (Hagerty, Ryan/USFWS, Public Domain)

Above right: A black-footed ferret peers out of a trap in 2007 just before it's transported to a facility for vaccination against canine distemper and fitted with a transponder chip. (Mulhern, Dan, Public Domain)

Black-footed ferret reintroduction in Meeteetse, Wyoming, in 2021. (Ryan Moehring/USFWS, Public Domain)

Above: African wild dog in Kenya, early 2020. (Liz Kalaugher)

Right: Harbour seal. (Peter Pearsall/ USFWS, Public Domain)

Above left: Hawaiian monk seal in marine debris at Papahanaumokuakea National Monument. (NOAA/NOAA Fisheries, Public Domain)

Above right: A sea otter holds onto a piece of ice while swimming. (Lisa Hupp/ USFWS, Public Domain)

Tasmanian devil.
(JJ Harrison (https://www.jjharrison.com.au/), CC BY-SA 3.0 <https://creativecommons.org/licenses/by-sa/3.0>, via Wikimedia Commons)

Above left: A Tasmanian devil in a collection tube. (Elizabeth Murchison, Transmissible Cancer Group, University of Cambridge)

Above right: A Tasmanian devil shows its impressive teeth. (Photo by Michael Jerrard on Unsplash)

Above left: Mussels on the hull of an abandoned ship. Non-abandoned ships transport mussels around the world. (LBM1948, CC BY-SA 4.0 <https://creativecommons.org/licenses/by-sa/4.0>, via Wikimedia Commons)

Above right: Artist's impression of the 'founder dog' that first gave rise to CTVT between 4,000 and 8,500 years ago. (Transmissible Cancer Group, University of Cambridge)

A fire salamander in Germany. (Photo by Bernd Dittrich on Unsplash)

Above: African clawed frog (*Xenopus laevis*), which were traded worldwide for use in early pregnancy tests and may have helped spread chytrid fungus. (Muffet, CC BY 2.0 <https://creativecommons.org/licenses/by/2.0>, via Wikimedia Commons)

Right: A mountain chicken frog. (Photo via goodfreephotos.com)

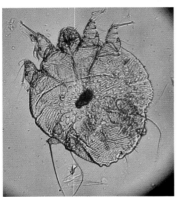

Above: A red fox. (Photo by Tavis Beck on Unsplash)

Left: A *Sarcoptes scabiei* mite, which causes mange. (Kalumet, CC BY-SA 3.0 <http://creativecommons.org/licenses/by-sa/3.0/>, via Wikimedia Commons)

Below: A San Joaquin kit fox. (Peterson, B., US Fish and Wildlife Service, Public Domain)

American crow. (Pearsall, Peter, USFWS, Public Domain)

An *Aedes albopictus* mosquito, which was one of the two main mosquito species to transmit West Nile Virus in New York to other animals. (James Gathany/CDC, Public Domain, via Wikimedia Commons)

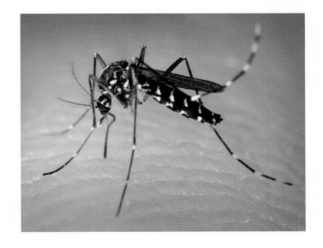

Northern muriqui in RPPN Feliciano Miguel Abdala. (Peter Schoen, CC BY-SA 2.0 <https://creativecommons.org/licenses/by-sa/2.0>, via Wikimedia Commons)

Brown howler monkey in RPPN Feliciano Miguel Abdala. (Peter Schoen, CC BY-SA 2.0 <https://creativecommons.org/licenses/by-sa/2.0>, via Wikimedia Commons)

Above left: Feliciano Miguel Abdala Natural Reserve (formerly Caratinga Biological Station) in Minas Gerais, Brazil. (Gilia Noel, CC BY 2.0 <https://creativecommons.org/licenses/by/2.0>, via Wikimedia Commons)

Above right: Golden lion tamarin. (Photo by Matt Flores on Unsplash)

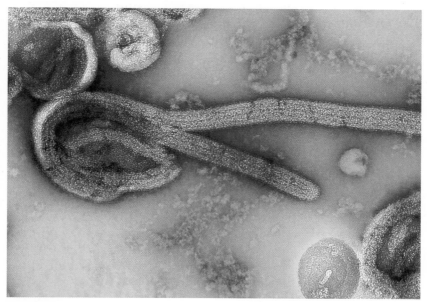

An electron microscopic image of the 1976 isolate of Ebola virus. (Photo by CDC on Unsplash)

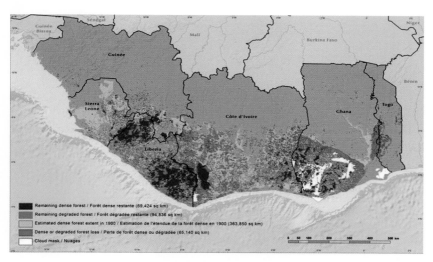

Upper Guinean Forest change from 1975 to 2013. The darkest green areas represent areas of dense forest that remained in 2013, pale green shows the estimated extent of dense forest in 1900, and red areas indicate forest loss between 1975 and 2013. (USGS, Public Domain)

An adult female mountain gorilla carries her infant in Bwindi Impenetrable National Park, Uganda, 2020. (Gorilla Doctors)

An infant mountain gorilla has a rope snare removed from his wrist during a snare rescue in Virunga National Park, DRC, 2020. (GorillaDoctors)

The infant mountain gorilla that was rescued from a snare fully recovered and thriving three months later, Virunga National Park, DRC, 2020. (Gorilla Doctors)

An Arctic skua near Noss before the avian flu outbreak. (Harvey Lloyd-Thomas)

Above: Northern gannets nesting on the cliffs at Noss before the avian flu outbreak. (Harvey Lloyd-Thomas)

Below: Colorized transmission electron micrograph of avian influenza A H5N1 virus particles (yellow) grown in Madin-Darby Canine Kidney (MDCK) epithelial cells. Microscopy by CDC; repositioned and recoloured by NIAID. (NIAID, CC BY 2.0 <https://creativecommons.org/licenses/by/2.0>, via Wikimedia Commons)

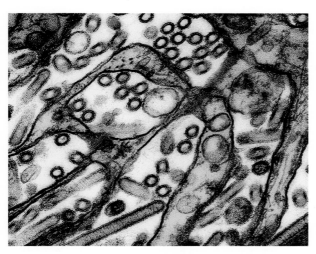

Ngeeti Lempate (Mama Grevy) on right with Grevy's Zebra Scout colleague Paatin Lebasha. (Liz Kalaugher)

Area in Samburu county near where an oil pipeline, road and rail route is slated for construction. (Liz Kalaugher)

Grevy's zebra near a disused boma in Samburu county, Kenya. (Liz Kalaugher)

Young Grevy's zebra in prime breeding territory in Samburu county, Kenya. (Liz Kalaugher)

12

Monkey Mix-up: Muriquis, Vanishing Forests and Yellow Fever

'And on the right hand of the sunlike throne
Would place a gaudy mock-bird to repeat
The chatterings of the monkey.'
                Percy Bysshe Shelley, *The Witch of Atlas*, 1824

Back in 1983 as a PhD student, Karen Strier headed to the Atlantic Forest in Brazil, an area that stretches – or used to stretch before we cleared more than 85 per cent of its up to 1.2 million square kilometres – along the length of South American coast that rises up from the south to meet the continent's easternmost point. Fresh from fieldwork with baboons in Kenya, Strier, who is now a primatologist at the University of Wisconsin-Madison, US, was keen to see how other non-human primates lived and worked, especially outside Africa and Asia, where primates had received more attention thanks to our interest in human evolution. Her research target? The Americas' largest but little-studied monkey, the northern muriqui – pronounced mor-EE-key – (*Brachyteles hypoxanthus*). At the time this woolly spider monkey, as it was also known, was 'only' endangered, though today it's critically endangered. In 2021, the IUCN assessed that at least two dozen populations of this primate have disappeared and only thirteen populations remain, with a total of fewer than a thousand individuals living in isolated fragments of forest in south-east Brazil.[1] Handily enough for Strier's research, the muriqui turned out to behave differently to most other primates – it's unusually egalitarian. 'I fell in love with the muriquis and the forest and Brazil and everything,' Strier says. 'That first fourteen months was like total exploration, everything was exciting and new. It was hard work to get the animals habituated and to learn how to recognise all the individuals in the group and to learn their routes through the forest and adapt to another culture and learn Portuguese and the whole thing. But I was a young PhD student, and it was all very exciting.'

Covered in light brown to gold medium-length fur, muriquis' bodies are nearly half a metre long and their prehensile tails even longer.[2] With long, thin limbs, pot bellies and a dark face freckled with pink markings and fringed by a band of white fur reminiscent of a face-framing hoodie, the northern muriqui lives mainly in trees, where it takes care, no doubt, that the branches can support it – males weigh more than 9 kilograms (20 lbs). This arboreal lifestyle makes studying the primate a challenge. Even finding the monkeys each morning was sometimes tricky. And once Strier had found them, she had to follow them. 'They can travel much more agilely through the trees than I could on the ground,' she says. 'I was dependent on being able to get through the vegetation.' If there wasn't a trail, Strier worked off-trail until she could hire someone from the local farming community to clear one for her. 'So I was covered in scratches from spiny plants, and sometimes you get trapped and have to cut your way out with a machete.' Fortunately for Strier, the muriqui's facial markings are unique to each individual. The only difficulty was getting close enough to recognize all 23 monkeys in her original study group. 'You follow them looking up,' she explains. 'It's a hilly habitat, so sometimes you can get more at eye level if you climb above them, but it's always at a distance with binoculars.' Each day, Strier followed the muriquis until they settled down to sleep then returned the next morning before they left. Which was fine until she had a day off and no longer knew where they were.

The patch of forest that Strier studied near Caratinga in the state of Minas Gerais was less than 10 square kilometres (4 square miles) in size. When I ask how the patch became so small, Strier says the real question is 'why was it protected?' Europeans cleared the region's forest for pasture or farming when they colonised Brazil; in aerial shots of the Atlantic Forest today, you typically see fields dotted with stands of trees, often on hills that were harder to clear and farm. But the new owner of this near-thousand-hectare patch of forest wanted to protect it. When he sold it in the mid-1940s, he extracted a promise from the purchaser, one Senhor Feliciano Miguel Abdala, that he'd protect it too. That land-owner kept his promise and later let Strier study the muriquis. When Senhor Feliciano died in 2000, his family made the area one of Brazil's natural heritage preserves – it's now known as the Reserva Particular do Patrimônio Natural (RPPN) Feliciano Miguel Abdala. The tract of forest sits there to this day, an island of trees and wildlife in a sea of pasture and coffee plantations. Thanks to birds and other animals dispersing seeds, the sites of selective logging have regrown and the forest edges have spread – there's more forest there today than in 1980.[3] In 2023, Strier and colleagues celebrated 40 years of field study at the site.[4]

FEVER THREAT

But in January 2017, as Strier travelled back to the forest after a semester of teaching, she didn't know what she'd find. The professor had listened to reports in the news and from colleagues about an outbreak of yellow fever, a viral disease that infects both humans and non-human primates. The epidemic, which began in November 2016, was the most severe in humans in the country in 80 years. When infected with yellow fever, some people experience no or mild symptoms, others suffer fever, chills, headache, nausea and vomiting. Most people recover from this stage, but the skin and eyes of the unluckiest turn yellow and they see a return of the fever, as well as abdominal pain, dark urine, more vomiting and perhaps even bleeding from their mouth, nose, eyes or stomach. Some 50 per cent of these unfortunate people die.[5] There's no treatment for the virus other than nursing care that relieves the symptoms. Between the end of 2016 and mid-2018, there were more than 7,000 notifications of human cases of yellow fever in the south-eastern Atlantic Forest region of Brazil.[6] A total of 744 people died, mostly in the states of Minas Gerais and Espírito Santo.[7]

The yellow fever virus, a member of the flavivirus family that consists of single-stranded RNA in an envelope, evolved in Africa some 3,000 years ago.[8] In 1927 it became one of the first viruses known to cause disease in humans when Dr Adrian Stokes discovered it in blood samples from a Ghanaian patient that had been delivered to him for investigation.[9] The Ghanaian patient, Asibi, recovered, but Stokes was infected,[10] probably during his studies of the virus in the lab,[11] and died. Yellow fever still kills on the continent today – in 2013 the disease took between 29,000 and 60,000 lives in Africa, where it's endemic in 34 countries.[12]

Also known as Bronze Jack or Yellow Jack, the disease reached the Caribbean and South America in the fifteenth or sixteenth centuries, most likely onboard ships associated with the slave trade.[13] Ships proved trouble for South America as well as for Hawaii's birds. Christopher Columbus himself may even have been the culprit, when his vessels resupplied in the Canary Islands to the west of Morocco and brought in infected eggs from the *Aedes aegypti* mosquito. It's unlikely that an ill human brought in yellow fever, as humans retain a high level of the virus in their blood for less than ten days, not long enough to outlast the one month it took to sail across the Atlantic.[14] Desiccated mosquito eggs, on the other hand, can survive several months,[15] having become infected in what's known as vertical transmission, when a female passes on the virus to the eggs in her ovary. Also known as the yellow fever mosquito, *Aedes aegypti*, whose Latin name comes from the Greek for 'unpleasant of Egypt',[16] bites by day and at dusk. As well as yellow fever, this insect transmits diseases such as dengue fever, chikungunya and Zika, plus two forms of avian malaria.[17] If you're a human, a non-human primate or a bird and you see a mosquito with white-banded legs and a lyre-shaped white marking on its thorax, do your darnedest to avoid it biting you.

At some point before or after it arrived in the Americas, where it's now endemic in thirteen countries,[18] this species began to depend on humans and to prefer living in villages, towns and cities over its original forests;[19, 20] Africa is now home to the newer human-friendly subspecies as well as the forest-dwelling version.[21] In South American cities, *Aedes aegypti*, along with the more recently arrived Asian tiger mosquito *Aedes albopictus* (which also strays away from cities), transmits yellow fever virus from human to human in what's known as the 'urban cycle'. Once in the Americas, yellow fever changed history; an epidemic among French troops in modern-day Haiti, then known as Saint-Domingue, dissuaded Napoleon Bonaparte from conquering land in the United States,[22] and an outbreak delayed construction of the Panama Canal.[23] The US cities of Philadelphia, New York and Boston all experienced serious outbreaks until the disease was eradicated from the US, thanks to people understanding – and acting on – the fact that it was spread by mosquitoes. In 1905, New Orleans experienced the very last epidemic.[24]

In South America, however, the virus lingers to this day. In this continent's forests, the virus gained new hosts: mosquitoes of the genus *Haemagogus* and the *Sabethes* genus, which lay their eggs in tree holes. In Africa, one of the main mosquito vectors of the virus had been the forest-dwelling *Aedes africanus*, which transmitted yellow fever between monkeys, in what's known as the sylvatic, or 'jungle', cycle. Whereas in Africa, monkeys are largely resistant to yellow fever – if they're infected they may grow sick but don't generally die and afterwards become immune,[25] in the Americas, monkeys are much more susceptible, which amplifies the virus and risks epidemics.[26] Once it arrived, yellow fever circulated in mosquitoes and monkeys in the Amazon and the Orinoco river basin in Venezuela and Colombia, sometimes flaring up and causing outbreaks in monkeys and, when humans entered the forest before returning to the city, humans too. The city of Rio di Janeiro documented a total of 19 outbreaks since 1849, each killing at least 1,000 people.[27] By the 1960s, through controlling *Aedes aegypti* in cities and vaccinating people en masse, Brazil thought it had managed to eliminate yellow fever.[28] The first yellow fever vaccine, a weakened version of the Asibi strain named after that first known yellow fever patient,[29] had become available for humans by the late 1930s; its inventor Max Theiler received a Nobel Prize for his work in 1951. But in 1963–64 and 1968–69, human epidemics of dengue, a disease caused by another flavivirus, showed that *Aedes aegypti* was on the rise once more.[30] Soon yellow fever had returned too. Meanwhile, deforestation and population growth increased the amount of contact between human populations and yellow fever virus circulating in the forest.[31] And it was this sylvatic cycle that was responsible for the 2016–17 yellow fever outbreak in Brazil,[32] which came perilously near to the major human conurbations of Sao Paulo – population more than 22 million – and Rio di Janeiro – population 12 million – in the country's South-east. Thankfully, this time the virus didn't make it back into the cities; Brazil hasn't seen an urban case of yellow fever since 1942.[33]

MURIQUI MYSTERY

At the time of this 2016–17 outbreak, scientists knew that howler monkeys were highly susceptible to yellow fever, but how muriquis would respond wasn't clear. Strier drove the near 300 kilometres from Vitoria to Caratinga together with long-term Brazilian collaborator Sérgio Mendes of the Universidade Federal de Espírito Santo and the Instituto Nacional da Mata Atlântica (INMA), and former student and now collaborator Carla Possamai of the Muriqui Instituto de Biodiversidade in Caratinga. 'We had heard from some of the local people about them finding along the road and in nearby forests evidence of dead animals,' Strier says. 'So we were really worried. And when we got there, the most noticeable thing was the forest was so quiet – we couldn't see or hear any howler monkeys.' Classified by the

IUCN as vulnerable,[34] the brown howler (*Alouatta guariba*) has a call like the venting of an angry steam train or a cross between a roar and a very long burp.[35] It's generally the male who howls, often to defend his mate or territory – these are no mellow muriquis. As he howls, he may stand, hands and feet curled round a branch, the tip of his tail coiled round a branch behind, as he thrusts his head forwards, mouth open wide above a shaggy beard.

But on this visit in 2017 the brown howlers weren't howling and the team couldn't see or hear any muriquis either. These monkeys have a range of calls; adult males' vocabulary includes a neighing sound, perhaps to keep the group together, while adult females may make staccato noises to maintain space between themselves and other females while they feed.[36] It wasn't time to panic quite yet – Strier's Brazilian students had just come back from their New Year's break and it often took a few days to relocate the muriquis after a gap. This time, however, instead of 324 animals in five groups,[37,38] which was about one-third of the northern muriqui's total population,[39] all the team could find was a small group of four or five animals near the research house. 'It felt almost like an abandoned area,' Strier says. 'The whole forest just felt different. And then Sergio found a dead howler monkey.'

A while later the researchers found the cranium of what they suspected was a muriqui and Strier became really concerned. The population here was the largest left in the world and Strier and her field team knew them as individuals. She called in a bigger team made up of her former students, who gave their time for free. Thankfully, although the researchers' fears about the skull bone they found turned out to be correct, this team managed to relocate most of the muriquis. 'It was like they had all gone off and everybody was kind of strange,' Strier says. Instead of these animals mostly being in their groups, Strier's team found 'a few here, others there, as if they were scattered and a bit lost'. In the six months since the last count, about 30 animals had disappeared, some 10 per cent of the population.[40] No one can be 100 per cent certain that these muriquis died of yellow fever, but the timing means it's very likely.

Despite ranging widely through a forest that they knew well and which had good trails, the researchers found the remains of only four muriquis. Yet the muriqui headcount was down by 40, implying that if you're in an area where you haven't tallied the primate population, the death count could be ten times higher than you would assume based on remains. 'If we were just going into the forest and we found four skeletons or craniums, we would never have known how many animals really died,' Strier says. 'When you read reports that hundreds of animals died, it's probably thousands.'

For the three other types of monkey in the RPPN Feliciano Miguel Abdala, the picture after the yellow fever outbreak was bleaker. The population of brown howler monkeys dropped from 522 in 2015 to just 70, a decrease of some 85 per cent.[41] Mendes' students collected over a hundred sets of howler monkey remains from the region around the field site. 'In many cases, these were entire groups,' Strier says. 'There'd be one howler monkey group that lived in a little tiny patch of forest, a few acres in size, and they would find the whole group or almost all the group. The students got there right away when the animals were still dying or in the process of dying. It was really horrible for them to see all that.'

Another of the reserve's monkey species, the buffy-headed marmoset (*Callithrix flaviceps*), looks a little like a clown – it has harlequin-style diamond black markings around its eyes and nose, and a cream face surrounded by apricot fur that extends into triangular fluffy tufts blossoming from either side of its head. But nothing about this animal's potential fate is funny. The species is under threat from hybridisation. Marmosets from other species, mainly common marmosets and black-tufted-ear marmosets, were taken for the pet trade and given alcohol by market traders to make them seem placid. Then, once they reached their new home and were scared and hungover, they seemed too wild to tame and were often released back into the wild, into areas already populated by other marmoset species.[42] And most – some 80 per cent – of these critically endangered buffy-headed marmosets lost their lives in the 2016 yellow fever outbreak; the population in the reserve declined from 85 individuals to 15–17.[43] 'We've become increasingly aware with these really solid databases, how disease can wipe out populations of primates,' Strier says. 'And because they're living in fragments, you can see it in their small populations. It contributes to the fragility of these places.' In a small area of forest, animals are likely to be crammed together at a higher density than they would have been before we destroyed most of their habitat, potentially making any disease spread faster. Carla Possamai didn't see any of the marmosets for months after the outbreak, even though she paced the trails, playing back their squeaking call.[44] In the before times, one group had regularly entertained tourists at a particular spot in the forest, but by 2021 it could take a couple of months to find any of these marmosets at all.

The buffy-headed marmoset is not to be confused with the buffy-tufted marmoset, black-tufted-ear marmoset, black-headed marmoset, white-headed marmoset or white marmoset. The world is home to 22 species of marmoset, but the only one I've come across was in the Peruvian Amazon, where three pygmy marmosets clung to the bark of a large tree at about head height. Looking incredibly delicate – they weigh some 100 grams, about as much as 25 teaspoons of sugar – and pretty much the length of a large handspan, they rotated their heads to peer back at us from flat doll-like faces with huge golden eyes, their pupils massive in the gloom of the forest. They were so small, so beautiful and so surprising that I could hardly breathe.

Back at the Feliciano Miguel Abdala reserve, the third species of non-human primate, black-horned capuchins (*Sapajus nigritus*), proved less susceptible than brown howler monkeys and buffy-headed marmosets, but more susceptible than muriquis – 40 per cent of these capuchins died and the population dropped from 377 to 226.[45] Non-human primates suffered elsewhere in Brazil too. Until this yellow fever outbreak, the endangered golden lion tamarin (*Leontopithecus rosalia*) had been something of a conservation success story. In the early 1970s, only a few hundred remained, in forest fragments around 80 kilometres outside Rio de Janeiro. Weighing just 600 grams and with long russet-gold fur that sweeps back in a mane around the bare skin of their faces, forming a regal quiff so long that it hangs down to the sides, a little like an extensively blow-dried celebrity, it's perhaps no surprise that these creatures were popular with zoos and the pet trade. But that, along with habitat destruction, led their numbers to plummet. A conservation programme that kicked off in 1984 re-introduced captive-bred animals into the wild, and by 2014 numbers had reached 3,700.[46] After the outbreak, numbers dropped by one-third, to 2,500.[47] For the tamarins, losses from what had been two of the largest populations meant conservation workers aiming to reconnect forest fragments containing at least 2,000 golden lion tamarins now had to link up eight patches of forest rather than three. Planting just one additional forest corridor was put at $138,216, so the total new costs were substantial.

PRIMATE PROBLEMS

By May 2017, the yellow fever outbreak had killed at least 5,000 non-human primates and would continue for another year.[48] This wasn't the first mass die-off of South America's monkeys. In 1951 a team found that numbers of Ecuadorian mantled howler monkeys on Barro Colorado

Island, which was created in 1914 when the Chagres river valley was flooded for the Panama Canal, had dropped by 48 per cent since they were monitored in 1933, most likely due to a recent outbreak of yellow fever.[49] In 1933, one C.R. Carpenter had seen a male mantled howler use his body and tail as a bridge for an infant to cross from one tree to another. But in 1951 the female that researchers watched almost continuously for eight hours 'was very slow and sluggish in all her movements, ate practically nothing and lagged more than four hours behind the other monkeys in the group progression'.[50] Howler monkeys in Brazil also suffered badly from yellow fever outbreaks in 1936, 1937, 1945, 1947,[51] 2008 and 2009.[52]

The most recent monkey troubles began in July 2014, when yellow fever broke out in non-human primates in the transition zone between the Amazon and the Cerrado, an area of tropical savannah to its south and east.[53] In 2016 the disease spread into the coastal Atlantic forest, where monkeys hadn't been exposed to the virus before and didn't have any immunity, then on into new regions in the south and south-east where people hadn't been immunised against yellow fever because the disease had never circulated there before. We don't know why the outbreak was so severe or how it spread so fast;[54] from early 2017, in the south-east the yellow fever virus dispersed on average roughly three kilometres per day.[55] Within a year the disease had reached four of Brazil's most populated states – Minas Gerais, Sao Paulo, Rio di Janeiro and Espírito Santo.[56] We also don't know how, when yellow fever reached the remnants of the Atlantic forest, it jumped from fragment to fragment.[57] Monkeys don't tend to travel large distances outside their home territory and *Sabethes* mosquitoes usually stay in or near the forest. Although *Haemagogus* mosquitoes could perhaps travel more than eleven kilometres,[58] particularly if rain has been scarce and females lack water-containing tree holes to lay their eggs, or if the numbers of animals to bite have dropped.

Humans, of course, travel further still. The most likely suspects for the spread between forest fragments are infected mosquitoes and infected humans who aren't aware they're ill; on this continent humans are much less susceptible than monkeys. Between 1940 and 2015, the human population in south-east Brazil grew 368 per cent, from 18,345 to 85,817, and rose from a density of 19.8 people per square kilometre to 92.9.[59] In addition, more people are travelling to and spending time in the forests, whether for work, eco-tourism, woods therapy, the 'back to nature' movement, neo-survivalism, or to live in the forest or visit a second home.[60] That exposes people to bites from infected mosquitoes more frequently, as well as degrading the habitat and perhaps forcing non-human primates to migrate to other forest patches, dispersing the virus. Vehicles and luggage may transport infected mosquitoes or eggs too.

Then there are the environmental factors. In the last 40 years, deforestation of Brazil's inland forests has sped up, enabled by big machines and use of fire, and forcing monkeys to relocate and live in altered environments and in overpopulated small patches of forest.[61] And agriculture has expanded in what's known as MATOPIBA territory, some 400,000 square kilometres in the states of Maranhao, Tocantins, Piaui and Bahia, where production of soybean and other grains has soared.[62] Monkeys whose habitat has suffered harm may be forced not only to cluster together in forest fragments at higher densities, but also to live closer to humans. As elsewhere, when people destroy swathes of forest, leaving only fragments, they put people and other primates in closer and more frequent contact, boosting the risks of disease transmission between species.[63] Loss of habitat and of wildlife (due to both the habitat loss and yellow fever outbreaks) mean mosquitoes have fewer non-human animal bodies to bite, so turn to humans.[64] What's more, the Asian tiger mosquito (*Aedes albopictus*), which reached Brazil in 1986,[65] can live in both human-modified environments and rural ones, potentially spreading the virus between the two. Perhaps Brazil is also a victim of its own success. In recent years, forest protection and regeneration programmes have boosted the numbers of both monkeys and mosquitoes.[66] Climate change may have played a role too. Strier favours the theory that when the rains returned after a two-year drought, dormant mosquito eggs hatched and mosquito numbers boomed. She and her colleagues felt like there were more mosquitoes at the time. But we don't fully know why the 2017 outbreak was so lethal, spread so fast and what its long-term consequences will be for non-human primates, including their social behaviour.[67] As monkeys and humans spend more time in each other's vicinity, there's a real risk that yellow fever may re-enter Brazil's cities and their outskirts, potentially affecting millions of people.

VACCINATE FOR VICTORY?

So what can we do to stop yellow fever in its tracks? The solutions for humans are easier to implement than those for non-humans; it's much easier to explain to a human than a non-human that vaccination will benefit them. A global vaccination programme aims to stop epidemics of yellow fever in humans by 2026;[68] in Brazil in 2018 the government began vaccination of 23.8 million humans, using one-fifth the normal dose of vaccine due to shortages, in 69 cities and towns in Rio di Janeiro state and Sao Paulo state.[69] Although many people feel safe outside the forest and are reluctant to be vaccinated.[70]

In cities it's also possible to control mosquitoes without harming the ecosystem. Fortunately, vaccinating people is a win for humans and a win for wildlife, as it could cut the number of asymptomatic humans unwittingly helping yellow fever spread into forest fragments.[71] Vaccination is an option for some of the more susceptible animals too, for whom scientists are trialling versions of the human vaccine adapted for other primates. In 2021, scientists began vaccinating golden lion tamarins in Brazil, enticing them into cages with recorded tamarin calls and bananas before sedating them.[72] By February 2023, the team had treated more than 300 animals and discovered that the vaccine had created antibodies in 90–95 per cent of the tamarins. By August 2023, the population had bounced back to 4,800 individuals.[73] 'Those are animals that are relatively easy to catch,' Strier says. 'They're small and you can lure them into a cage and they can be vaccinated. And that seems to be having a positive impact on them.' Howler monkeys, however, are much bigger and heavier and so are harder to handle. The plan is not to trap them in order to vaccinate them, but to vaccinate these monkeys when they've been trapped for other management purposes, such as translocations – deliberately moving them to another place – or reintroductions. 'That's one way to protect the most vulnerable species from yellow fever,' says Strier. 'But of course, we're all very concerned about what are the next outbreaks – what other diseases could potentially affect them? Even apart from diseases, these small populations are still really vulnerable.'

During outbreaks in some parts of Brazil, people added to monkeys' woes by harassing and killing them in fear that they were transmitting yellow fever, even in areas without any cases.[74] During the 2008–09 outbreak, some 40 per cent of dead howler monkeys tested may have been killed in this way rather than dying from the disease.[75] And human aggression in Rio Grande do Sul led the local red howler monkey to be listed once again as an endangered species.[76] But in 2016 and 2017, Strier and her team didn't see this happening near RPPN Feliciano Miguel Abdala. Perhaps this is a sign that campaigns such as 'Protect our guardian angels' have worked,[77] by educating people that howler monkeys are victims rather than perpetrators and should be appreciated for often providing the first clue of an outbreak, enabling people with access to vaccines to get themselves vaccinated; in Brazil vaccinations are available for free as long as you can get to the clinic.[78]

What of the other species at the reserve? The buffy-headed marmosets are starting to reproduce and the population may be recovering. The picture for the brown howler monkey isn't yet clear. Monitoring is still underway and Possamai hasn't yet been able to discern if howler babies are surviving, or if the social groups are restructuring and becoming larger as smaller groups merge; it's tricky to work out if you're not capturing the animals or manually marking them. 'We've been focusing on the marmosets because less was known about them,' Strier explains.

These days, Strier is seeing a higher death rate in muriquis than before the 2016 outbreak of yellow fever. She thinks this may be due to the dilution effect; now that so many brown howlers have disappeared, mosquitoes infected with the yellow fever virus are more likely to bite a muriqui. The previous higher biodiversity 'diluted' the effect of the disease. Not that the muriquis are showing signs of yellow fever. 'Maybe they get infected with yellow fever and it doesn't cause them to haemorrhage, but it makes them weaker and more vulnerable to other disease,' Strier says. 'We just don't know enough about their immunology.'

The sudden disappearance of many of the muriquis whose lives and habits she'd got to know changed Strier's views about how far to go with conservation. 'It had a profound influence for me on my perception of the fragility of our conservation efforts,' Strier says. 'When I give talks, I sometimes joke about how I felt smug because the forest was protected, and the muriqui population was growing and in good shape, and they had survived the drought. We had some scares with forest fires but they didn't impact the part of the forest that the muriquis used.' Everything was going OK. When yellow fever hit, however, 10 per cent of the muriquis disappeared within just six months. And the situation was even worse for other primates. 'To lose so many howler monkeys, to feel the loss of all that wildlife and energy, had a huge impact on me,' Strier says. 'It really impressed me how we can't relax our efforts and that these forest fragments are super fragile and super vulnerable.'

From being a staunch proponent of non-interventionist methods such as creating natural corridors so that animals can move themselves, Strier now feels that more hands-on techniques such as captive breeding or translocating animals are appropriate. 'In these really critical populations, we don't have that many options,' she says. And here Strier's research can help. Forty years ago, we knew most about non-human primates from studies of baboons and macaques, where young females remain in the groups they were born in. When Strier wrote up her dissertation back in the 1980s,[79] she realised that muriquis were different to these other primates – the males and females are the same size and tolerate one another, they hug for minutes at a time, and the males mate in full view and don't fight over, or try to dominate, females.[80] And female dispersal is much more common in primates than we'd thought. Strier discovered that female muriquis transfer to new social groups to find a mate at about five to seven years old, either alone or with a buddy, and often during the rainy season, from October to April. 'Sometimes they move directly into another group,' Strier explains, 'sometimes they shop around and they can take a year or longer to find the group they want to live in.' Knowing this helps with introducing a new female to a captive breeding programme or with translocations, when conservation workers deliberately move animals to another group in order to widen the gene pool. The more these processes resemble life in the wild, the more they're likely to succeed. Other research on muriqui diet and behaviour has helped maintain healthy populations in captivity. 'As a scientist, it's always super-rewarding when you can see the results of your scientific research have a real application as well,' Strier says. 'That motivates me even more to keep the project going because I think we can still contribute a lot.'

Today the muriquis at RPPN Feliciano Miguel Abdala are effectively living on an island that nobody visits, with nowhere to go. A few females have moved to tiny patches of neighbouring forest on other people's land, but options for such dispersal are extremely limited now that 85 per cent of the Atlantic Forest is gone. Fragments of forest suffer from the edge effect, since a greater proportion of their trees are exposed to wind and the elements, which can cause tree loss around the forest perimeter, shrinking the forest even further. And many species of tree in the Atlantic Forest no longer have populations large enough to be viable, whether through deforestation, selective logging, lack of seed dispersers, or climate change. Monkeys themselves often help vegetation spread and regenerate; with fewer monkeys in the forest after the outbreak, there may be knock-on effects for seed dispersal. The smaller the fragment of forest, the more of a concern this would be. 'There's lots of explanations but we're looking at a different kind of ecosystem,' Strier says. 'What you see in this region and certainly in fragments may not ever look again like the original Atlantic Forest.' That said, in this region of Brazil, use of land for agriculture and pasture is declining. At RPPN Feliciano Miguel Abdala, the forest is expanding through natural regeneration around the edges.

As the years pass, an isolated population in a small fragment of habitat is likely to lose its genetic diversity. And animals that are more genetically similar are less able to adapt to environmental changes or the arrival of a new disease. Primates are arguably better off than other animals in the face of change as they have more options for adjusting their behaviour – they can be more flexible in their diet than, say, hummingbirds, which need flowers to provide nectar. Whereas muriquis can choose to eat leaves, fruits, vines, flowers, bark, nectar, seeds and soil.[81] The brown howler monkey, however, may be too accommodating for its own good; some scientists think this flexibility may, by enabling individuals to live successfully in fragmented habitats, mask the longer-term need for their protection in larger tracts of forest to prevent genetic drift – genetic losses that happen by chance – and other problems of isolated populations.[82] 'Whether we're looking at species that are just hanging on to the end, or whether we're looking at animals that are able to survive and persist for a long time, we don't know,' Strier explains. 'But certainly living in isolated populations is not the ideal situation for any organism.'

Currently, Strier's conservation priority is building a corridor that connects the muriqui population at RPPN Feliciano Miguel Abdala and their neighbours more than 50 kilometres away at RPPN Mata do Sossego, in an even smaller patch of land of just 1.8 square kilometres (0.7 square miles). 'For these small forest fragments, the big emphasis has to be in connecting them into one another so that the animals have the opportunity to increase their population sizes, creating more habitat through those corridors, and then also through the opportunity of gene flow through the populations,' Strier says. RPPN Feliciano Miguel Abdala is home to one of the largest sub-populations of northern muriquis in the world, and they live in multiple groups so are probably in better demographic shape than their 28 or so neighbours living in one social group. Although that smaller population did benefit from fresh genes when conservation biologists translocated a female from another site. While 'Strier's muriquis' are potentially more robust as a population, they live in an area that's hotter and drier, so they're at greater risk from climate change. 'I see the value of the corridor as a way to provide these muriquis with a place to go so that they'll not only encounter other muriquis, but they'll also have access to more water and cooler temperatures, which might be better for their survival,' Strier says. Warmer summers could also, perhaps, see yellow fever reach new areas and speed up development of the virus.[83] Wetter weather could increase the number of mosquitoes too.

For her study population of muriquis in Brazil, Strier hopes that in twenty years' time there'll be more, better-connected habitat and the existing habitat will be improved, through natural regeneration as well as assisted regeneration such as tree planting. That could go some small way towards reversing our fragmentation of the forest that's put the animals at more risk of disease. Whatever we can do to boost the muriqui population will help protect against future outbreaks of other diseases – it's only pure fluke that muriquis aren't as susceptible to yellow fever as brown howler monkeys and buffy-headed marmosets. And there are plenty of other insect-transmitted diseases that affect primates in Brazil and about which we know very little, such as phlebovirus, orthobunyavirus and alphavirus.[84] Despite all this, Strier remains an optimist for the region's non-human primates because she's seen a small, isolated population of muriquis increase more than six-fold in size. 'We still have to do everything we can, though, to give the resilience of nature, including primates, a chance,' she says.[85]

13

Bats for Bushmeat: Plantations, Eco-tourism, Apes and Ebola

'What an ugly beast is the ape, and how like us.'

Cicero, 106–43 BCE

When you first see mountain gorillas (*Gorilla beringei beringei*) in the wild, it feels like discovering a sibling that you didn't know you had, according to Kirsten Gilardi, a wildlife health vet at the University of California, Davis, US. 'There's something compelling about the fact that we're so closely related to them,' Gilardi says. 'Then to be in their natural environment, seeing how they live in their family groups in a beautiful forest and interacting with each other in very human and very kind ways, it's a special experience. Hard to describe, honestly.'

To enable people to get this close, the gorillas are habituated – they've been taught over many months that humans don't pose a threat to their lives. That helps researchers to study them, vets to treat them and small groups of tourists to visit them. Crucially, the money that tourists pay helps manage and protect the parks where these gorillas live. When I re-enact the experience for myself via YouTube, I see a silverback rest on his side, limbs in the recovery position, knees bent, eyes shut in a grassy clearing among the trees of Rwanda's Volcanoes National Park.[1] The silver fur this group leader developed at the age of twelve or thirteen glints across his strong shoulders and shot-putter arms. Around the side of his face bristles dark fur like a Victorian gentleman's whiskers, and his hairline sits low beneath a high domed forehead. A few immature silverbacks, younger males, known as blackbacks, and females fold vegetation into their mouths and chomp it with their back teeth, sometimes bending their heads to the floor to eat something straight from the ground. Arms fold across pot bellies as the animals sit like hunched-up Buddhas. Amber eyes fringed with long lashes stare peacefully from wrinkled dark faces, their nostrils so wide they span almost the distance between the pupils.[2] Some sit and scratch between their shoulder blades with fleshy grey fingers. Others lie sprawled on their backs, mouths hanging open as they snooze alone or as another gorilla picks their fingers through the fur on their stomach. Some follow their leader in lying on their sides, a youngster perched on their shoulder. Other youngsters, woolly furred, knuckle-walk then whirl their arms around and spin their bodies in circles before wrestling each other in play fights – pushing, yanking, chasing, rugby tackling, rolling on the ground, jabbing at foreheads, baring sharp canines. The smallest cling to their mothers' chests, while one youngster, the fur on his head standing bolt upright like a punk hairstyle, lies on the ground in the crook of his mother's arm. He sticks his hand in her eye and she patiently pushes his tiny arm away.

No need for YouTube for Gilardi, who is also executive director of Gorilla Doctors, an organisation that provides veterinary care to eastern gorillas – mountain gorillas and their less thick-haired close cousins Grauer's gorillas (*Gorilla beringei graueri*). The roots of the organisation lie with famous researcher Dian Fossey, who began studying these gorillas in 1967 in the Virunga Mountains. Over time, Fossey noticed that walking upright or standing made the gorillas in her new study area nervous, so she knuckle-walked to get close to them and chewed sticks of celery to draw them even closer.[3] She learned to identify the gorillas by their noseprints – the wrinkles that, like our fingerprints, are pretty much unique to the individual.[4] And she realised that the animals were dying of their injuries after becoming trapped by snares set by poachers for wildlife like antelope;[5] traditionally, local people don't hunt mountain gorillas for meat, but adults have been killed when their infants were taken as pets,[6] and also poached directly.[7] By the mid-1980s, fewer than 300 mountain gorillas remained in the Virunga Mountains.[8] Fossey wanted the surviving apes to receive veterinary care to remove the snares, and was growing concerned that contact with humans was giving the non-human primates illnesses – the yellow fever that we heard about in the last chapter is by no means the only disease that humans, other apes and monkeys share.[9] Gorilla, chimp and human bodies are so similar biologically that it's easy for diseases to jump between us – when pathogens transmit from humans to other animals they're known as 'reverse zoonoses' or anthroponoses (short for zooanthroponoses), although some argue that the term zoonosis (often considered short for anthropozoonosis) should cover both directions – non-human animal to human, and human to other animal.[10] Fossey turned out to be right about the threat of gorillas catching respiratory illnesses from humans. But she was murdered in December 1985 and didn't see a different disease, one that's not respiratory but equally, if not more, deadly loom large for mountain gorillas: Ebola virus disease, also known as Ebola haemorrhagic fever.

EBOLA EMERGES

Most of us know of Ebola as a human disease that raged across West Africa from December 2013 until 2016, tragically killing more than 11,000 people.[11] I, for one, am terrified of it. In humans, Ebola causes high fever, abdominal pain, and bloody vomit and diarrhoea. You can die within 72 hours of first feeling ill as the virus invades your liver cells, immune cells and the cells that line your blood vessels.[12] The genus takes its name from the Ebola River in the Democratic Republic of the Congo (DRC), which flows near the location of one of the earliest known outbreaks in humans, back in 1976. Before this outbreak was stopped in its tracks by quarantine measures and better sterilisation of medical equipment – part of its spread was due to injections with germ-harbouring needles at a local hospital – it killed 280 of the known 318 people it infected.[13] As well as people, Ebola has also killed thousands of chimps and gorillas. We first realised great apes were succumbing to Ebola in November 1994 when researchers studying the behaviour of a group of 43 western chimpanzees (*Pan troglodytes verus*) in Côte d'Ivoire, on the southern coast of West Africa, found eight group members dead and noticed that others were missing.[14] The team observed two of the chimps appearing ill before they died – they showed signs of stomach pain, lethargy and not wanting to eat; it looked like nearly all the chimps who became infected didn't survive. In total, a quarter of the group disappeared in just a few months from the rainforest of Côte d'Ivoire's Taï National Park. Lab tests identified the virus as being from the Ebolavirus genus; this version is now known as the Taï Forest species. The same group of chimps had experienced a similar outbreak in 1992, though the cause of death wasn't identified.[15] Before both waves of the 1994 outbreak, the chimps had hunted and eaten a red colobus monkey; the chimps that had eaten more meat had a higher risk of catching Ebola. Red colobus monkeys are vegetarian and spend most of their time in the canopy, where they may have picked up the Ebola virus from bat secretions such as faeces, urine and saliva.[16] Alternatively, the chimps may have come into direct contact with infected bat bodily fluids when they spent days feeding from a fig tree.[17]

Several species of bat are under suspicion as the reservoir host for Ebola, including the hammer-headed fruit bat (*Hypsignathus monstrosus*), singing fruit bat (*Epomops franqueti*) and little collared fruit bat (*Myonycteris torquata*).[18] It's also possible that insectivorous bats or even an animal that's not a bat is the reservoir host.[19] To date, we have only found fragments of the virus' RNA in the liver and spleen of these fruit bats,[20] never a whole 'live' virus, though Gilardi is optimistic that we'll pin this down soon. It's likely Ebola viruses had been circulating among bats in Africa's forests for some time – 'aeons' according to Gilardi – before we identified them. They'd probably been spilling over into people too, and making them ill, for a long time before we realised.[21] Bats are a uniquely perfect group for evolving viruses, Gilardi believes. Firstly, they're among the oldest mammals on the planet, so bat viruses have had a long time – tens of millions of years – to co-evolve alongside them. Bats host almost all viruses without suffering ill effects.[22] Bats, it seems, may be protected against becoming ill from viruses by their evolution of flight, whose stringent requirements meant that, almost as a side-effect, they developed an immune system that counteracts viruses but limits inflammatory responses that can damage the bat itself.[23] This may be why bats are so incredible in other ways – they have a low incidence of cancer and make up 18 of the 19 mammal species that, when adjusted for body size, live longer than humans (the other is the naked mole-rat).[24] The longest-living bats we've recorded survive an average 3.5 times longer than non-flying mammals of the same size; we have records of five species of bats living longer than 30 years in the wild.[25] Finally, as Gilardi explains, many types of bat roost in large groups, letting viruses circulate easily, and many bats are migratory, taking their viruses with them – and shedding them into the environment – as they move around the planet.

Apes don't hunt bats but could encounter fruit bats roosting in the trees where they feed, or insectivorous bats in tree holes where the apes are seeking honey, insects or water. 'Ebola outbreaks could happen anywhere, anytime, under any circumstance,' Gilardi says. 'But it does require that close contact between a bat that's shedding the virus at the time of contact, and that virus making it onto a mucous membrane of a human or an [non-human] animal.'

However it happened in the 1994 outbreak, once chimps were infected with Ebola, whether live or dead, they would have been able to spread the virus to other chimps through direct contact. A researcher who performed a postmortem on one of the chimps also became infected, but no other humans did.[26]

At the time of the outbreak, deforestation of the Guinean forest zone had altered rainfall and temperature patterns and West Africa was experiencing a prolonged drought. (According to estimates, the upper Guinean forest extended at least 360,650 square kilometres before 1900, across parts of Guinea, Sierra Leone, Liberia, Côte d'Ivoire, Ghana, and Togo. By 1975, satellite images indicate that 84 per cent of that forest was gone, and a further 28 per cent of the forest remaining in 1975 had disappeared by 2013, some 15 per cent of it converted for agriculture, including plantations.)[27] What's more, refugees from the civil war in Liberia had doubled the local population and the park was feeling the pressure of illegal crop plantations and additional poachers. This chimp group lived in the north-western edge of the park, now just two kilometres from farmland and broken forest.[28] Perhaps these changes in the climate and environment altered the behaviour of the bats harbouring the Ebolavirus without suffering ill effects, causing the virus to spill over into the much more susceptible chimps. According to wildlife medicine specialist Gary Wobeser of the University of Saskatchewan, Canada, diseases are like weeds in that both thrive in disturbed environments.[29] Each year, the world loses 7 million hectares of forest – an area the size of Portugal – and humans have altered 75 per cent of the earth's terrestrial surface.[30] Some believe that the combination of rapid human population growth and deforestation in West Africa contributed to the unprecedented scale of the 2014 outbreak in people.[31]

Outbreaks of Ebola have often taken place during, and especially towards the end of, the dry season (which runs from December to May). The dry season is when bats breed and come into closer contact with each other to compete over females and to mate; pregnancy can alter the female's immune system and other mammals may feed on infected birthing fluids, blood and placental tissues.[32] In addition, as the dry season progresses, food resources become more scarce. This could force fruit bats, duikers (a type of antelope), chimps and gorillas to cluster together in or under the few trees that still bear fruit. And that would put animals at greater risk of coming into contact with secretions from infected bats, for example by foraging on partially eaten fruits or on fruit pulp that the bats spit out.[33] People cutting down trees and using the land for new purposes could also reduce fruit resources and bring animals into closer contact,[34] as could forest disturbances such as fire and drought.

This 1994 outbreak was the first known case in wild great apes. The first known Ebola cases in non-human primates were discovered in 1989 in a quarantine facility in Reston, West Virginia, US,[35] when cynomolgus monkeys, also known as crab-eating or long-tailed macaques (*Macaca fascicularis*), were found to be infected.[36] Some, but by no means all, of the ailing monkeys had a fever and haemorrhaged from their skin and internal organs. Some died from the illness and others were killed – their bodies exhibited enlarged kidneys and spleen, and sporadic organ haemorrhages.[37] These South East Asian natives, headed for a lifetime sentence in a laboratory, had probably picked up a different species from the Ebolavirus genus, Reston Ebolavirus, at a primate export centre in Manila in the Philippines. We don't know for sure how the animals encountered the virus, but the export centre was built on a former fruit orchard and the animals may have been exposed to fruit bat excreta.[38] This species of Ebola is unusual in coming from Asia; all other species have been found chiefly in Africa. It's also unusual in that it doesn't (yet?) infect humans. The Ebola genus of viruses contains six species that we know of, all members of the filovirus family of single-stranded RNA viruses with a fine, filament-like structure. The Taï Forest species and the species of Ebolavirus from the 1976 outbreak near the Ebola river, now known as the Zaire Ebolavirus after the name of the DRC at the time, are the only species that we know have made wild great apes sick. So far. Humans may suffer from the Zaire, Taï Forest, Sudan and Bundibugyo species of the virus.

GABON GOES ON

At the tail end of 1994, around the same time as the Côte d'Ivoire chimp deaths, eastern chimpanzees, gorillas and humans in Gabon on the West coast of Central Africa suffered an outbreak of the Zaire strain of Ebola. People living in gold-panner encampments in Minkebe Forest in the Minkouka area far in the north-east of the country saw dead chimps and dead western lowland gorillas (*Gorilla gorilla* ssp. *gorilla*) in the forest around the same time as humans started to die. In the Minkebe and Mwagne forests, a colossal 90–98 per cent of gorillas and chimps disappeared.[39] Deaths of non-human great apes accompanied all three human outbreaks of Ebola in north-eastern Gabon from 1994 to 1996.[40] Around

40 kilometres south of the gold-panner encampment, in the village of Mayibout 2, in February 1996, people first became infected when they butchered a chimp that a group of children had brought back from the forest after their dogs killed it.[41] It's illegal to hunt great apes in most protected parks, 'but great apes range well outside of protected areas, and they're ranging in forest where people also live and are subsistence hunters', Gilardi says. While all mountain gorillas live in protected areas, other gorilla species are not so lucky. Some of these apes live in protected areas but many others don't, especially in the vast forests of the Congo Basin.

Twenty people ate chimp meat and eighteen became infected,[42] passing the virus to other humans, perhaps while these new victims nursed them or washed their dead bodies in the traditional manner; a total of 34 villagers died over the next month.[43] 'In our whole history we have always eaten dead animals found in the forest,' Gustave Mabaza, a Gabonese anthropologist told *The Guardian* newspaper.[44] Mbaza spent time explaining to communities near forests why Ebola meant they shouldn't touch or eat dead animals. 'Now, for the first time ever, food gathered from the forest is seen to be dangerous,' he added. 'That changes the whole relationship between people and the forest. It has always been part of people's lives, all that is good comes from the forest. Now it is not.' The third of these human epidemics began in July 1996, some 160 kilometres south of Mayibout 2. The first human patient was most likely a hunter in a forest camp in the Booué area. Two hunters died, one of whom was said to have killed several mangabeys, a type of monkey;[45] a dead chimp infected with Ebola was found in the forest nearby at around the same time.[46]

Overall, between 1995 and the year 2000, biologists reckon that three-quarters of western lowland gorillas in six protected areas were lost to Ebola virus disease;[47] western lowland gorillas live mainly in Cameroon, Central African Republic, Republic of the Congo (ROC), Equatorial Guinea and Gabon. In part because of these Ebola losses, the IUCN upgraded the species to critically endangered.[48] After poaching, the disease is the second biggest cause of declining western lowland gorilla numbers.[49] In central chimpanzee (*Pan troglodytes* ssp. *troglodytes*) populations too, infectious disease, especially Ebola, is the second major driver of decline, after poaching.[50]

Chimps and gorillas, as far as we know, then remained free from Ebola until a second cluster of outbreaks between 2000 and 2005, in a largely intact area of forest straddling north-eastern Gabon and the north-western ROC.[51] In 2001, in the Zadié region of Gabon, villagers and researchers noticed that the forests were littered with an unusually high number of dead gorillas, chimps and duikers.[52] It's rare to find carcasses in the tropics – many animals are hunted and eaten by predators, and it can take just a month for the body of a gorilla to decompose, so this development was concerning. Most of the bodies were gorillas and within two hours' walking distance of villages. It's possible that hundreds or even thousands of gorillas died unnoticed in the 3,000 square kilometres of sparsely populated forest here. Following at least six different hunting incidents, Ebola broke out in humans too in both Gabon and neighbouring ROC. A total of 97 people died before the epidemic was declared over in May 2002.[53]

Soon afterwards, in late 2002 and early 2003 at the Lossi Gorilla Sanctuary in north-western ROC, eight groups of western lowland gorillas that researchers had monitored almost every day for ten years disappeared. A total of 143 individuals that had lived in a 35 square kilometre region of the sanctuary were never seen again.[54] Across some 5,000 square kilometres of the sanctuary, between 90 and 95 per cent of the western gorillas died during the outbreak, a total of at least 5,000.[55] The team found dead chimps too. In 2004, 377 western lowland gorillas from the Lokoué clearing in nearby Odzala-Kokoua National Park suffered a similar fatality rate, although solitary males, who made up 8 per cent of the population, fared slightly better – only 77 per cent of them died.[56] But 97 per cent of the group-living females and infants, who would have been essential to rebuilding the population, perished. The outbreak spread to humans and killed 29 people.[57] Some areas of the park lost all their gorillas – the animals probably died in their thousands before the outbreak petered out in early 2005.

At Lokoué clearing, the gorilla population began to recover within a decade; numbers fell to 40 then remained stable for six years before adult females migrated in, formed new breeding groups and gave birth.[58] But models indicate that the gorilla population in Rwanda's Virunga Mountains would need up to 131 years to recover completely,[59] and the genetic impacts on disease-struck gorilla populations could potentially carry on for centuries.[60]

In all, between 1992 and 2005, great apes probably suffered seven Ebola outbreaks.[61] Then the virus went quiet, for non-human apes, at least. No ape deaths from Ebola have been confirmed since, though it's possible that a human outbreak in Boende, DRC, in 2014 was due to contact with an infected primate carcass.[62] Despite this lull in chimp and gorilla deaths, by 2013, when the human outbreak in West Africa began, far more apes had died of Ebola than people – thousands of apes compared to roughly 1,500 humans.[63]

Even after 2005 the virus was still circulating in gorillas. Tests on gorilla faeces – it's much easier to pick up some dung than take a blood sample or swab from an animal that may be as strong as twenty men – showed that roughly 10 per cent of apes tested in 2005, 2007 and 2008 had antibodies to Ebola, indicating that they'd survived the disease.[64] We're now also finding people with antibodies against Ebola who didn't show clinical signs. There are probably Ebola viruses we haven't described yet. The USAID Emerging Pandemic Threats Predict Project, a US-funded surveillance programme for viral pathogens in wildlife that could pose a risk to people, found the Ebola Bombali virus in a bat before it had infected a non-bat species or caused any kind of illness.[65] 'Maybe it never will,' Gilardi, who helped with the discovery, says. 'Maybe it's one of those Ebola viruses that's not as pathogenic.' But other, pathogenic strains of Ebola continued to break out in people.

THE GREATEST APE?

Multiple Ebola outbreaks in humans have been linked to contact with wildlife infected by bats, mainly dead gorillas and chimps, plus some duikers, too.[66] In some outbreaks, Ebola may have been transmitted to humans by domestic dogs that ate the remains of infected wild-life.[67] Dogs don't show clinical signs of Ebola but have been found with antibodies against the infection. In the worst human outbreak of Ebola to date, however, humans perhaps caught the virus from a bat directly, without a non-human primate, antelope or dog acting as intermediary. The epidemic, which raged across Guinea, West Africa, in 2014 and infected more than 28,600 people, killing 11,325,[68] was 100 times larger than any of the previous 21 known outbreaks in humans, and the first to reach towns and cities.[69] It began in the small village of Meliandou in Guéckédou, which is surrounded by oil palm plantations and forest, in December 2013. Not many non-human primates live in this area.[70] Gorillas don't live in this region and chimps are not plentiful; there's no evidence of chimp infections. Some people believe the outbreak started when a two-year-old boy played in a hollow tree housing a colony of Angolan free-tailed bats, then died from the virus.[71]

BACK TO THE MOUNTAINS

In 2018, Ebola again broke out in humans via an unknown route, in the conflict-torn eastern DRC. For humans, this was the second largest Ebola outbreak in history, with 2,287 deaths.[72] This time, the disease came right up to the northern boundary of DRC's Virunga National Park, dangerously close to mountain gorillas. The subspecies had been doing well thanks to the veterinary care set in motion by Dian Fossey, which eventually evolved into the Gorilla Doctors organisation that Gilardi leads.

Mountain gorillas had never encountered Ebola, but in their western lowland gorilla cousins the disease had a mortality rate of up to 98 per cent and reduced the global population by one-third.[73] When you have 300,000 western lowland gorillas that's still a massive impact, but the dwindling mountain gorilla population here – just a few hundred by the end of Fossey's time – was much more vulnerable to extinction by a disease. For Gilardi, it was a 'big scare' that Ebola was 'so darn close to mountain gorilla range'. This time, the mountain gorillas were lucky – they don't live near the edge of the park and Ebola only came within 80 kilometres.[74] 'Once the outbreak zone came close, the park stopped human entry, which was the single most important thing to do – to just not have people going back and forth from their homes and villages into the park,' Gilardi says. But people live in and access the park illegally to hunt and harvest charcoal, water and food. 'That was a constant threat,' Gilardi says. 'It was a huge worry.' Fortunately, the human outbreak subsided and Gilardi's team didn't find any evidence that any of the mountain gorillas had been infected.

If Ebola gets close again, there's a plan. A vet vaccinated against Ebola – a human vaccine for the Zaire strain was first trialled in 2015, in the later stages of the 2013–16 human epidemic – and wearing full personal protective equipment would take a sample from any sick or dead gorillas believed to have the virus.[75] Then park staff would likely minimise contact between different groups of gorillas and consider vaccination of the animals against Ebola. (Ironically, development of a vaccine for chimps was delayed when the use of chimps in labs in the US was banned in 2015, possibly the first time the chimps would have been guinea pigs for their own species' benefit.)[76] If the disease were to strike even just a single individual, computer simulation models indicate that it would spread widely and the mountain gorilla population would plummet. On the Virunga Massif, the models indicated that less than

20 per cent of the population – some 112 to 125 of the original 607 – would survive 100 days after the first infection.[77] But vaccinating at least half the habituated gorillas (some 44 per cent of the population) within the first three weeks could mean more than half the gorilla population survives, even though at least 75 per cent of unvaccinated animals would be infected. 'The range countries for the mountain gorilla are all aware now of the fact that vaccination of a subset of the population would save a lot of lives,' says Gilardi. 'So that's a conversation we're in right now about how to approach that.'

Since the gorillas are habituated to humans, it would be relatively easy for a vet to get close enough to dart a few of the apes with vaccine; the human vaccine also works on apes.[78] But then the gorillas would realise something was up and would move as far away as possible. (In Bwindi Impenetrable National Park in Uganda, gorillas have learned to walk backwards when they see someone carrying a dart gun to hide their backs, where the dart needs to land.)[79] The only way to vaccinate most of the animals would be to have several vets darting at once. 'So it would be a huge effort,' Gilardi explains. 'That's why it's not a decision that anybody would make lightly.'

COVID CONCERNS

The mountain gorilla escape from Ebola was a big relief, until Covid-19 came on the scene. Though Covid-19 did lead to an operational change for the better. Eco-tourism is a mixed blessing. Tourists are understandably fascinated by the opportunity to spend time up close to their relations, and the money they pay for the experience helps protect the apes and the park, including all its other animals. 'This is beautiful, intact forest that is incredibly important for capturing and delivering water downslope from the parks,' Gilardi says. 'There's so much human pressure all the way around – in some parts of the area, there's a thousand people per square kilometre. If the governments didn't have a reason to protect the forest for the mountain gorillas, they'd be under tremendous pressure to allow people to remove that forest and plant crops.' The human population surrounding the mountain gorillas has the highest density in continental Africa.[80] (For comparison to Gilardi's 1,000 people per square kilometre figure, the population density of Cornwall, a rural county in the far South-west tip of England, is 161,[81] the US state of New Jersey has an average of 488 people per square kilometre,[82] and for some of the most crowded parts of central

London the same figure is roughly 15,700.)[83] But the very presence of tourists also puts the apes at risk, from respiratory diseases such as colds, flu, parainfluenza, human metapneumovirus, adenovirus, measles and Covid-19, as well as polio.[84] 'Because we are so closely related, we know that gorillas are susceptible to human pathogens,' Gilardi says. Roughly a quarter of mountain gorilla deaths are due to respiratory illness; it's second only to trauma as a cause of death in this species.[85] And respiratory illness makes many more gorillas ill; between just under half to all of the animals may grow sick in such outbreaks.[86]

Researchers and park workers can pose a risk to gorillas too. On the one hand, habituation helps these staff observe the gorillas from up-close and lets them see when gorillas are injured or sick and need help from the Gorilla Doctors vets. On the other hand, it puts the gorillas at risk of disease. To mitigate the risk, people, whether tourists, researchers or park employees, are not supposed to enter the park if they have any signs of illness such as fever or sore throat. But people who have paid thousands for a trip may be reluctant to self-declare and miss out, and in any case, the measure doesn't stop the asymptomatic spreading their germs. Habituation can put the gorillas at risk in other ways too. The process is stressful and may make the gorillas more vulnerable to attacks. In 2009 an adult female mountain gorilla from Bwindi Impenetrable National Park died when she roamed outside the park and a farmer threw a rock at her for eating his crops. If this mother hadn't been habituated, she wouldn't have let anyone get close enough to kill her with a rock. Her unweaned infant died later.[87]

In total, some 44 mountain gorilla groups – roughly 73 per cent of the animals in the Virunga Massif and 50 per cent of the mountain gorillas in Bwindi Impenetrable National Park – are used to humans, who visit in groups of up to ten.[88, 89] Researchers and tourists now visit the same groups each day in Rwanda and Uganda, so between 264 and 880 people are near these gorillas each day, or up to 321,200 people a year. For no other species of great ape does such a significant portion of its total world population spend a part of its day, every day, in close proximity to people.[90] The IUCN guidelines are for people to stay at least seven metres away but the gorillas don't know this and may approach much closer, or thick vegetation and difficult terrain may make seven metres unachievable. 'The fact that mountain gorillas are being visited by people every single day makes us feel obligated to treat them when they may be injured or ill because of their contact with people,' Gilardi says. 'Wildlife veterinarians are typically mostly focused on population health versus individual animal health. But with the Gorilla Doctors, we train the park workers how to recognise signs of illness or injury in individual gorillas, and then we can get into the park and take care of those animals.'

When Covid-19 struck, the team was extremely concerned and park managers made mask use mandatory for all those nearing gorillas – previously it had been compulsory only in Virunga National Park in DRC. They also increased the distance people should keep between themselves and a gorilla from seven to ten metres. 'Knock on wood, we were collecting either faecal samples or even direct nose or throat swab samples from mountain gorillas with respiratory illness during the outbreak, but never confirmed Covid-19 as a cause,' Gilardi says. In fact, the respiratory health of the gorillas in Volcanoes National Park in Rwanda even improved during the Covid-19 pandemic – in the five years before, the gorillas suffered an average of 5.4 outbreaks of respiratory illness in their family groups each year, but this fell to 1.6 outbreaks each year after the pandemic was declared.[91] 'That could be attributed to fewer people in the park because tourism was halted, but also to masks,' Gilardi says.

The new mask rules have continued in both Rwanda and Uganda. (Virunga National Park is currently closed to visitors due to the insecurity in the region.) Gilardi says this compulsory mask-wearing is a big relief. 'We were always advocating for those basic prevention measures, but it took the pandemic and the fact that the entire world was wearing masks and practising social distancing for the parks to realise this was acceptable and people will be willing to wear them,' she says. 'The gorillas don't care when people show up and have a mask on half of their face. So, despite the global tragedy for humans, it was a silver lining for gorilla tourism.'

Another way to improve the health of mountain gorillas is managing human excreta inside the park – there are no latrines and human waste litters the forest, exposing the gorillas to pathogens via another route.[92] Improving the health of the local human population through vaccination programmes for measles and other diseases could help too;[93] the non-governmental organisation Conservation through Public Health works with human communities in Uganda and the DRC with the aim of keeping both gorillas and their human neighbours healthy.[94] In the late 1980s, human measles likely caused an outbreak of respiratory illness in mountain gorillas in Volcanoes National Park.[95] And scabies mites can also reach the gorillas from humans or their livestock, while free-ranging feral dogs pose a risk of transmitting diseases such as rabies.

MOUNTAIN RISING

Despite all these threats, today mountain gorillas are a rare success story. They now number 1,063, split among the central Africa countries of Rwanda, the DRC and Uganda. Some 604 mountain gorillas live in the Virunga Massif, in Rwanda's Volcanoes National Park, Virunga National Park in DRC, and Mgahinga Gorilla National Park in Uganda.[96] The other 459 members of the species live some 25 kilometres to the north in what's possibly the most pleasingly named park ever – Bwindi Impenetrable National Park in Uganda.[97] The mountain gorilla is the only great ape whose numbers in the wild are increasing, a result that Gilardi ascribes to 'extreme conservation measures' – efforts that don't just minimise negative human influences but seek to increase positive influences.[98] The veterinary medicine she and her colleagues provide to the mountain gorillas explains almost half of the annual growth rate of the habituated gorilla population.[99] By rescuing, for example, a young female trapped in a snare and treating her injuries, vets keep her alive and increase her opportunity to contribute a lifetime's worth of reproduction. From 1967 to 2008, the population of unhabituated gorillas in the Virunga Massif declined by 0.7 per cent each year, whereas numbers of habituated gorillas increased by just over 4 per cent each year, to 339.[100] Each group of habituated gorillas benefits from intensive conservation measures, including veterinary treatment for snare injuries, respiratory disease and other life-threatening conditions, and guarding against poaching during daylight hours. As well as the guards who spend daylight hours near gorilla groups, the Virungas currently have more than 50 field staff per 100 square kilometres to protect gorillas, a number more than 20 times the global average.[101] These staff remove more than 1,500 snares each year. Over the 40 years of the study, humans killed 26 habituated gorillas, a total of 12 per cent of all the known deaths. Three gorillas were caught in snares set for other animals, 15 were shot by militia groups (although this number would probably have been higher had eco-tourism not been so valued), and 8 were killed by villagers or poachers for reasons including capture for the pet trade, to stop crop raiding, or for bushmeat.[102] In total, 16 habituated gorillas died from respiratory disease, whether transmitted from humans or not; vets monitored 17 outbreaks of respiratory disease that affected more than 245 habituated gorillas, treating 42 gorillas, of which 36 recovered. Treatment for snare wounds was provided 42 times and all but one gorilla survived.[103]

Another factor behind the mountain gorilla population growth is that all mountain gorillas live in national parks. 'Mountain gorillas are unique,' Gilardi says. 'That's one of the reasons that they're a conservation success story – they all live in protected forest.' In 2018, the IUCN downgraded mountain gorillas from critically endangered to endangered.[104]

But Ebola could emerge from a bat at any time, or an eco-tourist or other human could give the gorillas a different disease. 'That's a constant threat that's not going to go away,' Gilardi says. 'And if anything, that threat's going to grow because more and more people are living in that area, and we're learning more and more about the kinds of viruses that could transmit from people to gorillas.' The world is changing in other ways too. Models based on the distribution of bat species suspected to be reservoir hosts for Ebola indicate that the disease could emerge across most countries in Central and West Africa.[105] The warm and wet conditions that fruit bats probably prefer are expected to become more widespread as climate changes,[106] which could increase Ebola prevalence in areas to the north, south and east of the current bat species' range. And human population growth will not only increase contact between humans and wildlife, but also force wildlife into closer contact with one another as forests become degraded and converted to plantations. For now, though, mountain gorillas are faring relatively well in well-protected habitat, a fact for which Gilardi says she is incredibly grateful. 'They're big vegetarians living in a gigantic salad bowl in an open-air environment, living extremely healthy lives, and that really helps them be as resilient as possible in the face of disease,' Gilardi says. Enabling other wild animals to live in well-protected, intact habitat would guard them against diseases, and other conservation threats, too.

14

Big Farma: Industrial Farming and Bird Flu

'As we still ascend from shelf to shelf, we find the tenants of the tower serially disposed in order of their magnitude: gannets, black and speckled haglets, jays, sea hens, sperm-whale birds, gulls of all varieties – thrones, princedoms, powers, dominating one above another in senatorial array'

Herman Melville, *The Encantadas*, 1984

In May 2010, from our holiday croft in Mainland, Shetland, my friends and I take the ferry to the island of Bressay and walk for an hour across its short grassy turf dotted with wildflowers to another ferry station on the east coast. After a three-minute hop in an inflatable boat to the island of Noss and another trek, we reach our goal, the Noup of Noss, otherwise known as Noss Head, where the cliffs soar 181 metres above the sea.

As we walk, great skuas – or 'bonxies' – wheel threateningly overhead. With brown-speckled plumage, stubby tails, short necks, white-flashed wings and hooked beaks, they look strangely otherworldly, as if they have faded into life on Earth from another place and might yet fade away. These ground-nesters attack humans who inadvertently, or not, stray too close to their eggs. But this time it didn't come as close as the lesser black-backed gull that swooped millimetres from my head when it was nesting on a roof near my home in suburban Bristol. I was lucky that this bird only dived at me once – my silver-haired neighbour learned to take a brolly with her even in fine weather to fend them off. The birds on Noss leave us alone, however. Puffins sit at the top of the cliff, their bright orange feet looking strangely like 'reduced' stickers, and the bright arcs of their red, blue and yellow beaks and the sculpted white planes of their cheeks pointing out to sea.

Suddenly, it's as if we've been transported into a wildlife documentary. When we peer over the edge of the turf down at the grey rock and ledges of the Noup, the smell, sound and sight hit us all at once. The tang of seabird guano mixed with saltwater assaults our nostrils, the noise is like a soundtrack made by the special effects department – a cacophony of squawks and yelps and yaks and squeals from gannets ('solans' in Shetland-speak), guillemots ('longvies'), fulmar ('maalies') and more. It's almost too loud to hear each other speak, not that we can talk for the first few moments in any case. We're all too stunned to say anything except 'Wow'.

The cliff to our left is a wall of birds. Some perch on ledges in small hollows, others seem to stand on an area less long than their feet. Everywhere streaks of white guano run down the rock. It's like a vertical Noah's ark – pairs of one species just inches away from other species, while single birds wait for their partner to return with food. The matronly, solid white bodies of the gannets, one of the UK's largest seabirds at a hefty 3 kilograms (6.6 lbs) or so, stand out against the grey rock as they make calls like a trilling quack. Their blue-ringed pale-blue eyes, kohl-dark stripes projecting front and rear, shine bright above pale beaks sectioned with black lines as if engineered from triangles of steel. Guillemots stand like penguins, their bright white chests contrasting with their dark grey heads and shoulders; from time to time the birds stoop to check on the white eggs speckled with grey that lie precariously near the cliff edge in front of their long webbed feet. Fulmars swoop in with dark-smudged eyes, legs dangling akimbo beneath pristine white undersides, their wings silver-grey, and their beaks and tails seemingly sheared off at the ends to leave extra space on the rocks. Birds come and go in a constant stream as we watch, the gannets plunging straight into the sea, black-tipped wings held out to the side, 'elbows' bent downwards, and their long white necks, topped with yellow-dusted heads, outstretched. Other birds head further out in their search for fish.

SKUA SENTINEL

Eleven years later, in the summer of 2021, great skuas from Shetland and St Kilda, an island in the Outer Hebrides, off the other side of Scotland, began to die from H5N1 flu.[1] Flus of the A species, including avian flu and all flus that have caused human pandemics such as the one in 1918, are named according to the proteins that project from their spherical outer envelope – haemagglutinin (H or HA) and neuraminidase (N or NA). To date, we know of eighteen types of H and eleven of N,[2] though some are found only in flu viruses that infect bats. All four species or types of flu (species is a fuzzy concept when it comes to viruses), influenza A, B, C and D, come from the family Orthomyxoviridae. Influenza B viruses mainly infect humans and together with Influenza A viruses are responsible for seasonal flu outbreaks in people. The C species only causes mild illness in humans (it can also infect pigs) and the D species mainly affects cattle, though may spill over into other animals.[3]

This summer outbreak in seabirds in Shetland was odd, as avian flu usually broke out in winter in waterfowl like ducks and geese (though deaths in waterfowl, particularly Barnacle Geese, did happen that winter).[4, 5] During the following summer, northern gannets grew sick from H_5N_1[6] as well as skuas – the first time that these birds had ever suffered from an H_5 flu.[7] In a newspaper photo from July 2022, a dead gannet lies on the shore of a Scottish beach, its feathers bedraggled and its head hanging limp.[8] Other seabirds and shorebirds, the group that includes sandpipers, plovers, gulls and auks, also died,[9] in an outbreak of H_5N_1 that would become the worst ever recorded in Europe.[10] A National Trust worker from the Farne Islands, off Northumberland, to the south of Scotland, said of the islands' bird cliffs that 'instead of the smell of guano, it's the smell of death'.[11] Skuas, gannets and guillemots were hit particularly hard. In total, thousands of all three species died; numbers at great skua colonies declined by up to 85 per cent, while up to 25 per cent of birds at gannet colonies succumbed.[12] Since more than half the world's northern gannets, some 220,000 pairs, nest on UK coasts, these losses were globally significant.[13]

DUCKING DISEASE

The 2.3.4.4b strain of H_5N_1 flu devastating the Scottish seabirds had first started to emerge 25 years earlier and thousands of miles away, in domestic geese in the Guangdong province of China.[14] Humans were essentially midwives for this new form of the disease.

In Bristol, for a while in the 1990s or 2000s, on the canal dug by Napoleonic prisoners of war to divert the tidal river away from the harbour, lived a white duck adorned by an unusual array of bouffant feathers. As I walked to the train station alongside this New Cut, I'd think, *Ah, it's the duck with the silly head.* Later, I discovered that friends more sophisticated than me had named this duck Einstein. I've not seen Einstein for well over a decade, but back then he would have had a relatively good chance of surviving bird flu. Dabbling ducks like Einstein, especially mallards, are the prime suspects for being this flu's natural reservoir. Wild waterbirds have lived alongside this virus for generations; it used to infect only their lower gut. When a bird with flu defecates or regurgitates as it swims, it emits virus particles into the water that other birds then ingest as they feed and dive. The flu virus has a very high affinity for water, where it spreads faster and more efficiently, explains Valentina Caliendo of the Dutch Wildlife Health Centre at Utrecht University in the Netherlands, a vet who did a PhD on bird flu. What's more, birds near water tend to congregate in large groups, enabling the virus to spread to more individuals. And sometimes wild birds pass the virus on to domestic species. When infected with this type of flu, birds tend to show either no or mild symptoms, such as an increased temperature, watery faeces or, in the case of chickens, a reduced production of eggs. Both wild and domestic birds could easily cope with these kinds of clinical signs.

But then, in poultry in several places at several times over the years, something changed. Type A flu viruses consist of a single strand of RNA in eight segments, all packed into a particle roughly 100 nanometres in diameter. 'The virus has a very simple genetic constitution,' says Caliendo. 'It does not have long genomes that have to be decoded by the host cell in a certain way. Things happen very fast when the virus infects a cell, and the virus can mutate easily.' Helping that fast mutation is the viru's RNA structure, so there's no double-stranded check of its genetic code as it replicates, or its slim, single-stranded formation.[15] The viru's segments can also jumble up as it replicates. If a cell is infected with two different strains of bird flu, the segments mix and match in a process like those put-the-right-head-with-the-right-(or wrong)-body-and-legs kids' books. More formally, it's known as reassortment.[16] In combination, these two factors – easy mutation and reassortment – mean flu viruses can change fast.

I've not been a fan of chickens since one pecked a hole in my tiny wellington boot when I was five or so, trying to feed its flock grain on the smallholding belonging to my grandparents' neighbours. Chickens seem scary when you're only a few feet tall and surrounded. I don't, however, think chickens deserve to be crammed together in massive sheds without ever seeing daylight or fresh air, and where the ventilation fans may circulate germs. Earth is currently home to 50 billion chickens, more than there are wild birds, and many of these number are genetically similar.[17] Several times in domestic birds held captive in conditions even more crowded than those on a sea cliff, the virus became more lethal. Now it infected birds' whole systems, not just their guts, and it killed. 'Basically, when the virus was introduced into commercial poultry farms, it found its ideal environment and its ideal hosts,' Caliendo says. 'Around that time, the first mutation happened in the virus and it became very lethal for chickens. The highly pathogenic form of avian influenza can kill a chicken within a few hours.' One of the early tests to see if a flu virus was low or highly pathogenic was to intravenously inject it into chickens – injection makes the infection progress faster. If a high proportion of the chickens died relatively fast, the flu was due to highly pathogenic avian influenza virus, or HPAIV for short.[18] (The term 'highly pathogenic' relates to the effect on poultry, not humans. Sometimes humans die of low pathogenic strains of bird flu, sometimes we don't grow sick from strains that are deadly to birds.)[19]

Our high-density poultry farms had bred not just new domestic birds and new eggs, but new, lethal forms of a disease; both the H5 and H7 subtypes of flu developed highly pathogenic forms. H7 HPAIV may be responsible for the reports of 'fowl plague' in poultry in Europe from 1878 onwards.[20] And sometimes farmed chickens, turkeys, ducks and geese passed on their HPAIV woes to wild birds. The first flu outbreak known to kill wild birds was in 1961, when the disease struck down hundreds, if not thousands of common terns in South Africa.[21] The birds suffered from an H5N3 strain that may have been related to the highly pathogenic version of H5N1 that broke out in Scottish chickens in 1959,[22, 23] perhaps transported to South Africa by a tern migrating from Europe.[24] Or the outbreak may have been conncted to the ostrich industry.[25] After this, wild birds stayed free of mass flu deaths for decades.

IN FLEW FLU

Until, that is, a few years after a new strain of H5N1 flu emerged in China in 1996,[26] the first of this subtype to be seen in Asia. First found in domestic waterfowl in Guangdong province in the South of the country, this lineage was dubbed A/goose/Guangdong/1/1996 (H5N1), or GsGd for short, after the species, place and year in which it was discovered, along with an A because it's influenza A, H5N1 to denote the subtype and a 1 for the strain number. It's thought to have come from H5 flus in wild migratory birds.[27] In 1997 a similar strain of H5N1 broke out in poultry in Hong Kong, its spread between poultry perhaps facilitated by a change in the NA gene.[28] This H5N1 flu jumped to humans and killed them for the first time, making eighteen people ill, six of whom died.[29] The authorities took dramatic action, mandating culls of all poultry at markets and all chickens on farms in Hong Kong.[30] The HK-97 lineage of virus that caused the deaths went extinct,[31] though similar lineages survive to this day; by spring 2021, more than 860 people around the world had contracted H5N1 and 455 of them had died, a death rate of more than 50 per cent.[32]

Meanwhile, two versions of flu were running in parallel in this region – a low pathogenic version in wild birds and a highly pathogenic version in poultry. 'The first step was the establishment of these very crowded poultry farms where the virus just found the perfect spot to mutate,' says Caliendo. In many cases, highly pathogenic flu spread from farm to farm as workers accidentally transmitted the virus on contaminated equipment, or captive poultry were transferred without sufficient biosecurity. 'In a certain way it was a man-made disease,' Caliendo says, 'and the fact that then the virus escaped the farms to infect wildlife, that's also because of poor biosecurity containment.' Many consider that biosecurity in Asia is a bit looser than in the rest of the world. Not only that, but many small-scale farmers keep chickens alongside ducks that range freely on ponds and rice paddies and mingle with wild birds. In the absence of refrigeration, people in this region buy fresh meat at traditional wet markets where all sorts of species are kept alive together, including wildlife. 'This [Gs-Gd highly pathogenic flu] virus was introduced into nature, and then it becomes modern history,' Caliendo adds. This time when the highly pathogenic version of flu entered wild birds, it mixed with the low pathogenic form, creating several new strains, as did mixing of highly pathogenic and low pathogenic flus in domestic birds. In essence, the flu virus passed from wild birds to poultry, which cooked up a more dangerous version then passed it back into the wild, where it learned some new tricks – how to spread better between wild birds and on into another group.

In Hong Kong in 2002, wild migratory birds died from H5N1. The sporadic cases were the first known deaths due to avian flu (of any type) in wild aquatic birds since 1961.[33, 34] And matters soon worsened. By 2003, one particular new mix-and-matched strain had become dominant.[35]

In late 2003 and 2004 it broke out in poultry in China and other countries in South-east and East Asia,[36] including Vietnam and Thailand, which were both particularly badly hit, Cambodia, Indonesia, Japan, Korea, Lao People's Democratic Republic (PDR) and Malaysia.[37] By this time, H5N1 had grown even more virulent for poultry and millions of chickens in South-east Asia died, either from the disease or the culls that aimed to stop flu in its tracks. The virus also killed wild birds and mammals, including humans and cats, both domestic and big.[38] In 2004 a total of 147 tigers and 2 leopards in zoos in Thailand that were fed infected poultry carcasses died of H5N1,[39] and the following year several Owston's civet, an endangered striped and spotted cat-like creature with a thin pointed face and long tail, died from H5N1 at a conservation centre in northern Vietnam.[40] The animals had not been fed poultry and the source remained unknown, although villages nearby had seen unexpected poultry deaths. And this version of flu was here to stay. Whereas outbreaks of highly pathogenic avian influenza (HPAI) in poultry in other regions have each time developed from a low pathogenic form caught from wild birds, in Asia since 1996 highly pathogenic H5N1 viruses have circulated continuously.[41]

Over time the virus grew more deadly and better at spreading. In 2005 a virulent form of H5N1 broke out in waterbirds at Lake Qinghai in western China, and more than 6,000 birds died, mostly bar-headed geese,[42] in the first mass outbreak of H5N1 in wild birds.[43] Before this die-off, highly pathogenic flu had been unusual in free-ranging wildlife. The problem with this outbreak was that the birds had come from far and wide and, if well enough, would migrate back, taking their germs with them. 'All the textbooks will tell you if a bird gets a highly pathogenic avian influenza virus, it'll drop dead,' says Ab Osterhaus of the Tierärztliche Hochschule Hannover, Germany, who we met through his work on morbilliviruses in seals. 'And dead birds don't fly.' But Osterhaus and colleagues found that not all duck species die when infected with highly pathogenic flu, and chances are that some would be well enough to migrate.[44] Not that anyone believed him at first. 'It is so funny,' he says. 'When you try to publish these things, people say it's impossible.'

Humans are not entirely innocent of spreading bird flu, either; legal movements of poultry and duck meat, and trade at live bird markets, may also spread the disease.[45] Fighting cocks smuggled into Malaysia probably introduced highly pathogenic flu to the nation. And staff at international borders have intercepted illegally transported birds infected with H5N1 at least three times, including doves in Malaysia, crested hawk-eagles smuggled into Brussels airport in Belgium from Thailand,[46] and ducks in Taiwan.[47]

But it does seem that in the autumn of 2005, when the birds migrated from Qinghai Lake to Africa, the Middle East and Europe, they spread this virulent strain of H5N1 into wild birds and poultry across the globe. In Europe, the virus first reached Russia and south-eastern nations, then spread to the north and west during the winter of 2006.[48] In wild birds, flu's worst excesses took off in the winter and early spring of that year.[49] More than 700 dead mute swans, whooper swans, common pochards and tufted ducks in Europe tested positive for flu; the actual number of deaths was probably higher.[50]

NEW 'N'S

Thanks to flu's ability to reassort, over time H5 viruses from poultry and wild birds swapped genes when birds became infected with two different types of flu at the same time, a highly pathogenic subtype and a low pathogenic subtype, say, or two low pathogenic subtypes. Since emerging in 1996, the H5N1 GsGd lineage has split into ten major clades (groups that share a common ancestor). And since 2008, the sub-clade 2.3.4.4 from clade 2 has frequently reassorted with low pathogenic strains, acquiring N2, N5, N6 or N8 proteins. Collectively, these viruses are known as H5NX.[51] In the winter of 2014, H5N8 spread to Europe with waterfowl migrating from eastern Asia to summer breeding grounds in the north, and then birds migrating down into Europe. From these summer breeding grounds H5N8 flu also spread east across to North America, the first recorded time that a virus with an HPAI H5 from the GsGd lineage had crossed the Bering Strait.[52] After H5N8 mixed with local flus, the resulting H5N2 killed more than 50 million poultry.

Back on the other side of the Atlantic, when common teal and Eurasian wigeon migrated from their Eastern breeding grounds to winter in Europe for the 2016–17 season, they brought with them, scientists suspect, a strain of H5N8 that was more harmful to ducks and killed more wild birds than the 2014–15 outbreak.[53] This H5N8 subtype struck the Netherlands, where Caliendo is based, for the first time and affected mainly species of duck. 'For common teal and Eurasian wigeon, it was a pretty bad year with many dead birds,' Caliendo says.

The following winter it was the turn of H5N6 to hit the Netherlands, UK and Germany.[54] Soon, the H5N8 and H5N6 subtypes had outcompeted the original H5N1. In the Netherlands, however, after the 2016–17 season, flu cases largely dropped off until October 2020 when new cases of H5N8 bird flu appeared. Again, Eurasian wigeons were the first species found with infections and are prime suspects for importing the virus. This time around, while some wigeons still showed flu symptoms and died, the birds did not experience mass mortality, probably because they had some residual immunity from the previous outbreaks.[55] Though immunity to flu in wild birds doesn't last long.

The wigeons may have been responsible for what happened next too. These ducks use grassland more than other ducks, spending time feeding on this habitat in large flocks. If infected, they excrete flu virus onto the grass, putting geese, who share the wigeons' liking for grassland, at risk of infection. This time around, much higher numbers of barnacle and greylag geese suffered from flu and, as the outbreak continued, their mortality skyrocketed. 'We believe that one of the reasons that the outbreak could fuel so much was because for the first time, there were lots of animals from the same species that could really amplify the infection,' Caliendo says. 'In the Netherlands, there are quite large populations of wild geese, so that really overloaded the environment with virus.'

Other bird species in the Netherlands and elsewhere also suffered from flu for the first time, including scavengers and predators of dead or infected birds, such as birds of prey like buzzards, peregrines and short-eared owls,[56] and corvids (members of the crow family).[57] These species hadn't previously been found suffering from low pathogenic avian flu.[58] Infected birds may look depressed, cough, sneeze, have watery eyes, lack coordination and balance, have head and body tremors, exhibit drooping wings, have necks contorted into awkward positions, be paralysed, and lie on their backs paddling their feet in the air.[59,60] Or they may look absolutely fine or, at the other end of the scale, be dead. Flu's harms vary depending on the species and the individual. H5N1, for example, birds in at least 11 of the 27 avian orders, including: gulls and shorebirds; storks, herons, flamingos and egrets; pigeons and doves; eagles, kites and vultures; cranes; passerines (perching birds); pelicans and cormorants; parrots and parakeets.[61]

SEASONAL SURPRISE

For Caliendo, the flu outbreak's biggest surprise was the change in the way the virus transmits. Until 2020, avian flu in Europe's wild birds was seasonal, breaking out when birds migrated into the region in the autumn or spring. But at time of writing bird flu is present in wild birds all year round in the Netherlands and elsewhere in Europe – it's enzootic and persists at a fairly low, consistent level.[62] Just nine years earlier, in 2011, HPAI H5N1 was considered enzootic only in poultry in Bangladesh, China, Egypt, India, Indonesia and Vietnam.[63] Caliendo says that now there is a high load of flu in the environment, it's easier for birds to get infected. And resident species such as barnacle geese, some of which have stopped migrating to the Arctic north for the summer,[64] maintain the infection locally all year round.

The year 2020 saw another change in avian flu too. The H5N8 sub-type reassorted in wild birds and H5N1 bounced back, this time with a new version of the H gene known as clade 2.3.4.4b and with an N1 gene adapted to transmit in wild birds. The first of these wild-bird-adapted HPAI H5N1 viruses were identified in Europe during the fall of 2020 and spread into Africa, the Middle East and Asia before heading even further afield.[65] And then there were the mammals.

MAMMAL MALADY

In 2021, H5 infections were discovered in mammals for the first time since 2005–06, including in carnivores such as red foxes, harbour seals and grey seals, as well as humans.[66] Readers in the UK may remember the case in December 2021 of a man who caught highly pathogenic H5N1 from the Muscovy ducks he was keeping as pets in his house, though thankfully, unlike the ducks, many of which died, the man remained asymptomatic.[67,68] In May 2021 a couple of red fox cubs were seen act-ing aggressively near Groningen in the Netherlands and were tested in case they had rabies.[69] But it was H5N1 bird flu, which had made them blind. 'We suspect that these foxes were fed by the mother or themselves on infected wild birds,' says Caliendo. Dog-walkers in the area were advised to keep their dogs on leads to stop them coming into contact with infected birds.[70] In the Netherlands, polecats, stone mar-tens, badgers and otters also became infected with H5N1.[71, 72] In the UK, otter, red fox, common dolphin, harbour porpoise, harbour seal and grey seal all died from this subtype.[73] In mammals, flu generally infects the respiratory tract, which spans from nose to lungs. It varies in its severity but often causes pneumonia and may sometimes kill. Highly pathogenic H5N1 flu, however, may infect mammals' whole systems and cause severe infections.[74] In some cases, H_5N_1 has been found in carnivore's brains, rather than their lungs.[75] The 2020–23 panzootic of H_5N_1 flu infected more than 48 mammal species in 26 countries, including nations in North and South America. It hit more mammal species in more countries than previous waves of infection.[76] It infected 12 species in 10 countries in Asia, Europe and Africa between 2003 and 2019.[77] The earlier waves mainly affected terrestrial and semiaquatic mammals, whereas the panzootic also affected marine mammals, often causing mass deaths, for example more than 10,000 American sea lions in South America.[78]

So why were so many more mammals succumbing this time round? The answer may be relatively simple. So many birds were infected with highly pathogenic H_5 flu viruses in the Netherlands in 2020 that it was easy for mammals to feed on their carcasses. And some of the virus particles that mammals ingested had mutations that meant they could more easily adapt to and replicate in mammal cells.[79] Mammals run at lower temperatures than birds, and one of these mutations helps the virus deal with this relative cold. And to infect the receptors on the surfaces of mammalian, rather than bird, cells, the flu virus needs an extra mutation. 'The final step would be that a virus can directly transmit from a mammal to a mammal without the intermediate step of the birds in between,' Caliendo says. 'And that would be when the virus would be highly functional for mammal species too.' To get to that point, the virus would only need a handful of mutations. As a mammal, this sends a chill down my spine.

It's happened in the past. Both human and swine flu likely originated in birds many moons ago. In humans, seasonal flus circulate each autumn and winter, sometimes A type, sometimes B. Over the years we've built some immunity by catching different versions. To stop seasonal flus killing older people and those with impaired immune systems, scientists predict the strains that will emerge each winter and prepare vaccines. But we can't predict when and which type A flus will jump to us from a bird, perhaps via another mammal. The flus we've never experienced before and have very little immunity to. If these flus are able to transmit human to human, they may become pandemic. In 1918 the 'Spanish flu' killed 50 million people worldwide.[80] (Nobody knows where this flu originated, but Spain's newspapers were first to report it as the country was neutral in the First World War and not censoring its media.)[81] Pigs fell ill around the same time as humans; it's still under debate whether the H_1N_1 virus behind the pandemic, which likely originated in a bird, switched from pig to human or from human to pig. Pigs are a perfect host for flu viruses to mix and match in – pigs catch avian and human flu as well as swine flu, and are known as a 'mixing vessel'.

Since 1918, the world has experienced a further three flu pandemics. The H2N2 virus behind the 'Asian flu' pandemic in 1957 and the H3N2 that caused 'Hong Kong' flu in 1968 both originated at least partially in birds. The H1N1 swine flu pandemic in 2009 arose when different strains of flu that had been circulating in farmed pigs for at least ten years reassorted; they contained segments that came from birds some thirty years earlier.[82] The next flu pandemic could kill 200 million people,[83] an order of magnitude more than we lost to Covid.[84]

Every now and then, poultry workers catch bird flus of the H5, H6, H7, H9 and H10 subtypes from their charges. They may themselves die, and have occasionally also passed bird flu on to their close human contacts.[85] But as far as we know, the chain of transmission has never yet extended beyond this. In flus more adapted to mammals, including swine flu and seasonal flu in humans, transmission tends to be via particles of virus breathed out into the air, rather than excreted in faeces. 'It is possible that this H5 virus has acquired that capacity of being less water-reliant and has found a way to spread in air,' Caliendo says. 'Then it's really scary because that's when we say, "OK, this is going to be the next pandemic".' Like Covid, such a flu virus could be transmitted from mammal to mammal, including from human to human, via the air.

Now is probably not a reassuring time to mention that in North-west Spain in October 2022, the H5N1 flu that broke out at a mink farm may have spread from mink to mink rather than via repeated infections from wild birds. While I was writing this chapter, in April 2024, news broke that a farmer in Texas had become the first human to catch H5N1 flu from another species of mammal, after picking up the virus from his cattle. Fortunately, he only suffered eye inflammation.[86] The disease had been confirmed in cattle for the first time only the previous week, in Texas, Kansas and Michigan,[87] after a 'mystery dairy cow disease' struck a few weeks earlier.[88] The cows recovered by themselves. Other farm workers, and barn cats, subsequently tested positive too. I have everything crossed that I don't need to update this section of the book anytime soon. Should that situation change, scientists have already created candidate vaccine viruses for H5 flus from the same clade 2.3.4.4b,[89] and I very much hope that either any vaccines we create from them provide good protection against H5N1 or that we are able to produce a more effective candidate vaccine virus fast. Of mammals in the wild, marine mammals and polar bears are likely most at risk from avian flu. 'Species that are really geographically confined and are already threatened because of other causes are the ones at higher risk of extinction,' Caliendo says. Although only one polar bear has so far been reported with H5N1 flu, in December 2023,[90, 91] Caliendo believes it's possible that many more polar bears are infected because it's so difficult to sample animals in extreme environments. The animal with a confirmed case died.

GOING GLOBAL

By the end of 2021, H5N1 was the predominant subtype in Europe, Asia, the Middle East and Africa,[92] and had hit the Shetland Isles where, as in the Netherlands that season, sea birds fared worst.[93] Also by the end of 2021, HPAI H5N1 had reached wild birds in Canada and the US,[94] perhaps imported from Europe by gulls.[95] These days, even birds on Antarctica catch this flu. Just before Christmas 2023, Argentinian researchers found Antarctic Skuas dead near Orcadas Station and suspected they had caught highly pathogenic flu.[96] Soon at least ten species of birds and mammals had been discovered with the infection, including gentoo penguins, wandering albatross, kelp gulls, Antarctic terns, fur seals and elephant seals, which suffered a mass die-off at the end of 2023. Patient zero, according to the virus' genetics, was likely a bird migrating in from South America. 'Once it was out, we got to the current situation where basically only one continent, only Australia, is free from highly pathogenic avian influenza in wild birds,' Caliendo says. And she believes highly pathogenic bird flu is also likely to reach Australia with infected migrating birds such as gulls.

So what can we do about avian flu? It's a science-writer cliché that every single news story about a recent discovery could end with 'More research is needed'. But when I ask Caliendo what we still need to know about flu, her answer has a twist. 'It's a virus that has been very well studied, although it keeps evolving,' she says. 'Everything that is new needs to be investigated. But I also think it's a spiral that could have been stopped before. So it's a bit of "what have we learned?" more than what we need to know. We can have lots of studies, but what for? We need to refine why we are doing all this research. If we really want to stop the infection, we have to go back to what we already know and what's the best way to use it. It's time to use that knowledge that we already have.'

If we were to take that approach, we would reconsider our meat consumption and whether we need to produce that much poultry, Caliendo says. 'Other things like migration of birds, I don't think we can do much about.' Today, instead of letting our chickens roam outdoors, we pack them into giant sheds where they breed diseases. Large-scale poultry farms are overly represented in the flu case numbers. Cramming 15,000 to 70,000 birds into a single shed, using ventilation fans that spread dust and germs as well as air, and letting bird waste fall into fishponds where wild birds feed, all help spread disease. Industrial farming also shifts animals between sites and across borders. And increasingly we farm giant sheds of pigs next to giant sheds of poultry, potentially enabling transfer of pig and avian viruses between the two. Perhaps because of this phenomenon, swine flus are now evolving more rapidly. And these 'pathogen factories' created by industrial-scale farming are harming health in other ways too.[97] Using antimicrobials to boost animal growth on industrial farms means we're breeding drug-resistant bacteria alongside our dinner, as well as flu viruses. The phrase 'reap what you sow' springs to mind.

Richard Kock, the vet we met through his work on rinderpest, says the solution is really easy. 'The trouble is the veterinary community is subsidised by the agricultural community,' he explains. 'Vets aren't going to stick their necks out because that's where they get their money from, and they don't shout against the poultry industry. It needs other people to do that.' In some ways, we all have direct control of this industry through our shopping and eating habits. And we could simply opt out by choosing not to buy meat or eggs from chickens, turkeys or other farmed birds. As fixes go, it's relatively simple. If you're struggling to cut down your meat consumption, every now and then you could look at the plastic shrink wrap around your chicken breast or pork chop and imagine it covered in mould and germs, a mini pathogen factory like the much larger version it's likely come from. That might do the trick. Or if you're a bird-watcher or marine mammal enthusiast, you could imagine corpses of your favourite creatures littering the beaches. If your meat is organic and free-range, the pathogen factory image is less relevant, but you could turn to land-use change and picture chainsaws chopping down rainforests instead. Raising animals for meat requires a lot more land than growing crops and that means less land for wildlife, and disrupted, unhealthy ecosystems.

Aside from the most effective, and logical, option of stopping farming poultry in this way, we could tweak any farming we continue with. Now that highly pathogenic bird flu is so well established in the wild, farmers would be wise to avoid locating their farms where there's a risk of wild birds approaching. In the Netherlands, poultry farms used to be located around water and wetlands, but now the guidelines recommend avoiding that. Farmers are also advised to reduce the number of poultry animals they keep per farm so that a single outbreak kills fewer birds. Lastly, there's the option of vaccinating captive birds, as Egypt and China do. Although the EU parliament and a number of nations have permitted flu vaccines on poultry farms for many years, farmers haven't wanted to vaccinate their animals for fear of reducing the chances of selling their meat and eggs. But, following several vaccine trials, Caliendo thinks it's likely that farmers in the Netherlands will start vaccinating their poultry. When vaccinating birds against flu, it's important to match the strain in the vaccine closely with the strain that's currently circulating. If you don't, the virus may be able to spread silently – without obvious symptoms – through your poultry, evolving as it goes.[98] Zoos in some countries are also licensed to vaccinate their endangered birds but it's not feasible to vaccinate wild birds by injection on a massive scale, although if the vaccine came in aerosol form that would be another matter. A few special cases, such as wild California condors in the US, are receiving injections after the US approved their use in a bird for the first time in 2023.[99, 100]

Other solutions for slowing the emergence of new flu outbreaks include banning mink farms; strong biosecurity at poultry farms; surveillance to keep an eye on how the virus is evolving and which species it's infecting; stopping the release of millions of game birds, such as pheasants and partridge bred in captivity for the autumn shooting season, that could catch flu and act as a reservoir,[101,102] especially given recent findings of links between infected poultry premises and the abundance of game birds in the area;[103] stopping hunting and bird ringing; preventing human access to wetlands to stop human exposure and stop humans spreading the virus elsewhere via their shoes or clothing; disinfecting vehicles, boats and other gear used at infected sites; manipulating water levels of small managed wetlands to create increased heat or ultraviolet light levels that will more rapidly kill viruses in surface water;[104,105] and communicating with the public to stop humans and their pets being exposed to bird flu, for example by handling dead birds, droppings or feathers, and by pets eating dead birds or raw meat.[106] Protecting lakes and wetlands could also stop wild waterfowl giving avian flu back to poultry, probably by keeping poultry farms away from these waters and by tempting the wild birds away from human-dominated landscapes and into protected natural habitats, a study in China found.[107]

After outbreaks in wild birds, we can reduce the spread of flu to both birds and mammals by removing as many bird carcasses as possible. In outbreaks in the Netherlands, the wetlands became like large cemeteries, according to Caliendo. Following concerns about biosecurity and possible infections of humans by the large numbers of dead birds lying on the ground, the authorities invested manpower and resources into carcass clean-up. In 2022, in colonies of sandwich terns in France, Belgium, Denmark, Germany, Poland, Sweden, the Netherlands and the UK, researchers discovered that removing dead flu victims improved the other birds' chances of survival.[108] It looks like taking away the bodies helps contain the spread of the infection, to mammals and to other birds.[109] However, it's not clear if removing carcasses is always the best option. 'Entering colonies to remove carcasses could bring benefits but also comes with the risk of disturbing birds, which could increase stress and bird movements, and potentially the risk of transmission,' says conservation scientist Susie Gold of the UK's RSPB (Royal Society for the Protection of Birds). 'Early on in an outbreak, removing carcasses could stop the spread; however, once there's already large numbers of infected birds, there's so much virus present in the environment that it's not clear if removing dead birds would reduce the overall impact.'

In the UK, the ban on sandeel fishing that came into force in April 2024 in Scottish and English sections of the North Sea could take the pressure off seabirds by keeping their food sources intact, though they must still contend with disturbance and accidental killing by the fishing industry, plastic pollution, climate change and invasive species like mice and rats on islands.[110] 'A lot of what we can do is just try and build population resilience,' says Gold. 'The more colonies you have and the healthier they are, hopefully the better able to withstand disease outbreaks.'

On Noss so many gannets died during the 2022 outbreak that the water at the bottom of the breeding colonies was, with characteristic Scottish dark humour, dubbed 'gannet soup'.[111] In an attempt to stop the virus from spreading, tourists were no longer allowed to visit the island and stand near the cliffs as we did all those years before.[112] The same went for other birding hotspots in Scotland.[113] Around the world, three-quarters of gannet colonies, a total of 40 in Iceland, the UK, Ireland, Canada, Germany, Norway and the English Channel, recorded unusually high mortality in the outbreak.[114] Like most seabirds, gannets live a long time – the UK record is 37 – and reproduce slowly, each year generally producing just a single chick.[115] So their population could take decades to recover from the flu deaths, especially as the birds are also at risk from fishing operations and plastic pollution.[116]

If I went back to Noss, I might notice, as well as reduced numbers of seabirds like skuas and gannets, that some of the gannets had irises that weren't their usual pale blue-grey, but instead were completely black or mottled,[117,118] although they still retained the distinctive pale blue rings around their eyes. Researchers studying gannets on Bass Rock off the east of Scotland found this change in some of the healthy birds that they discovered had flu antibodies in their blood, indicating they had recovered from the disease.[119] We don't yet know how flu caused this change in iris colour or whether it affects the birds' vision, survival or the likelihood that they'll reproduce successfully.[120] But the immunity, at least, is a positive sign. The eyes of some terns that survived flu turned cloudy.[121]

'The evolutionary theory is that seabirds are long-lived species, so they should put a lot of energy into immunity,' says Gold, 'which hopefully should protect them quite long-term once they've seen a virus. But at the moment we don't know how long immunity to avian influenza will last in these species.' When I spoke to Gold in June 2024, she said the UK had been quiet for flu so far, with no positive reports since April, perhaps because a lot of birds have immunity or because the virus has become less transmissible or less pathogenic (although there's currently no evidence for this). But Europe had a few positive cases and the US had many. In the UK, as this immunity wanes over time or the population of new birds who have never met the virus grows, flu could break out again. 'The hope is that this was a unique event,' Gold says, 'but I think realistically we could keep on seeing outbreaks in future years or in any populations which haven't seen the virus before. It's really difficult to predict what will happen, but I think the virus hasn't gone away.'

That's what immunity is. That's our main hope, for now – that in any future outbreaks increasing numbers of birds, as gannets, Dutch Eurasian wigeons and sandwich terns seem to have, develop immunity to avian flu, and the amount of virus excreted into the environment drops. 'We hope that nature will self-contain the infection risks and the disease,' Caliendo explains. 'If fewer birds are infected, also fewer mammals will be infected.' Most mammals, so far, have caught flu from birds. But Caliendo says it's very difficult to predict precisely how things will evolve. 'This is a virus that has surprised us in many ways.' It's also difficult to predict whether we'll use what we already know to stop more flu strains emerging from our high-density industrial-scale farms. That's our only other hope.

15

Changing Our Stripes: Protecting Ourselves and Other Wildlife

'At the approach of danger there are always two voices that speak with equal power in the human soul: one very reasonably tells a man to consider the nature of the danger and the means of escaping it; the other, still more reasonably, says that it is too depressing and painful to think of the danger.'

Leo Tolstoy, *War and Peace*, 1867
Translated by Louise and Aylmer Maude

Ngeeti Lempate lives in a manyatta, a small collection of homes surrounded by a thorn-bush fence. After passing through the gap in the fence, I sit under a tree with her. Also known as Mama Grevy, as she was one of the very first Grevy's Zebra Scouts, Lempate squats on the ground while I perch on a white plastic chair because the Kenyans politely insisted. The earth is red here. When I get home I find red dust in the crevices of my watch and between the lenses of my glasses and the frame. Soon after I begin asking my questions via Damaris Lekiluai from the Grevy's Zebra Trust, who kindly translates them into Samburu, it starts to rain, an event so rare here that schoolchildren chant 'Rain, rain, come again' rather than the 'Rain, rain, go away, come again another day' I sang in Britain. We relocate to Lempate's hut some twenty feet away. It's in the traditional style, built by making a line of vertical sticks in the ground, weaving others horizontally in between then adding a roof, these days using plastic sacking rather than cow hides. The rain patters on the roof of sticks and plastic, but inside it's dark, cosy and dry. A granddaughter, who's maybe three or four, stares at me and grabs at my pen. We compromise on her having the part with the ink but not the lid.

While Lempate talks I hear the gentle tinkling of her traditional beaded necklace that's tassled with thin metal discs about a centimetre in diameter. As a Grevy's Zebra Scout, Mama Grevy tells me, she gets up at 5am and does her 'woman's work' duties like preparing food for children who are going to school, cleaning calabashes ready for milk, milking the goats and cows, preparing tea, cooking food and sweeping the house. By 8am she's out on patrol looking for Grevy's zebras. When Lempate sees wildlife, she notes it down and records its coordinates – she had to learn on the job and tells me she never imagined she'd be able to write in a notebook or operate a GPS machine. She notes the locations of tracks and signs such as droppings too.

The role has changed Lempate's life both by teaching her about wildlife, she says, and by providing her with income – the community put her forward for the post in part because she was a widow, in part because they thought she was a quick learner and in part because they trusted her. The role has changed life for Grevy's zebras too. These days the community reports to Grevy's Zebra Scouts when they see sick zebras, or emaciated zebras during a drought. The scout then contacts the Grevy's Zebra Trust, which sends out a vet or arranges a drop-off of hay. Community members will now even share grass with wildlife, and volunteer to do the hard task of removing the invasive species *Acacia reficiens* (false umbrella thorn), which stops grass and other plants growing beneath its branches. People used to live freely with wildlife but didn't take care of them; now they're taking more responsibility. And the zebras are surviving better.

Once the herders realised that chasing Grevy's away from water sources and the best grass to save these resources for their cattle harmed the zebras, particularly the youngsters, they changed their ways.[1] And the Grevy's thrived – they no longer had to wait to drink until herders had left the area to secure their cattle behind a thorny fence overnight, safe from lions. Grevy's zebra numbers stabilised in 2015, with infants and juveniles making up nearly 28 per cent of the group, a promising sign for population growth. A few humans changing their behaviour made a big difference to wildlife survival. This zebra is a rare success story in the field of conservation and has, for now, ceased slipping towards the void.

ONE DIRECTION

The Grevy's is lucky on the disease front too – it's relatively hardy thanks to its loose social structure and small-group living. Yet the lives of Grevy's, humans and livestock are closely intertwined. As indeed are the lives of all animals – human and non-human alike. That's why we must embrace One Health. To protect us all from disease, we need people with expertise across the fields of human, other animal and wildlife health, as well as conservation and environmental protection. Neil Vora of Conservation International compares this approach to the difference between the US and European healthcare systems – many European countries have far more general practitioners per head of population than the US, and better health outcomes. 'Generalist doctors help treat a patient holistically, not just on particular issues,' he says. 'We need to be thinking more about that for how we deal with health on this planet.'

Over the next few decades, humans' damage to the environment may boost the chances that there's a pandemic in any given year up to three-fold.[2] As we've seen in earlier chapters, the way humans farm poultry has created more virulent versions of flu that have spilled into wild birds and people, leading to human pandemics in the past and likely in the future too. A disease that sprang up when we domesticated cattle spilled back into wildlife across Africa, killing countless wildebeest, buffalo, kudu, warthogs, cows and more, as well as humans by destroying their livelihoods. Humans transported yellow fever from Africa to South America, where, spread by transported mosquitoes, it impacted people and monkeys. I could go on, but it's time to talk, with Vora's help, about how to stop this.

'Very little good came out of Covid,' he says, 'but at least now at the highest levels of government we have world leaders talking about the need to do a better job with pandemics. This is something that experts have been calling for for decades because it's not a question of *if* a pandemic is going to occur, it's a matter of *when*.' If you include Covid-19, since 1918 the world has seen at least six viral pandemics – H1N1 or 'Spanish' flu in 1918, H2N2 flu in 1957–58, H3N2 flu in 1968, H1N1 flu in 2009 and HIV from 1981 onwards.[3] If we continue to harm nature, by encroaching on the lives of other animals, spreading pathogens around the world, destroying habitat and ecosystems, and warming the climate, the frequency of major epidemics and pandemics is only going to increase; the annual probability of an extreme epidemic could triple in coming decades.[4] At the time of writing, mpox had just been declared a Public Health Emergency of International Concern, and bird flu was showing signs of becoming better at transmitting from mammal to mammal – passing between cows and farmers in the US.

GROWING HEALTH

Vora grew interested in public health because his dad caught smallpox as a child in India. Dr Vora Senior was fortunate to survive a disease that kills some 30 per cent of those infected, although he still bears its scars. 'When I was growing up, [my dad] told me about the incredible public health workers around the world, particularly in the sixties and seventies, who vaccinated people and eventually eradicated the disease,' Vora says. The last naturally occurring cases of smallpox broke out in 1977 and the disease was formally announced as eradicated in 1980, the first – and, so far, only – human virus ever to be eradicated. His father's experiences led Vora to train in medicine and public health. As a result, he would help during the two biggest Ebola outbreaks in Africa, run Covid contact tracing in New York City, and most recently come to work at Conservation International promoting the prevention of pandemics, a role that enables him to combine his passions for public health and conservation. 'These spillovers [of diseases into people] are happening because of what humanity is doing to nature and how we are treating [non-human] animals,' Vora says. 'And spillovers are becoming more and more common because of those pressures.' In the last decade alone, the world has seen major outbreaks of Ebola in the African nations of Guinea, Liberia, Sierra Leone, DRC and Uganda, Zika in the Americas in 2015, Covid-19, and mpox, in Europe, the Americas and Africa from 2022 onwards. All these viruses originated in wildlife.

As we've seen throughout this book, ongoing pressures such as globalisation, habitat loss and climate change are likely to increase the rate of emergence of wildlife diseases too.[5] According to conservation biologist and disease ecologist Johannes Foufopoulos of the University of Michigan, US, diseases emerging in humans and wildlife go hand in hand. Although the two problems don't overlap completely, the same processes that encourage spillover of wildlife pathogens into humans also aid the transmission of human germs to wildlife and the jumping of wildlife pathogens into other, more susceptible wildlife species. Most of the public debate about emerging diseases, especially after Covid, has been from the perspective of human health. In the long term, however, Foufopoulos thinks the impact on wildlife is much more dramatic. 'It exceeds most people's narrow view of the world, which is "I'm just interested in human health, in the things that I can get",' he says. 'As ecologists, we need to step back, understand what the bigger picture is and then inform society about what really is going on beyond this immediate narrow scope. This is such an important topic and it's also complex because we live in a big, complicated, interconnected world where things we really don't think connect to each other actually do.' Species extinction is 'bad for everyone – it's bad for the planet, and it's bad for society in the long term'.

Globalisation mixes up species by introducing them to new areas both deliberately and by mistake,[6] forcing animals to meet pathogens they've never encountered before and have little immunity to. Habitat loss makes it easier for humans, livestock or dogs and their pathogens to encroach on wildlife, it may force wildlife to crowd together in smaller spaces, enabling disease to spread more easily, and it causes animals stress, suppressing their immune systems.[7] Climate change, meanwhile, mixes up species when they respond to the changing conditions in different ways and at different rates – some will move towards the poles and up mountains faster than others, while some may not be able to adapt in time. None of this is good for wildlife health. According to Foufopoulos, putting species in contact that weren't in contact before by, for example, intruding into rainforest habitat or changing the climate, 'stir[s] the pot in all directions', increasing emerging diseases for humans and wildlife alike. So in most cases, reducing zoonotic pathogen spillover into humans will also help stop new diseases emerging in wildlife. Although wildlife is at risk in a few additional ways, including translocations of animals into new areas for conservation purposes or to restore an ecosystem, many drivers are the same. 'We have the saying "it's very difficult to separate a parasite from its host",' Foufopoulos adds. Wildlife managers try and ensure the individuals to be moved are healthy by quarantining them beforehand, but it can be very hard to tell if a wild animal is infected – they've generally evolved to look healthy so that predators don't single them out. So translocated individuals can take their pathogens with them to their new home. Artificial feeding sites, captive breeding programmes, feral populations of escaped domestic or captive animals, and the pet trade also put wildlife health at risk more than they do human health. For example, pets imported from another country that are then released or escape put native wildlife at risk of meeting new diseases. 'Once something has been introduced – once the genie is out of the bottle, it's almost impossible to put back in,' Foufopoulos says. 'Look how hard it was to get Covid-19 under control even while having a vaccine at hand and having good ways to communicate with the patients. Imagine a system where you have Covid-19 except no vaccine and your patients hide from you and you can't talk to them.'

CAN'T SEE THE WOOD

So far, we've looked at the emergence of individual wildlife diseases and ways to eradicate, control or mitigate them once they've emerged. One way to help wildlife deal with a newly emerged disease is to reduce other stressors by protecting habitat, stopping hunting, reducing pollution and so on. Which could feed back into reducing disease emergence in the first place. 'It is better to prevent than to cure,' the philosopher Hippocrates, who lived from 460–370 BCE, said in Greek. In the 1730s, Benjamin Franklin was more imperial and wrote that 'An Ounce of Prevention is worth a Pound of Cure'.[8] For human pandemics, part of the 'cure' is 'preparedness and response', approaches that are essential but almost inevitably can't develop vaccines and treatments and make them widely available fast enough to prevent many deaths. (Though we could do a lot better on that front too.) Prevention, on the other hand, stops no end of trouble. 'When you prevent an outbreak from happening in the first place, everyone benefits everywhere,' says Vora. 'This is fundamentally an issue of equity because we've seen time and time again that the tools of response, whether we're talking about medicines or vaccines or even information, are inequitably distributed.' Prevention also stops pathogens that have jumped into humans from another species of animal from spilling back into other wildlife. And there's an economic benefit. 'Primary prevention costs a fraction of the cost of cures,' concluded one scientific paper.[9] Primary prevention aims to limit viral spillovers from happening at all.[10] Hippocrates and Franklin would have been proud.

So what can we do to enact this primary prevention and protect ourselves from diseases emerging in the first place? Fortunately for wildlife, given humans' selfish ways, the solutions for both human and wildlife health – like the problems – are linked. The biggest driver of emerging wildlife diseases is habitat modification and environmental degradation.[11] And habitat destruction in the form of clearing and degradation of tropical and subtropical forests is a big factor behind the emergence of infectious diseases in humans. In 2013 tropical forests lost more than a hundred trees per second, a figure that is mind-boggling.[12] The only way I can picture this is as a domino-toppling session with dominoes spread densely across the board and a single finger push felling more than three packs simultaneously. With that finger jabbing every single second, in a minute it's knocked over more than 214 packs of dominoes (a standard pack contains 28 dominoes if you'd like to check my maths) and in an hour flattened nearly 12,900 packs. You don't need to be a parent supervising games night to know that's unsustainable.

'Saving rainforest is critical,' Vora says. Deforestation creates new edges where humans and domestic animals start interacting with wildlife. And it destroys biodiversity, particularly the species that evolved to live only in the rainforest – the more generalist species, the rats, the bats, can often live alongside people. Bats, rodents and primates are most diverse in tropical and subtropical forests and tend to carry an unusually high proportion of zoonotic viruses.[13] 'So that loss of biodiversity can increase risk,' Vora adds. 'Reason number three is that deforestation stresses animals out and makes them more likely to get infected and to shed pathogens. Just like when we're stressed, we're more likely to get sick.'

As well as increasing the chances that human and non-human animal diseases emerge, habitat destruction is the single biggest cause of extinction for terrestrial biodiversity.[14] 'Guess what?' Vora says. 'When we invest in tropical forests [to prevent pandemics], we're also mitigating climate change, and we are also protecting biodiversity. It's like a three for one deal, a win-win-win situation.' Plus, if we needed any more incentives, protecting tropical forests also protects the indigenous peoples who live there.

For the first two decades of this century, deforestation was most extensive in the Amazon basin, West and Central Africa and South East Asia.[15] Cutting deforestation of the Amazon could be particularly important for preventing pandemics and protecting wildlife. It's one of the world's most biodiverse regions, particularly for bats and primates, and people here often eat non-human primates and large rodents.[16] What's more, cities in the Amazon have limited capacity for containing infectious epidemics. In the first wave of the Covid-19 epidemic in 2020, the Amazonian cities of Iquitos, Peru, and Manaus in Brazil experienced the world's highest infection and death rates.[17]

Brazil has already demonstrated we can turn this around. In 2023, President Luiz Inácio Lula da Silva (commonly known as 'Lula') almost halved forest loss in the Brazilian Amazon,[18] and he has pledged to end deforestation by 2030.[19] Under Lula's predecessor, Jair Bolsonaro, who was president from 2019 to 2022, deforestation rates soared and illegal gold miners invaded indigenous territory, destroyed forest, polluted rivers with mercury and human waste, and made Yanomami people sick from malaria and STDs, and short of food by contaminating fish and scaring away prey animals with mining machinery.[20] Before this, including during

Lula's previous term, from 2003 to 2011, Brazil had made more effort; by 2014 deforestation had declined 70 per cent.[21] Achieving this cut involved recognising indigenous territory, expanding protected areas, enforcing protected areas and other laws, monitoring deforestation with satellites, putting market restrictions on illegal landholdings, restricting credit for municipalities with high deforestation rates, and paying small farmers for ecosystem services,[22] as well as intervening in beef and soy supply chains.[23] To preserve our health and nature's, we must do more of this protection, not just in the Amazon but in the world's other tropical and subtropical forests too. It will need funds, from the countries with rainforest within their borders and from the nations who will benefit and are in best shape to assist. On average between 2000 and 2014, the Brazilian government spent $1 billion each year on forest conservation.[24] Norway committed $1 billion to the Amazon Fund between 2009 and 2019,[25] and in January 2023 Germany promised funding of at least $38 million.[26] Perhaps it's time for other nations to step up too, to protect us all.

CLIMATE VISION

Traditionally, the Grevy's zebra is highly significant in Samburu culture, Lempate tells me as we shelter from the rain in her hut. When the rains fail and drought strikes, people follow the zebras to find pasture for their cattle. And Grevy's are visionary – if you hear a Grevy's making noise at a dried-out water pan, it's a sign that rain is nearing. In recent years, however, five consecutive rainy seasons failed and, after one of the worst droughts in Kenya for nearly 40 years, the Grevy's Zebra Trust cancelled the Great Grevy's Rally for 2022. As climate changes, droughts may worsen and drier ground may make outbreaks of anthrax more likely, as Grevy's scrabble around for the sparse grass and inhale spores from the dust, then succumb to one of the few diseases able to kill them.[27]

Slowing deforestation would protect the climate by preventing the release of the carbon stored in trees and soil, stopping it joining the carbon we've pumped into the air by burning fossil fuels. We could gain more than one-third of the climate mitigation we need by 2030 to stabilise warming to well below 2°c if we conserved, restored and improved our management of forests, wetlands and grasslands.[28] Cutting climate change would feed back into disease protection, for wildlife and humans alike: it would stop disease vectors like mosquitoes moving into new territories, and stop animals being forced to move towards the poles and up mountains, taking pathogens into new areas with them, and causing the movers stress and perhaps reduced food intake. Not to mention reducing yet another pressure that can push species towards extinction, by slowing the rise in 'natural' disasters like floods and droughts, limiting harm to the economy and reducing risks to human health from heat extremes, premature births, floods, mental distress and more. We'd be crazy not to reduce our carbon emissions. And cutting deforestation is buy one solution, get two or three free.

COMMUNITY CHEST

Funding health clinics for communities near rainforests could help keep trees in place. Vora previously sat on the board of Health in Harmony, an NGO that does just this, as well as providing training on alternative livelihoods and educating young people. The organisation always asks the local community what it needs, but it's often all too easy to tell other people what to do without listening. Vora has a story that illustrates the dangers: just before dinner on the first night of his visit to Samburu County in Kenya for a meeting on conservation and health, Vora was outside and heard a thud: a bat had fallen out of the air and landed on the grass. The Samburu guide taking care of the visitors picked the bat up and moved it to the side, out of the way. 'Being a bat-borne disease expert, I got worried that, *Wow, this guy might get exposed to some pathogen,*' Vora says, 'because was this bat sick? Is that why it fell out of the air? So I told him, look, go wash your hands. You can get exposed to something.' The guide duly did as he was advised. The next day brought another incident before dinner. One of Vora's colleagues looked down at her foot and saw a tick. 'That same Samburu guide reached out, ripped the tick off her foot with his fingers and was about to crush it,' Vora recalls. 'My colleague and I said, "Hold on! Don't do that because this tick might have a virus in it, such as Crimean Congo haemorrhagic fever virus or some other dangerous pathogen. It's better to put the tick down and crush it on the ground so you don't get exposed to anything."' Again, the local guide complied. But then he looked at the two researchers and playfully told them that he'd been picking up bats and ripping off ticks from cows since he was a kid and had never got sick. 'I had a huge realisation in that moment about my arrogance,' says Vora. 'Here I am, a New Yorker telling this guy who's lived on this land his whole life how to interact with the wildlife, and if you put me out in that Samburu

County for fifteen minutes, I would probably die, definitely within an hour or two, whereas he can survive for days and days.' Vora believes it's important to come into such conversations with humility and respect for the expertise and lived experience of local communities, and to come up with joint solutions together; conservation has in the past made huge mistakes in excluding people. 'People are part of the landscape and indigenous peoples are the best custodians of protected areas,' he says. 'People need to be front and centre, otherwise this is all going to fail.'

In its project near Gunung Palung National Park in Indonesian Borneo, Health in Harmony consulted with the local community from 2005 to 2007 before building the clinic people requested. The clinic let patients pay for its services with handicrafts, seedlings or manure that was then used in conservation; in the past they might have paid with cash from selling wood they'd logged illegally. In areas where illegal logging had slowed, the community received a discount for medical services.[29] Over a decade, these investments reduced deforestation by around 70 per cent,[30] while more than 28,000 people received healthcare,[31] and infant mortality dropped by two-thirds.[32,33] To put that a way that somehow brings it home more clearly, for every three babies that would have died in the past, two of them now survived. 'We've been sold some type of dystopic vision of the world that is human or planet, not both,' Vora says. 'That is not true. Indigenous communities have shown us this for thousands of years. We have to remember that we are a part of nature, that people and planet don't need to be in conflict, there's a better path forward possible.'

As one route to this path, we can build disease considerations into our planning processes by making sure that the environmental impact assessments for construction and other projects include the project's impact on the health of humans and wildlife,[34] and protect habitat as necessary.

COUNTERCULTURE

We could also support ecological countermeasures that will mitigate some of the harm we've already done.[35] Proposed to prevent human pandemics, such countermeasures would also protect wildlife health. They involve, among other things, conserving and restoring nature at key sites used by reservoir hosts so that they suffer less disturbance and come into less contact with humans, and mitigating the risks for humans and wildlife that are in most danger. Damaging an ecosystem risks breaking down the first three barriers to the spread of a disease into a new species and its development into a pandemic or panzootic. It can mix up species so that their distributions overlap more than they otherwise would, cause resevoir hosts stress and reduce their food intake so that their immune systems work less well and they shed more virus, and alter the new species' exposure to pathogens by disrupting its behaviours and interactions. (The fourth and fifth barriers depend on biology – the fourth hinges on whether the virus can bind to the new animal's cells and how effective its immune system is against this particular germ, and the fifth depends on whether the pathogen in that new animal can amplify, be excreted and then transmit to other animals.)[36] Ecological countermeasures address all three of the first barriers by aiming to return reservoir hosts back to their original habitat, cause them less stress, and minimise changes in behaviour that make humans or other new species of animals more likely to be exposed. The measures also tie in nicely with the UN's current Decade on Ecosystem Restoration.

For a reservoir host like a bat, for example, ecological countermeasures involve protecting where the animals eat and where they roost, and protecting people and livestock at risk by minimising livestock interactions with bats and their excreta, designing bat-proof houses and safely excluding bats from existing houses, protecting food such as fruit trees and water supplies from bats, keeping urban expansion away from large wildlife populations, and using trees that don't attract fruit-eating bats in parks. In landscapes that are already degraded, protecting where bats eat may mean restoring and connecting key food sources that sustain bats when resources are scarce, for example during winter or the dry season, and when they're pregnant or lactating and need extra food.

To implement these countermeasures, teams of ecosystem health workers, including forestry, wildlife, veterinary, medical and public health officers, could work together locally, as well as providing environmental education and liaising with larger government One Health teams.[37] Effectively a new class of workers in the health sector, these ecosystem health workers could also consult with the indigenous people and local communities who have lived in balance with nature for thousands of years.[38] 'As my colleague Jonathan Jennings likes to say, our current design of health systems is around a single species, *Homo sapiens*, but that's not going to cut it,' Vora says. 'We cannot protect our health without taking care of the planet and taking care of [all] animals.'

TRADING PATHOGENS

In June 1987, ten plains zebras from Namibia's Etosha National Park were imported to a safari park near the Spanish capital Madrid. During August, several horses at the park died. The prime suspects were poisoning by rattlepod or water hemlock plants, or insecticides that had got into their feed. Then horses on nearby farms became ill too and the true culprit was identified – African Horse Sickness,[39] caused by a virus transmitted by midges. The zebras didn't seem ill as they're a natural reservoir for the disease and don't tend to become sick, but they passed it on to the domestic horses, who weren't so lucky.

The zebras probably became infected during their three weeks' quarantine in Namibia, in open pens accessible to midges – not the best design. Controlling the outbreak involved slaughtering 146 equids at the safari park and neighbouring farms, vaccinating horses in nearby provinces, and banning the movement of horses throughout the country. Horse breeders in Spain lost income estimated at $20 million.[40] But ten zebras are a stripy speck on the face of the colossal amount of wildlife shipped around the world each year, whether for zoos and safari parks, human consumption, the pet trade, the fur trade, medicines or ornaments. In 2011, some 6 million tonnes of non-human animals each year were extracted from the Congo and Amazon basins, equivalent in weight to roughly 8 million cows.[41] Around the world, hunting threatens more than 300 land mammals in developing countries with extinction, including elephants, pangolins, primates and armadillos, with larger mammals generally more at risk.[42] A quarter of all mammal species are traded, including a high percentage of rodents, bats and non-human primates, which host a high diversity of viruses that can jump to humans.[43]

One of the biggest global importers of wildlife, including for the exotic pet industry, is the US; tens of millions of individual animals are imported to the US each year.[44] The transit conditions, lack of health screening and the warehouses that store animals before and after import, often alongside other species they wouldn't normally encounter, are similar to those at live animal markets.[45] All are likely to help spread diseases.[46] Most of these species, such as corals, fish, reptiles and amphibians have a relatively low risk of passing on infections to humans. But they could, as we've seen in this book with amphibians and zebras, make related species sick. The US Fish and Wildlife Service (FWS) inspects all legal shipments of wildlife to make sure they comply with the Convention on International Trade in Endangered Species of Wild Fauna and Flora (CITES) regulations that control the trade of 36,000 animals and plants. The FWS also routinely tests for diseases, but only a few, including highly pathogenic avian flu in some poultry, psittacosis (a bacterial infection) in parrots, and foot and mouth disease, caused by a virus, in hooved animals;[47] traditionally we've chiefly been interested in wildlife diseases that can infect us or our livestock.[48] Many countries don't test imported wildlife for diseases at all. And, as far as we know, no country tests for unknown or novel pathogens.[49]

We know even less about illegal shipments of wild animals and their products or those that don't cross national boundaries. Yet even legal shipments have led to trouble. The trade in lab animals imported Ebola Reston virus from the Philippines to the US, and the pet trade in pouched rats whisked the monkeypox virus now known as mpox from Ghana to Texas in 2003.[50] The trade in amphibians spread the deadly chytrid fungus worldwide. And I could go on. Despite these proven dangers, no one is fully responsible for preventing animal diseases spreading via trade. CITES says that monitoring pathogens in the wildlife trade is not within its mandate.[51] Meanwhile, the World Organisation for Animal Health (WOAH) focuses mainly on infectious disease threats to livestock from the trade in wildlife. If a disease threatens profits from trading livestock, WOAH lists it as notifiable and health professionals must report suspected cases to the authorities.[52] While the organisation has the power to list a disease that threatens wildlife through environmental sources, it rarely does. That said, WOAH did list the amphibian chytrid fungus because of its threat to the trade in amphibians.[53]

Given funding, WOAH and CITES could work together more closely – the pair signed a memorandum of understanding in 2015 'to protect CITES-listed species and conserve biodiversity by ensuring the efficient implementation of surveillance and disease control measures needed to protect [non-human] animal and human health worldwide'.[54] Researchers propose that the money for a substantial increase in WOAH's operating budget, which was $35 million in 2018,[55] could come from governments, consumers or businesses in the wildlife trade such as fashion houses, pet sellers and aquarium traders. Traders could be required to obtain a permit to import any species of wildlife, not just those listed on CITES.[56]

Around the world, regional wildlife enforcement networks (WENs), such as those in the EU, Caribbean and southern Africa, bring together CITES, customs and police authorities from neigh-bouring countries, in an attempt to regulate the wildlife trade. These networks don't currently monitor human or other animal health, and have suffered from lack of financial and political support. Perhaps that should change – throwing money at the problem could make a big difference. We could beef up WENs to deal better with the disease threat from all types of wildlife trade, legal or illegal.[57] The oldest, ASEAN-WEN for countries in South East Asia, has been in action since December 2005 and has an annual budget of just $30,000, a far cry from the half a million to a million dollars researchers estimate such networks need each year to be effective.[58] Given funding, other organisations attempting to fight the illegal wildlife trade could battle against diseases too. Providing CITES, WOAH and national agen-cies with enough money to research, monitor and enforce restrictions on the wildlife trade would reduce its risks and lower the chances of diseases spreading into new species, including humans.[59] The funds could also pay for vets to conduct surveillance in high-risk locations.[60] Many parts of Africa have 2 vets per 100,000 people, whereas Spain, with 2 vets per 1,000 people, has 100 times this figure.[61]

One of the most important policies for diseases and the wild-life trade is currently enshrined in the Convention on Biological Diversity's Kunming Montreal Global Biodiversity Framework, which was agreed in 2022. Target five in the framework is to 'ensure that the use, harvesting and trade of wild species is sustainable, safe and legal, preventing overexploitation, minimizing impacts on non-target species and ecosystems, and reducing the risk of pathogen spillover, applying the ecosystem approach, while respecting and pro-tecting customary sustainable use by indigenous peoples and local communities'.[62] Since countries are now developing national plans that align with this global framework, some will adopt measures to address the risk of spillover at wildlife markets and through the wildlife trade.

Scientists have also proposed that, as part of pandemic prevention, these national plans have targets to reduce deforestation, close or regu-late wildlife markets, and increase surveillance at the interface between humans, domestic animals and wildlife, whether at mink farms in the US, wet markets in Asia, or highly deforested areas in South America.[63] In emerging disease hotspots, this surveillance could be provided by clinical care for humans and domestic animals that would pick up patho-gens as they begin to spill over from wildlife.[64]

BAN BAN

Drastic times need drastic measures. In January 2020, after Covid-19 emerged, the Chinese state temporarily banned all wildlife trade until the end of the epidemic.[65] The following month the nation permanently banned consumption – and so effectively the food trade – of all terrestrial wildlife,[66] except a handful of farmed species with low health risks and established farming techniques,[67] including Sika deer, red deer, caribou, llama, guinea fowl, pheasant, partridge, common mallard and ostrich. 'Wildlife consumption is neither needed for subsistence of local communities nor essential for Chinese diets,' wrote one team of academics.[68] The Chinese state began wildlife farming in 1954 then, after the reforms of 1978 put a stop to collective farming and state control, millions of peasant households started to farm wildlife in the 1980s.[69] Ironically, it was a disease that recently led more people to farm, hunt and eat wildlife, when African swine fever broke out in Chinese pigs in 2018.[70]

In Kenya several people told me about wildlife they'd eaten in the past, before it was illegal here.[71] Guinea fowl ('tastes like chicken'), spur fowl, python, mice (although one person translated this as 'big mice' so it may have been rat), crocodile and, once, Grevy's zebra. During the first years of the Covid-19 pandemic, many called for those living near forests around the world to stop eating, and trading, bushmeat. But Vora doesn't think that such a ban is the answer. For a start, people need the nutrition – unlike in China, in many places supporting yourself does depend on eating wildlife. These people should have a right to harvest wildlife, Vora says, especially as they may have cultural reasons. Instead of calling for a ban, we can make eating bushmeat safer by giving people information, such as to avoid eating animals they find dead. 'I've seen bushmeat hunting in the Congo Basin,' adds Foufopoulos. 'Tricky. It's a huge risk, especially I would say bat hunting and eating bats – you're playing with fire there because bats carry so many diseases.' In rural settings, eating wildlife is in any case relatively safe for people elsewhere – any infections that result and start to spread from human to human are likely to peter out due to the low population density. But once a pathogen reaches a town or city, it can spread more quickly. 'There is really no reason why live wild birds and mammals should be sold in urban settings,' Vora says. 'That is definitely something that needs to be strictly regulated, even shut down.' In today's world, people can fly from one city to another on another continent within 24 hours, taking any newly acquired infections with them. 'We all live in an interconnected world, and a health threat anywhere is a health threat everywhere,' Vora says.

FUR NEVER

China still allows trade in wildlife for fur, exhibition, pets, research and medicine,[72] although medicine sourced from wildlife was removed from China's basic medical insurance coverage in 2019, making such medicines more expensive for consumers.[73] The new regulations on wildlife consumption and farming allow for farming of mink, silver fox, blue fox and racoon dog for non-edible uses, presumably fur. Yet raccoon dogs were one of the mammals infected by SARS-CoV in wet markets in Guangdong before the outbreak in 2003, and farmed mink in Denmark became infected with Covid-19 and transmitted it back to people. Avian flu also raged through a mink farm in Spain. Whether for fur, pigs or poultry, and perhaps also now cattle, high-density animal farms help flu genes mix and match to create new strains that jump to wildlife and humans. To reduce demand for fur, we could insist, by law, that fur in garments be identified with the species name and country of origin. Campaigns for corporate social responsibility, and social pressure against wearing fur from factory-farmed animals, could also help.[74] Or perhaps it's simply time to ban the fur trade? In farms for meat, dairy and eggs, we must improve biosecurity, and ensure animals are less tightly crammed together into these 'pathogen factories'.[75] And we could simply eat less meat. 'If we could reduce our meat consumption on the planet by three quarters, which still means there's a lot of livestock, it would make a huge difference in terms of risk,' says Richard Kock of the UK's Royal Veterinary College. 'It would help reduce greenhouse gases very significantly and we will release land – we'll probably be able to recover most biodiversity.' To get there, Kock thinks we need to involve more psychologists, social scientists and behaviourists 'because the natural scientists, we've been saying things for decades and decades, but nobody listened'.

Human nature is also a concern of Rebecca Henderson of Harvard University, US, an economist who is 'reimagining capitalism for the climate crisis'. (As we heard above, what's good for the climate is also good for wildlife, and human, diseases.) So far, business and government efforts to cut carbon emissions have made huge strides but progress hasn't gone far or fast enough, Henderson explained to delegates at a hybrid meeting in Oxford.[76] It's not just a technical and policy issue, it needs something more and it's time for humans to shift from being *Homo economicus* – competitive, seeing humans as economic instruments and finding meaning in consumption and status – to being *Homo reciprocans*, cooperative, seeing humans as beings that need respect and finding meaning in flourishing and connection. Activities like tai chi, meditating, spending time in nature, singing together, dancing, poetry and more could help us make this shift, Henderson explained. This was not something I expected to hear from an economics professor, who went on to add that we need to talk about what it really means to be a successful human being and get to a point where status lies in doing good in the world rather than owning yachts. We must dematerialise growth in rich countries, expanding economies while using fewer resources, and support growth elsewhere so that everyone reaches a decent standard of living. For those expecting economists to talk purely numbers, although Henderson's many figures proved she can do this too, perhaps we should move on to some costings.

SICKNESS BENEFIT

The estimated cost for primary prevention of pandemics – stopping new diseases emerging and jumping into humans, rather than stopping them spreading once they've emerged – is around $20 billion a year. That $20 billion price tag sounds like a lot, yet this amount is one-twentieth or even less of the statistical value of the human lives lost each year since 1918 to viral diseases that have spilled over from other animals, and less than a tenth of the economic productivity wiped out each year.[77] The Covid-19 pandemic alone cost millions of lives and trillions of dollars.[78] The economics of preventing pandemics are like insulating the roof of your house if you live somewhere cold – an outlay at the start saves you lots of money in the long-term. Both are no-brainers, even before you consider the loss, grief, fear and suffering the Covid pandemic brought us.

The $20 billion a year total comes from adding up the median cost estimates of $1.53–9.59 billion to halve deforestation, $19 billion to close China's wildlife farming industry, $476–842 million to reduce spillover from livestock into humans, $250-750 million on wildlife trade surveillance, $120–340 million on viral discovery and $217-279 million on early detection and control.[79] Viral discovery would help us know what's out there in bats and other wild animals. Though others feel that, rather than aiming to build a database of every single virus in existence, it's better to let the viruses themselves show us which are most dangerous by setting up clinics near forests and other potential disease spillover hotspots and checking people's blood for the subset of viruses that have proved they can jump to humans.[80] No matter how much effort we put into primary pandemic prevention, it's possible that diseases will still spill over, so we'll need early detection and control – belt and braces.

No one, as far as I'm aware, has costed up ways to stop new diseases emerging for wildlife, although preventing pandemics will, as a side-benefit rather than the main goal (but let's not look a gift-animal in the mouth), also protect wildlife. Foufopoulos questions whether it's even ethical to talk about cost when it comes to saving a species that's evolved over millions of years from human destruction. 'Once you start talking about what's the value of a species and going extinct, how much is it?' he asks. 'Is this really an issue of money? If we apply the same idea to human life, would you be willing to be saved if your life is worth this much? But if it's not [worth] that much, then we'll let you die or go extinct. We live in a capitalist society, so we tend to put a dollar value to everything, but we need to also be aware of what it means to think of everything in monetary terms.' Because the world loses more than money when we lose species – biodiversity has an immense value that we tend not to be aware of. 'Just because bees don't send you an invoice when they pollinate your flowers or your crops, it doesn't mean what they do doesn't have value,' Foufopoulos says. 'We typically realise that once the species has gone extinct. Businesses like to ignore that because what they are interested in is externalising the costs – essentially having society pick up the costs of them producing their product cheaply. So if a species goes extinct, let society deal with that.' Even though no one has looked at the cost benefit of protecting wildlife from emerging pathogens and disease outbreaks, Foufopoulos argues that it's very much worth it, 'number one, because wildlife extinction is expensive for society, and number two, the same expensive actions for the most part that prevent wildlife disease are also the same actions that will protect human health'.

Nature is overlooked as a solution, Vora believes. For example, conserving, restoring and better managing forests, wetlands and grasslands could provide more than 30 per cent of the emissions reductions we need to reach our climate goals, but receives less than 5 per cent of overall climate funds.[81,82] 'Same thing is happening on the health side – we ignore nature,' Vora says. 'Nature is not the solution to everything, but nature is critical, a necessary condition for a better future. We need to start investing in nature now and that will save costs down the road.'

So where might the money for this investment come from? As a result of Covid-19, the World Bank set up the Pandemic Fund. 'It's shocking that there wasn't such a fund before, but what's also shocking is that this fund struggled to raise money,' Vora says. The fund has a relatively modest target, given what's needed, of raising $10 billion a year for pandemic activities, particularly in lower income countries. 'They have struggled to raise more than just a few billion dollars in the first year,' Vora says. 'We just got out of the Covid pandemic. Yet countries are still being so stingy and not investing in pandemic prevention, preparedness and response.'

At time of writing, the World Health Organization was co-ordinating negotiations on a pandemic agreement that were due to be completed by the end of 2024, a delay from the original May deadline. It remains to be seen whether this agreement's articles on One Health and primary pandemic prevention will get to stay and how strong they will be – to date, much of the focus has been on preparedness and response once a pandemic has already emerged.[83] Equity has been a major sticking point during the negotiations, with poorer countries concerned that the drafts didn't go far enough and richer countries unwilling to budge. Yet equity is crucial for everyone because infectious diseases often take advantage of socio-economically marginalised groups of people. And such diseases know no borders. Some people bitterly resent the agreement itself. 'A number of highly influential voices in social media and elected officials have spread misinformation on what this agreement is about,' Vora says, '... for example, that this agreement will force a country to lose its sovereignty. That is not true.'

TAKING ACTION

Back in Kenya, the Grevy's Zebra Technical Committee negotiated a rerouting of a planned road, rail and oil pipeline away from part of the prime zebra breeding territory. Yet how much better if we didn't spend an estimated $1.8 trillion a year – yes, that's trillion, a thousand billion – on subsidising environmentally damaging industries?[84] We could also strengthen our laws to protect wildlife from such industries and the diseases that disrupted ecosystems bring. In 2019 the Pacific Island state of Vanuatu proposed to the International Criminal Court that ecocide – destroying the environment – becomes a crime in peacetime as well as during war. The crime would be held against individuals, not businesses, but it could make CEOs think twice before they lead their company into hacking down forests and mining and drilling and creating plantations in wild landscapes. The next step is for the International Criminal Court to admit and adopt the amendment, which needs a two-thirds majority from 123 nations. At the time of writing, momentum was building: in August 2021, France included ecocide in its Climate and Resilience Act, and in 2024 the European Council voted to criminalise cases 'comparable to ecocide'.[85] Perhaps one day the Grevy's and all other wildlife – and humans – will be protected by legislation that ensures people take the environment into account in their decisions. 'One thing that you learn as an ecologist is that the world is interconnected and things that are just beyond your borders actually can affect you,' says Foufopoulos. 'Covid, for example, came into the US from this 'outside' world. Whether we like it or not, we live in an interconnected planet and it behoves us, especially the affluent countries who don't live hand to mouth and have the luxury of using resources to prevent problems, to step back and say, "How can we make sure this becomes a better world, how can we pre-vent these things on a global scale?"' Foufopoulos sees this movement to Stop Ecocide as exactly along these lines – 'the realisation that what happens in the Amazon is actually our business as well, and eventually this is going to come home to roost, these things that we ignore, so it is wise to start thinking on a big scale'.

So far, these solutions – protecting habitat, particularly forests, slow-ing climate change, and regulating trade and farming – have mainly involved high-level policy action. But what can you do as an individual, assuming you're not a policymaker or researcher, should you feel so moved? Once you've informed yourself about the problem, which you already have by reading this book, the number one action, Vora believes, is to vote if you can 'for elected officials who believe in science instead of elected officials who deny climate change and deny the tragedy of Covid'. These officials will need to care about investing in nature too; you might like to write to them and make your views known. 'Action is the antidote to despair,' Vora says. 'Get involved – you will feel more hopeful. We can't buy into this delusion that it is too late to act, that it's out of our control. We can vote, we can get politically engaged, we can lead community-based efforts. There's so much stuff that we can do.' I'd like to suggest an action too – if you're able, please support a wildlife conservation charity, particularly one that works with local communities.

Every little helps, though the real change must come at a societal level. Vora compares this to public health. 'If you leave the onus on the individual to stop smoking, they're not going to,' he says. 'It's very hard to stop smoking. But you see massive benefits when you start implementing societal changes – implementing huge levels of taxation on cigarettes and putting big labels on packages about the health risks of smoking. That's a systems-level, societal-level change that's far more effective than shaming an individual and expecting an individual to stop smoking.'

Asked what the picture might look like in ten years' time, Foufopoulos says it's hard to predict because so much depends on what we do today and tomorrow. If we as a society step up to the challenge and take measures that might be expensive or difficult, we're going to have less of a wildlife disease problem. 'If there's one thing we've seen over the last few years, it's that management and policy matters,' Foufopoulos says. 'Conservation actions can change the world.' It's time to act more. 'If you look at all the trends like the number of outbreaks in the last hundred years, what you see is an exponential rise,' Foufopoulos adds. 'I'm afraid it's going to get worse but that doesn't mean we can't do things to make things better.'

If we carry on as we are, without a doubt there will be more pandemics. I, for one, don't want to experience another. And I also don't want to see wildlife grow sick and disappear. 'Not everyone believes in the intrinsic value of nature,' Vora says. 'But we need nature to survive. That's a fact. I hope that people will realise that we are a part of nature and that having our governments invest in protecting nature is a good thing.' Protecting nature, wildlife and livestock must become part of public health – to save ourselves, we must pay attention to our bodies and stop thinking humans are separate from other animals and nature. There is no way to untangle ourselves from the web of life without catastrophic damage.

Before he started working on pandemic prevention in 2021, Vora was anxious about the future. But now he's never been more hopeful. 'I see all the people around me who are dedicating their life for climate action and preventing pandemics and protecting biodiversity,' he says. 'I see the solutions are all around us, and I look at the younger generation who are so active and aware, and I think that we are going to succeed.'

When I asked Mama Grevy about the future, she said she'd like wildlife to be taken care of as well as livestock and given the same priority.

She'd like people to work in unity, listen to each other, and be able to protect our land and our wildlife. She hopes that there is peace among people and that everything is OK through having one voice. That's what I hope for wildlife and people too. One voice and one health. The sooner we stop seeing ourselves as separate from wildlife and nature, the better. For everyone.

'Ashe oleng,' I say at the end of the interview, in my terrible Samburu accent. It's about the only word I know – 'thank you'. Mama Grevy chuckles warmly and says it back.

Endnotes

PREFACE

1.   Pedersen, A. B., Jones, K. E., Nunn, C. L., & Altizer, S. (2007). Infectious diseases and extinction risk in wild mammals. *Conservation Biology, 21*(5), 1269-1279.

2.   Diamond, J.M. (1984). 'Normal' extinction of isolated populations. In M.H. Nitecki (Ed.), *Extinctions*, 191-246, Chicago University Press.

3.   Stork, N. & Lyal, C. (1993). Extinction or 'co-extinction' rates? *Nature, 366,* 307. https://doi.org/10.1038/366307a0.

4.   McCallum, H., Foufopoulos, J., & Grogan, L. F. (2024). Infectious disease as a driver of declines and extinctions. *Cambridge Prisms: Extinction, 2*, e2. https://doi.org/10.1017/ext.2024.1.

5.   Leopold, A. (1933). *Game management*. Scribner.

6.   Quoted in Deem, S., Karesh, W. & Weisman, W. (2001). Putting Theory into Practice: Wildlife Health in Conservation. Conservation Biology, *15*, 1224-1233. https://doi.org/10.1111/j.1523-1739.2001.00336.x.

7.   WHO Director-General's opening remarks at media briefing. (August 7, 2024). Retrieved from https://www.who.int/director-general/speeches/detail/who-director-general-s-opening-remarks-at-the-media-briefing---7-august-2024.

8.   Jones, K. E., Patel, N. G., Levy, M. A., Storeygard, A., Balk, D., Gittleman, J. L., & Daszak, P. (2008). Global trends in emerging infectious diseases. *Nature, 451*, 990-993. https://doi.org/10.1038/nature06536.

9.   Riley, T., Anderson, N. E., Lovett, R., Meredith, A., Cumming, B., & Thandrayen, J. (2021). One Health in Indigenous Communities: A Critical Review of the Evidence. *International Journal of Environmental Research and Public Health, 18*(21), 11303. https://doi.org/10.3390/ijerph182111303.

10.   Ibid.

11.   The Manhattan Principles. *One World, One Health*. Retrieved from https://oneworldonehealth.wcs.org/About-Us/Mission/The-Manhattan-Principles.aspx.

12.   Plowright, R. K., Ahmed, A. N., Coulson, T., Crowther, T. W., Ejotre, I., Faust, C.L., '... &' Keeley, A. T. H. (2024). Ecological countermeasures to prevent pathogen spillover and subsequent pandemics. *Nature Communications, 15*(1), 2577. https://doi.org/10.1038/s41467-024-46151-9.

1. WILD HORSE CHASE

1.   Irving, A. T., Ahn, M., Goh, G., Anderson, D. E., & Wang, L. F. (2021). Lessons from the host defences of bats, a unique viral reservoir. *Nature*, 589 (7842), pp. 363–70. https://doi.org/10.1038/s41586-020-03128-0.

2.   Banerjee, A., Baker, M. L., Kulcsar, K., Misra, V., Plowright, R., & Mossman, K. (2020). Novel insights into immune systems of bats. *Frontiers in Immunology*, *11*(26). https://dx.doi.org/10.3389/fimmu.2020.00026.

3.   United Nations. (2024). World Population Prospects 2024. *United Nations Population Division*. Retrieved from https://www.un.org/development/desa/pd/world-population-prospects-2024.

4.   WWF. (2020). Living Planet Report 2020 – Bending the curve of biodiversity loss. R.E.A Almond, M. Grooten and T. Petersen (Eds). WWF, Gland, Switzerland. Retrieved from https://www.worldwildlife.org/publications/living-planet-report-2020.

5.   IPBES. (2019). *Global assessment report on biodiversity and ecosystem services of the Intergovernmental Science-Policy Platform on Biodiversity and Ecosystem Services*. E. S. Brondizio, J. Settele, S. Díaz, and H. T. Ngo (Eds). IPBES secretariat, Bonn, Germany. 1148 pages. Retrieved from https://doi.org/10.5281/zenodo.3831673.

6.   Secretariat of the Convention on Biological Diversity. (2020). *Global Biodiversity Outlook 5 – Summary for Policy Makers*. Montréal. Retrieved from https://www.cbd.int/gbo5.

7.   Rubenstein, D., Mackey, B. L., Davidson, Z., Kebede, F., & King, S. (2016). *Equus grevyi. The IUCN Red List of Threatened Species 2016*. Retrieved from https://dx.doi.org/10.2305/IUCN.UK.2016-3.RLTS.T7950A89624491.en.

8.   Grevy's Zebra Trust. (2023). *Annual Report 2023*. Retrieved from https://www.grevyszebratrust.org/annual-reports/2023-annual-report/#flipbook-df_30008/9/.

9.   Rubenstein, D., Mackey, B. L., Davidson, Z., Kebede, F., & King, S. (2016). *Equus grevyi. The IUCN Red List of Threatened Species* 2016. Retrieved from https://dx.doi.org/10.2305/IUCN.UK.2016-3.RLTS.T7950A89624491.en.

10.   Fabricant, J. (1998). The Early History of Infectious Bronchitis. *Avian Diseases*, *42*(4), 648–650. https://doi.org/10.2307/1592697.

11.   Vijayanand, P., Wilkins, E., & Woodhead, M. (2004). Severe acute respiratory syndrome (SARS): a review. *Clinical Medicine*, 4 (2), pp. 152–160. https://doi.org/10.7861/clinmedicine.4-2-152.

12.   Skowronski, D. M., Astell, C., Brunham, R. C., Low, D. E., Petric, M., Roper, R. L., ... & Babiuk, L. (2005). Severe acute respiratory syndrome (SARS): a year in review. *Annual Review of Medicine*, *56*(1), 357-381. https://doi.org/10.1146/annurev.med.56.091103.134135.

13.   Ge, X. Y., Li, J. L., Yang, X. L., Chmura, A. A., Guangjian, Z., Epstein, J. H., Mazet J.K., Hu, B., Zhang, W., Peng., C., Zhang ,Y. J., Luo, C.M., Tan, B., Wang., N. Zhu, Y., Crameri, G., Zhang, S. Y., Wang, L. F., Daszak, P., & Shi, Z. L. (2013). Isolation and characterization of a bat SARS-like coronavirus

that uses the ACE2 receptor. *Nature*, *503*(7477), 535-538. https://doi.org/10.1038/nature12711.

14.　Guan, Y., Zheng, B. J., He, Y. Q., Liu, X. L., Zhuang, Z. X., Cheung, C. L., ... & Poon, L. L. (2003). Isolation and characterization of viruses related to the SARS coronavirus from animals in southern China. *Science*, *302*(5643), 276-278.

15.　De Wit, E., Van Doremalen, N., Falzarano, D., & Munster, V. J. (2016). SARS and MERS: recent insights into emerging coronaviruses. *Nature Reviews Microbiology*, *14*(8), 523-534.

16.　Ramadan, N., & Shaib, H. (2019). Middle East respiratory syndrome coronavirus (MERS-CoV): A review. *Germs*, *9*(1), 35–42. https://doi.org/10.18683/germs.2019.1155.

17.　De Wit, E., Van Doremalen, N., Falzarano, D., & Munster, V. J. (2016). SARS and MERS: recent insights into emerging coronaviruses. *Nature Reviews Microbiology*, *14*(8), 523-534.

18.　Gorbalenya, A., Baker, S., Baric, R. S., de Groot, R., Drosten, C., Gulyaeva, A. A., ... & Ziebuhr, J. (2020). The species Severe acute respiratory syndrome-related coronavirus: classifying 2019-nCoV and naming it SARS-CoV-2. *Nature Microbiology*, *5*(4), 536-544.

19.　MRC - UofGlasgow Centre for Virus Research. (2021, August 9). Evolutionary Origins of SARS-CoV-2. [YouTube video]. Retrieved from https://www.youtube.com/watch?v=-lXFu4uXRXI.

20.　Lytras, S., Hughes, J., Martin, D., Swanepoel, P., de Klerk, A., Lourens, R., ... & Robertson, D. L. (2022). Exploring the natural origins of SARS-CoV-2 in the light of recombination. *Genome Biology and Evolution*, *14*(2), evac018.

21.　Temmam, S., Vongphayloth, K., Baquero, E., Munier, S., Bonomi, M., Regnault, B., ... & Eloit, M. (2022). Bat coronaviruses related to SARS-CoV-2 and infectious for human cells. *Nature*, *604*(7905), 330-336.

22.　https://www.nature.com/articles/s41586-022-04532-4.

23.　https://www.biorxiv.org/content/10.1101/2023.07.12.548617v1.full.

24.　Pekar, J. E., Lytras, S., Ghafari, M., Magee, A. F., Parker, E., Havens, J. L., ... & Lemey, P. (2023). The recency and geographical origins of the bat viruses ancestral to SARS-CoV and SARS-CoV-2. *bioRxiv*.

25.　https://www.biorxiv.org/content/10.1101/2023.07.12.548617v1.full.

26.　Worobey, M., Levy, J. I., Malpica Serrano, L., Crits-Christoph, A., Pekar, J. E., Goldstein, S. A., ... & Andersen, K. G. (2022). The Huanan Seafood Wholesale Market in Wuhan was the early epicenter of the COVID-19 pandemic. *Science*, *377*(6609), 951-959.

27.　Crits-Christoph, A., Levy, J. I., Pekar, J. E., Goldstein, S. A., Singh, R., Hensel, Z., ... & Débarre, F. (2023). Genetic tracing of market wildlife and viruses at the epicenter of the COVID-19 pandemic. *BioRxiv*. https://www.biorxiv.org/content/10.1101/2023.09.13.557637v1

28.   Alwine, J., Goodrum, F., Banfield, B., Bloom, D., Britt, W. J., Broadbent, A. J., ... & Yurochko, A. (2024). The harms of promoting the lab leak hypothesis for SARS-CoV-2 origins without evidence. *Journal of Virology*, e01240-24.

29.   Crits-Christoph, A., Levy, J. I., Pekar, J. E., Goldstein, S. A., Singh, R., Hensel, Z., ... & Débarre, F. (2024). Genetic tracing of market wildlife and viruses at the epicenter of the COVID-19 pandemic. *Cell*, *187*(19), 5468-5482.

30.   Huang, X. Y., Chen, Q., Sun, M. X., Zhou, H. Y., Ye, Q., Chen, W., ... & Qin, C. F. (2023). A pangolin-origin SARS-CoV-2-related coronavirus: infectivity, pathogenicity, and cross-protection by preexisting immunity. *Cell Discovery*, *9*(1), 59.

31.   Jones, K. E., Patel, N. G., Levy, M. A., Storeygard, A., Balk, D., Gittleman, J. L., & Daszak, P. (2008). Global trends in emerging infectious diseases. *Nature*, *451*(7181), 990-993.

32.   World Health Organization. (2015). WHO publishes list of top emerging diseases likely to cause major epidemics. *World Health Organization*. Retrieved from https://www.who.int/news/item/10-12-2015-who-publishes-list-of-top-emerging-diseases-likely-to-cause-major-epidemics.

2. A MAMMOTH PROBLEM

1.   Green, R. E., Krause, J., Briggs, A. W., Maricic, T., Stenzel, U., Kircher, M., & Pääbo, S. (2010). A draft sequence of the Neandertal genome. *Science*, *328*(5979), 710–722. https://doi.org/10.1126/science.1188021.

2.   Hendry, L. Who were the Neanderthals? *Natural History Museum*. Retrieved from https://www.nhm.ac.uk/discover/who-were-the-neanderthals.html.

3.   Hunt, C. O., Pomeroy, E., Reynolds, T., Tilby, E., & Barker, G. (2023). Shanidar et ses fleurs? Reflections on the palynology of the Neanderthal 'Flower Burial' hypothesis. *Journal of Archaeological Science*, 105822.

4.   Spikins, P., Needham, A., Wright, B., Dytham, C., Gatta, M., & Hitchens, G. (2019). Living to fight another day: the ecological and evolutionary significance of Neanderthal healthcare. *Quaternary Science Reviews*, *217*, 98-118.

5.   García-Diez, Marcos. (2022). 'Art': Neanderthal symbolic graphic behaviour. In F. Romagnoli, F. Rivals, S. Benazzi (Eds), *Updating Neanderthals*, 251–250. Academic Press. https://doi.org/10.1016/B978-0-12-821428-2.00009-3.

6.   Greenbaum, G., Getz, W. M., Rosenberg, N. A., Feldman, M. W., Hovers, E., & Kolodny, O. (2019). Disease transmission and introgression can explain the long-lasting contact zone of modern humans and Neanderthals. *Nature Communications*, *10*(1), 1-12.

7.   Butler, T. (2005). Demise of passenger pigeon linked to Lyme disease. *Mongabay*. Retrieved from https://news.mongabay.com/2005/11/demise-of-passenger-pigeon-linked-to-lyme-disease/.

8.   McGill, I. and Jones, M. (2019). Cattle infectivity is driving the bTB epidemic. *Veterinary Record*, *185*(22), 699-700.

9.   Woodroffe, R., Donnelly, C. A., Ham, C., Jackson, S. Y., Moyes, K., Chapman, K., ... & Cartwright, S. J. (2016). Badgers prefer cattle pasture but avoid cattle: implications for bovine tuberculosis control. *Ecology Letters*, *19*(10), 1201-1208.

10.   Tackling Bovine TB Together: Towards Sustainable, Scientific and Effective bTB Solutions. (2024). *Badger Trust*. Retrieved from https://www.badgertrust. org.uk/_files/ugd/15d030_3d8b1a74ac874a1d8b9abe959179b333.pdf.

11.   Macdonald, D. (2023). A Commentary on Current Policy, A preamble to Badger Trust's report 'Tackling Bovine TB Together: Towards Sustainable, Scientific and Effective bTB Solutions'. *Badger Trust*. Retrieved from https:// www.badgertrust.org.uk/_files/ugd/15d030_704f2b8094654fda99d30d8ced30 96b6.pdf.

12.   Ibid.

13.   Schuenemann, V. J., Avanzi, C., Krause-Kyora, B., Seitz, A., Herbig, A., Inskip,S.,... & Krause, J. (2018). Ancient genomes reveal a high diversity of *Mycobacterium leprae* in medieval Europe. *PLoS Pathogens*, *14*(5), e1006997. https:// doi.org/10.1371/journal.ppat.1006997.

14.   Rasmussen, S., Allentoft, M. E., Nielsen, K., Orlando, L., Sikora, M., Sjögren, K. G., ... & Willerslev, E. (2015). Early divergent strains of *Yersinia pestis* in Eurasia 5,000 years ago. *Cell*, *163*(3), 571-582.

15.   Ferreira, R. C., Alves, G. V., Ramon, M., Antoneli, F., & Briones, M. R. (2024). Reconstructing prehistoric viral genomes from Neanderthal sequencing data. *Viruses*, *16*(6), 856.

16.   Kennedy, J. (2024, May 30). Scientists have discovered a 50,000-year-old herpes virus – and perhaps how modern humans came to rule the world, *The Guardian*. Retrieved from https://www.theguardian.com/commentisfree/ article/2024/may/30/50000-year-old-herpes-virus-humans-dna-homo-sapiens-n eanderthals.

17.   Greenbaum, G., Getz, W. M., Rosenberg, N. A., Feldman, M. W., Hovers, E., & Kolodny, O. (2019). Disease transmission and introgression can explain the long-lasting contact zone of modern humans and Neanderthals. *Nature Communications*, *10*(1), 1-12.

18.   Green, R. E., Krause, J., Briggs, A. W., Maricic, T., Stenzel, U., Kircher, M., ... & Pääbo, S. (2010). A draft sequence of the Neandertal genome. *Science*, *328*(5979), 710–722. https://doi.org/10.1126/science.1188021.

19.   Dannemann, M., & Kelso, J. (2017). The contribution of Neanderthals to phenotypic variation in modern humans. *The American Journal of Human Genetics*, *101*(4), 578-589.

20.   Ibid.

21.   Zeberg, H., Pääbo, S. (2020). The major genetic risk factor for severe COVID-19 is inherited from Neanderthals. *Nature*, *587*, 610–612. https://doi. org/10.1038/s41586-020-2818-3.

22.   Zhou, S., Butler-Laporte, G., Nakanishi, T., Morrison, D. R., Afilalo, J., Afilalo, M., ... & Richards, J. B. (2021). A Neanderthal OAS1 isoform protects individuals of European ancestry against COVID-19 susceptibility and severity. *Nature Medicine*, *27*(4), 659-667.

23.   Arppe, L., Karhu, J. A., Vartanyan, S., Drucker, D. G., Etu-Sihvola, H., & Bocherens, H. (2019). Thriving or surviving? The isotopic record of the Wrangel Island woolly mammoth population. *Quaternary Science Reviews*, *222*, 105884.

24.   MacPhee, R. D. (2018). *End of the megafauna: the fate of the world's hugest, fiercest, and strangest animals*. WW Norton & Company.

25.   Ibid.

26.   MacPhee, R.D.E. & Marx, P.A. (1997). Humans, hyperdisease, and first-contact extinctions. In S. M. Goodman & B. D. Patterson (Eds.): *Natural Change and Human Impact in Madagascar*. Smithsonian Institution, pp169–217.

27.   McFarlane, D. A., MacPhee, R. D., & Ford, D. C. (1998). Body size variability and a Sangamonian extinction model for *Amblyrhiza*, a West Indian megafaunal rodent. *Quaternary Research*, *50*(1), 80-89.

28.   Wyatt, K. B., Campos, P. F., Gilbert, M. T. P., Kolokotronis, S. O., Hynes, W. H., DeSalle, R., ... & Greenwood, A. D. (2008). Historical mammal extinction on Christmas Island (Indian Ocean) correlates with introduced infectious disease. *PloS One*, *3*(11), e3602.

29.   Murchie, T. J., Monteath, A. J., Mahony, M. E., Long, G. S., Cocker, S., Sadoway, T., ... & Poinar, H. N. (2021). Collapse of the mammoth-steppe in central Yukon as revealed by ancient environmental DNA. *Nature Communications*, *12*(1), 7120.

30.   Nickell, Z. D., & Moran, M. D. (2017). Disease introduction by aboriginal humans in North America and the pleistocene extinction. *Journal of Ecological Anthropology*, *19*(1), 29-41.

31.   Brazier, R.E., Elliott, M., Andison, E., Auster, R.E., Bridgewater, S., Burgess, P., & Vowles, A. (2020). The River Otter Beaver Trial Science and Evidence Report. *The River Otter Beaver Trial (ROBT) Science and Evidence Forum*. Retrieved from https://www.devonwildlifetrust.org/what-we-do/our-projects/river-otter-beaver-trial.

32.   Doughty, C. E., Prys-Jones, T. O., Faurby, S., Abraham, A. J., Hepp, C., Leshyk, V., ... & Galetti, M. (2020). Megafauna decline have reduced pathogen dispersal which may have increased emergent infectious diseases. *Ecography*, *43*(8), 1107-1117.

33.   Ibid.

34.   Ibid.

35.   Goheen, J. R., Augustine, D. J., Veblen, K. E., Kimuyu, D. M., Palmer, T. M., Porensky, L. M., ... & Young, T. P. (2018). Conservation lessons from

large-mammal manipulations in East African savannas: the KLEE, UHURU, and GLADE experiments. *Annals of the New York Academy of Sciences*, *1429*(1), 31-49.

36.   Doughty, C. E., Wolf, A., & Field, C. B. (2010). Biophysical feedbacks between the Pleistocene megafauna extinction and climate: the first human-induced global warming?. *Geophysical Research Letters*, *37*(15).

37.   Richmond, D. J., Sinding, M. H. S., & Gilbert, M. T. P. (2016). The potential and pitfalls of de-extinction. *Zoologica Scripta*, *45*, 22-36.

3. THE CANARY IN THE HAWAIIAN CHAIN

1.   Huelani. Sweet Lei Mamo (Mamo is the safflower or false saffron from Asia). *HUAPALA Hawaiian Music and Hula Archives*, Retrieved from https://www.huapala.org/SW/Sweet_Lei_Mamo.html.

2.   Native Birds of Hawaii. Hawaii Department of Land and Natural Resources. Retrieved from https://dlnr.hawaii.gov/wildlife/birds/.

3.   Discovery Channel South East Asia. (2015, November 26). Male Singing to Female that Will Never Come – Racing Extinction. [YouTube video]. Retrieved from https://www.youtube.com/watch?v=5THqAY3u50Y.

4.   Shallenberger, R. (1975). Recording of Kaua'i 'ō'ō at Alakai Swamp, duet from 4:18 onwards. *CornellLab Macaulay Library*. Retrieved from https://macaulaylibrary.org/asset/6031.

5.   Native Birds of Hawaii. Hawaii Department of Land and Natural Resources. Retrieved from https://dlnr.hawaii.gov/wildlife/birds/.

6.   BirdLife International. (2019). *Melamprosops phaeosoma. The IUCN Red List of Threatened Species* 2019: e.T22720863A153774712. Retrieved from https://dx.doi.org/10.2305/IUCN.UK.2019-3.RLTS.T22720863A153774712.en.

7.   Navine, A.K., Paxton, K.L., Paxton, E.H., Hart, P.J., Foster, J.T., McInerney, N., Fleischer, R.C., & Videvall, E. (2023). Microbiomes associated with avian malaria survival differ between susceptible Hawaiian honeycreepers and sympatric malaria-resistant introduced birds. *Molecular Ecology*, *32*, 6659–6670. https://doi.org/10.1111/mec.16743.

8.   Pratt, T.K., Atkinson, C.T., Banko, P.C, Jacobi, J.D. and Woodworth, B.L. (2009). *Conservation Biology of Hawaiian Forest Birds: Implications for Island Avifauna*, New Haven & London: Yale University Press, p38.

9.   Pratt, H. D. (2002). *The Hawaiian Honeycreepers*. Oxford: Oxford University Press, p46.

10.   Pratt, T.K., Atkinson, C.T., Banko, P.C, Jacobi, J.D. and Woodworth, B.L. (2009). *Conservation Biology of Hawaiian Forest Birds: Implications for Island Avifauna*, New Haven & London: Yale University Press, p21.

11.   Ibid., p62.

12.   Native Birds of Hawaii. Hawaii Department of Land and Natural Resources. Retrieved from https://dlnr.hawaii.gov/wildlife/birds/.

13.   Pratt, T.K., Atkinson, C.T., Banko, P.C, Jacobi, J.D. and Woodworth, B.L. (2009). *Conservation Biology of Hawaiian Forest Birds: Implications for Island Avifauna*, New Haven & London: Yale University Press, p63.

14.   Ibid., p65.

15.   Herman, D. (2020). Shutting Down Hawai'i: A Historical Perspective on Epidemics in the Islands. *Smithsonian Magazine*. Retrieved from https://www. smithsonianmag.com/history/shutting-down-hawaii-historical-perspective-epide mics-islands-180974506/.

16.   Stannard, D.E. (1989). *Before the Horror: The Population of Hawaii on the Eve of Western Contact*, Hawaii: University of Hawaii Press.

17.   Pratt, T.K., Atkinson, C.T., Banko, P.C, Jacobi, J.D. and Woodworth, B.L. (2009). *Conservation Biology of Hawaiian Forest Birds: Implications for Island Avifauna*, New Haven & London: Yale University Press, p312.

18.   Warner, R.E. (1968). The Role of Introduced Diseases in the Extinction of the Endemic Hawaiian Avifauna, *The Condor*, 70, 2, 101-120. https://doi. org/10.2307/1365954.

19.   Pratt, T.K., Atkinson, C.T., Banko, P.C, Jacobi, J.D. and Woodworth, B.L. (2009). *Conservation Biology of Hawaiian Forest Birds: Implications for Island Avifauna*, New Haven & London: Yale University Press, p312.

20.   Ibid., p234.

21.   Dennis LaPointe, personal communication.

22.   van Riper III, C., van Riper, S. G., & Hansen, W. R. (2002). Epizootiology and effect of avian pox on Hawaiian forest birds. *The Auk*, *119*(4), 929-942.https:// doi.org/10.1093/auk/119.4.929.

23.   Ibid.

24.   Pratt, T.K., Atkinson, C.T., Banko, P.C, Jacobi, J.D. and Woodworth, B.L. (2009). *Conservation Biology of Hawaiian Forest Birds: Implications for Island Avifauna*, New Haven & London: Yale University Press, p234.

25.   Ibid., p237.

26.   Warner, R.E. (1968). The Role of Introduced Diseases in the Extinction of the Endemic Hawaiian Avifauna, *The Condor*, 70(2), 101-120. https://doi. org/10.2307/1365954.

27.   LaPointe, D. (2012). *Culex quinquefasciatus* (southern house mosquito). *CABI Compendium*. Retrieved from https://doi.org/10.1079/cabicompendium.86848.

28.   Warner, R. E. (1968). The Role of Introduced Diseases in the Extinction of the Endemic Hawaiian Avifauna, *The Condor*, 70(2), 101-120. https://doi. org/10.2307/1365954.

29.   Avian Pox. *Cornell* Wildlife Health Lab. Retrieved from https://cwhl.vet. cornell.edu/disease/avian-pox.

30. van Riper III, C., van Riper, S. G., & Hansen, W. R. (2002). Epizootiology and effect of avian pox on Hawaiian forest birds. *The Auk, 119*(4), 929-942. https://doi.org/10.1093/auk/119.4.929.

31. Dennis LaPointe, personal communication.

32. Pratt, T.K., Atkinson, C.T., Banko, P.C, Jacobi, J.D. and Woodworth, B.L. (2009). *Conservation Biology of Hawaiian Forest Birds: Implications for Island Avifauna*, New Haven & London: Yale University Press, p318 and 328.

33. Warner, R.E. (1968). The Role of Introduced Diseases in the Extinction of the Endemic Hawaiian Avifauna, *The Condor, 70*(2), 101-120. https://doi.org/10.2307/1365954.

34. Martínez-de la Puente, J., Santiago-Alarcon, D., Palinauskas, V., & Bensch, S. (2021). Plasmodium relictum, Parasite of the month, *Trends in Parasitology, 37*(4), 355-356.

35. Atkinson, C. (2009). *Plasmodium relictum. CABI Compendium*, Retrieved from https://doi.org/10.1079/cabicompendium.69051.

36. State of Hawaii Department of Health. Malaria. *State of Hawaii Department of Health, Disease Outbreak Control Division*. Retrieved from https://health.hawaii.gov/docd/disease_listing/malaria/.

37. Beadell, J. S., Ishtiaq, F., Covas, R., Melo, M., Warren, B. H., Atkinson, C. T., ... & Fleischer, R. C. (2006). Global phylogeographic limits of Hawaii's avian malaria. *Proceedings of the Royal Society B: Biological Sciences, 273*(1604), 2935-2944, quoted in Pratt, T.K., Atkinson, C.T., Banko, P.C, Jacobi, J.D. and Woodworth, B.L. (2009). *Conservation Biology of Hawaiian Forest Birds: Implications for Island Avifauna*, New Haven & London: Yale University Press, p237.

38. Atkinson, C. (2009). *Plasmodium relictum. CABI Compendium*, Retrieved from https://doi.org/10.1079/cabicompendium.69051.

39. Pratt, T.K., Atkinson, C.T., Banko, P.C, Jacobi, J.D. and Woodworth, B.L. (2009). *Conservation Biology of Hawaiian Forest Birds: Implications for Island Avifauna*, New Haven & London: Yale University Press, p234.

40. Atkinson, C. (2009). *Plasmodium relictum. CABI Compendium*, Retrieved from https://doi.org/10.1079/cabicompendium.69051.

41. Lowe S., Browne M., Boudjelas S., De Poorter M. (2000) *100 of the World's Worst Invasive Alien Species A selection from the Global Invasive Species Database*. Published by The Invasive Species Specialist Group (ISSG) a specialist group of the Species Survival Commission (SSC) of the World Conservation Union (IUCN), 12pp. Updated and reprinted version: November 2004. https://portals.iucn.org/library/sites/library/files/documents/2000-126.pdf.

42. Pratt, T.K., Atkinson, C.T., Banko, P.C, Jacobi, J.D. and Woodworth, B.L. (2009). *Conservation Biology of Hawaiian Forest Birds: Implications for Island Avifauna*, New Haven & London: Yale University Press, p234.

43. Ibid.

44. Benning, T. L., LaPointe, D., Atkinson, C. T., & Vitousek, P. M. (2002). Interactions of climate change with biological invasions and land use in the Hawaiian Islands: modeling the fate of endemic birds using a geographic information system. *Proceedings of the National Academy of Sciences*, *99*(22), 14246-14249.

45. Atkinson, C. T., Utzurrum, R. B., Lapointe, D. A., Camp, R. J., Crampton, L. H., Foster, J. T., & Giambelluca, T. W. (2014). Changing climate and the altitudinal range of avian malaria in the Hawaiian Islands – an ongoing conservation crisis on the island of Kaua'i. *Global Change Biology*, *20*(8), 2426-2436.https://doi.org/10.1111/gcb.12535.

46. Grandoni, D. (2023). A tiny Hawaiian bird was nearing extinction. Then the Maui fires came. *Washington Post*. Retrieved from www.msn.com/en-us/news/us/a-tiny-hawaiian-bird-was-nearing-extinction-then-the-maui-fires-came/ar-AA1fyz5u.

47. BirdLife International. Akekee. *BirdLife International Data Zone*. Retrieved from http://datazone.birdlife.org/species/factsheet/akekee-loxops-caeruleirostris/text.

48. Dadam, D., Robinson, R. A., Clements, A., Peach, W. J., Bennett, M., Rowcliffe, J. M., & Cunningham, A. A. (2019). Avian malaria-mediated population decline of a widespread iconic bird species. *Royal Society Open Science*, *6*(7), 182197. https://doi.org/10.1098/rsos.182197.

49. Martinez, D., Duran, E. M., & Bauer, N. (2020). Saving 'Ōhi 'a: a case study on the influence of human behavior on ecological degradation through an examination of rapid 'Ōhi 'a death and its impacts on the Hawaiian Islands. *Case Studies in the Environment*, *4*(1), 1104118.

50. Woodworth, B. L., Atkinson, C. T., LaPointe, D. A., Hart, P. J., Spiegel, C. S., Tweed, E. J., ... & Duffy, D. (2005). Host population persistence in the face of introduced vector-borne diseases: Hawaii amakihi and avian malaria. *Proceedings of the National Academy of Sciences*, *102*(5), 1531-1536. https://doi.org/10.1073/pnas.0409454102.

51. Atkinson, C. T., Saili, K. S., Utzurrum, R. B., & Jarvi, S. I. (2013). Experimental evidence for evolved tolerance to avian malaria in a wild population of low elevation Hawai 'i 'Amakihi (*Hemignathus virens*). *EcoHealth*, *10*, 366-375. https://doi.org/10.1007/s10393-013-0899-2.

52. Camp, R. J., LaPointe, D. A., Hart, P. J., Sedgwick, D. E., & Canale, L. K. (2019). Large-scale tree mortality from Rapid Ohia Death negatively influences avifauna in lower Puna, Hawaii Island, USA. *The Condor*, *121*(2), duz007. https://doi.org/10.1093/condor/duz007.

53. Restoring biodiversity. *Pacific Rim Conservation*. Retrieved from https://pacificrimconservation.org.

54. Kaua'i Forest Bird Recovery Project. *Kaua'i Forest Bird Recovery Project*. Retrieved from https://kauaiforestbirds.org.

55. Birds, Not Mosquitoes – So the Forest Birds Thrive. *Birds, Not Mosquitoes.* Retrieved from https://www.birdsnotmosquitoes.org/.

56. Weston, P. (2024, June 21). Millions of mosquitoes released in Hawaii to save rare birds from extinction. *The Guardian.* Retrieved from https://www.theguardian.com/environment/article/2024/jun/21/mosquitoes-hawaii-rare-bird-honeycreeper-malaria-wolbachia-bacteria.

57. LaPointe, D. A., Hofmeister, E. K., Atkinson, C. T., Porter, R. E., & Dusek, R. J. (2009). Experimental infection of Hawaii amakihi (*Hemignathus virens*) with West Nile Virus and competence of a co-occurring vector, *Culex quinquefasciatus*: Potential impacts on endemic Hawaiian avifauna. *Journal of Wildlife Diseases, 45*(2), 257-271.

4. MEASLY MIGRATION

1. Cerulli, E. (1923). Una raccolta Amarica di canti funebri. *Rivista Degli Studi Orientali, 10*(2/4), 265–280. http://www.jstor.org/stable/41863602.

2. Spinage, C.A. (2003). *Cattle Plague: A History.* Springer.

3. Ibid.

4. Düx, A., Lequime, S., Patrono, L. V., Vrancken, B., Boral, S., Gogarten, J. F., … & Calvignac-Spencer, S. (2020). Measles virus and rinderpest virus divergence dated to the sixth century BCE. *Science, 368*(6497), 1367–1370. https://doi.org/10.1126/science.aba9411.

5. Ibid.

6. Ibid.

7. Spinage, C.A. (2003). *Cattle Plague: A History.* Springer.

8. Scott, G. R., The Murrain Now Known As Rinderpest, *Newsletter of the Tropical Agriculture Association,* UK, *20* (4) 14-16 (2000). https://www.queensu.ca/academia/forsdyke/rindpsto.htm.

9. Pastoret, P. P. (2006). Rinderpest: A general introduction. In Thomas Barrett, Paul-Pierre Pastoret, William P. Taylor (Eds), *Rinderpest and Peste des Petits Ruminants.* Academic Press. https://doi.org/10.1016/B978-012088385-1/50031-9.

10. FAO. A centuries old challenge in Rome. Retrieved from https://www.fao.org/fileadmin/user_upload/newsroom/docs/Lancisi.pdf.

11. WOAH. Eradicating rinderpest: moments in time. Retrieved from https://www.woah.org/app/uploads/2021/06/rinder-timeline-final.pdf.

12. Pastoret, P. P. (2006). Rinderpest: A general introduction. In Thomas Barrett, Paul-Pierre Pastoret, William P. Taylor (Eds), *Rinderpest and Peste des Petits Ruminants.* Academic Press. https://doi.org/10.1016/B978-012088385-1/50031-9.

13. Thaddeus Sunseri, personal communication.

14. McVety, A. K. (2018). *The rinderpest campaigns: a virus, its vaccines, and global development in the twentieth century.* Cambridge University Press.

15.　WOAH. Eradicating rinderpest: moments in time. Retrieved from https://www.woah.org/app/uploads/2021/06/rinder-timeline-final.pdf.

16.　Sunseri, T. (2018). The African Rinderpest Panzootic, 1888–1897. In *Oxford Research Encyclopedia of African History*. Retrieved from https://oxfordre.com/africanhistory/view/10.1093/acrefore/9780190277734.001.0001/acrefore-9780190277734-e-375.

17.　Spinage, C.A. (2003). *Cattle Plague: A History*. Springer.

18.　WOAH. Rinderpest. Retrieved from https://www.woah.org/en/disease/rinderpest/

19.　Ibid.

20.　Spinage, C.A. (2003). *Cattle Plague: A History*. Springer.

21.　Sunseri, T. (2018). The African Rinderpest Panzootic, 1888–1897. *Oxford Research Encyclopedia of African History*. Retrieved from https://oxfordre.com/africanhistory/view/10.1093/acrefore/9780190277734.001.0001/acrefore-9780190277734-e-375.

22.　Ibid.

23.　Rowe, J. A., & Hødnebø, K. (1994). Rinderpest in the Sudan 1888-1890: the mystery of the missing panzootic. *Sudanic Africa, 5*, 149–178. http://www.jstor.org/stable/25653249.

24.　Normile, D. (2008, March 21). Driven to Extinction, *Science, 319*, 5870. https://doi.org/10.1126/science.319.5870.1606.

25.　Sunseri, T. (2018, April 26). The African Rinderpest Panzootic, 1888–1897. *Oxford Research Encyclopedia of African History*. https://oxfordre.com/africanhistory/view/10.1093/acrefore/9780190277734.001.0001/acrefore-9780190277734-e-375.

26.　Ibid.

27.　Spinage, C.A. (2003). *Cattle Plague: A History*. Springer.

28.　Ibid.

29.　Tiki, W., & Oba, G. (2009). Ciinna – The Borana Oromo narration of the 1890s great rinderpest epizootic in north eastern Africa. *Journal of Eastern African Studies, 3*(3), 479-508.

30.　Sunseri, T. (2018, April 26). The African Rinderpest Panzootic, 1888–1897. *Oxford Research Encyclopedia of African History*. https://oxfordre.com/africanhistory/view/10.1093/acrefore/9780190277734.001.0001/acrefore-9780190277734-e-375.

31.　Ibid.

32.　Ibid.

33.　Ibid.

34.　Spinage, C.A. (2003). *Cattle Plague: A History*. Springer.

35.　Barfield, D. (2020, April 17). Royal Veterinary College Podcast Episode 60: Prof Richard Kock. *Royal Veterinary College*. Retrieved from https://www.rvc.ac.uk/research/podcasts/60-prof-richard-kock.

36.   Baird, J.L. (1900). *Foreign Office* 1/39, 20 May 1900, via Spinage, C.A. (2003). *Cattle Plague: A History*. Springer.

37.   IUCN SSC Antelope Specialist Group. (2017). *Tragelaphus derbianus* ssp. *derbianus*. *The IUCN Red List of Threatened Species 2017*: e.T22056A50197188. https://dx.doi.org/10.2305/IUCN.UK.2017-2.RLTS.T22056A50197188.en.

38.   East, R. (1990). *Antelopes: global survey and regional action plans. Part 3: West and Central Africa*. Gland: IUCN. ISBN 2-8317-0016-7. Retrieved via https://portals.iucn.org/library/node/8976.

39.   Sunseri, T. (2018). The African Rinderpest Panzootic, 1888–1897. *Oxford Research Encyclopedia of African History*. https://oxfordre.com/africanhistory/view/10.1093/acrefore/9780190277734.001.0001/acrefore-9780190277734-e-375.

40.   Spinage, C.A. (2003). *Cattle Plague: A History*. Springer.

41.   Richard Kock, personal communication.

42.   Sunseri, T. (2018). The African Rinderpest Panzootic, 1888–1897. *Oxford Research Encyclopedia of African History*. https://oxfordre.com/africanhistory/view/10.1093/acrefore/9780190277734.001.0001/acrefore-9780190277734-e-375.

43.   Richard Kock, personal communication.

44.   Thaddeus Sunseri, personal communication.

45.   Sunseri, T. (2018). The African Rinderpest Panzootic, 1888–1897. *Oxford Research Encyclopedia of African History*. https://oxfordre.com/africanhistory/view/10.1093/acrefore/9780190277734.001.0001/acrefore-9780190277734-e-375.

46.   Schmitz, O. J., Wilmers, C. C., Leroux, S. J., Doughty, C. E., Atwood, T. B., Galetti, M., ... & Goetz, S. J. (2018). Animals and the zoogeochemistry of the carbon cycle. *Science, 362*(6419), eaar3213. https://doi.org/10.1126/science.aar3213.

47.   Sunseri, T. (2018). The African Rinderpest Panzootic, 1888–1897. *Oxford Research Encyclopedia of African History*. https://oxfordre.com/africanhistory/view/10.1093/acrefore/9780190277734.001.0001/acrefore-9780190277734-e-375.

48.   Thaddeus Sunseri, personal communication.

49.   Ibid.

50.   Spinage, C.A. (2003). *Cattle Plague: A History*. Springer.

51.   Ibid.

52.   Ibid.

53.   Thaddeus Sunseri, personal communication.

54.   Sunseri, T. (2018). The African Rinderpest Panzootic, 1888–1897. *Oxford Research Encyclopedia of African History*. https://oxfordre.com/africanhistory/view/10.1093/acrefore/9780190277734.001.0001/acrefore-9780190277734-e-375.

55.   Ibid.

56.   Talbot, L. M., & Talbot, M. H. (1963). The Wildebeest in Western Masailand, East Africa. *Wildlife Monographs, 12*, 3–88. http://www.jstor.org/stable/3830455.

57. Ibid.

58. Richard Kock, personal communication.

59. Ibid.

60. Tounkara, K., & Nwankpa, N. (2017). Rinderpest experience. *Revue Scientifique et Technique, 36*(2), 569-578.

61. Roeder, P., Mariner, J., & Kock, R. (2013). Rinderpest: the veterinary perspective on eradication. *Philosophical Transactions of the Royal Society of London. Series B, Biological Sciences, 368*(1623), 20120139. https://doi.org/10.1098/rstb.2012.0139.

62. IUCN SSC Antelope Specialist Group. (2017). *Tragelaphus derbianus* ssp. *derbianus. The IUCN Red List of Threatened Species 2017*: e.T22056A50197188. https://dx.doi.org/10.2305/IUCN.UK.2017-2.RLTS.T22056A50197188.en.

63. Kock, R. A., Wambua, J. M., Mwanzia, J., Wamwayi, H., Ndungu, E. K., Barrett, T., ... & Rossiter, P. B. (1999). Rinderpest epidemic in wild ruminants in Kenya 1993-97. *Veterinary Record, 145*(10), 275-283.

64. Spinage, C.A. (2003). *Cattle Plague: A History*. Springer.

65. Ibid.

66. Ibid.

67. Kock, R. A., Wamwayi, H. M., Rossiter, P. B., Libeau, G., Wambwa, E., Okori, J., ... & Mlengeya, T. D. (2006). Re-infection of wildlife populations with rinderpest virus on the periphery of the Somali ecosystem in East Africa. *Preventive Veterinary Medicine, 75*(1-2), 63-80.

https://doi.org/10.1016/j.prevetmed.2006.01.016.

68. Myers, L., Metwally, S., Marrana, M., Stoffel, C., Ismayilova, G. & Brand, T. (2024.) Global Rinderpest Action Plan – Post-eradication, Second edition. Rome, FAO and WOAH. https://doi.org/10.4060/cc9269en.

69. BBC News. Largest world stock of animal-killing virus destroyed by UK lab. (June 2019). Retrieved from https://www.bbc.co.uk/news/science-environment-48629469.

70. Düx, A., Lequime, S., Patrono, L. V., Vrancken, B., Boral, S., Gogarten, ... & Calvignac-Spencer, S. (2020). Measles virus and rinderpest virus divergence dated to the sixth century BCE. *Science, 368*(6497), 1367–1370. https://doi.org/10.1126/science.aba9411.

71. Barfield, D. (2020, April 17). Royal Veterinary College Podcast Episode 60: Prof Richard Kock. *Royal Veterinary College*. Retrieved from https://www.rvc.ac.uk/research/podcasts/60-prof-richard-kock.

72. IUCN SSC Antelope Specialist Group. (2023). *Saiga tatarica. The IUCN Red List of Threatened Species 2023*: e.T19832A233712210. https://dx.doi.org/10.2305/IUCN.UK.2023-1.RLTS.T19832A233712210.en.

73. Ibid.

74. Fereidouni, S., Freimanis, G. L., Orynbayev, M., Ribeca, P., Flannery, J., King, D. P., ... & Kock, R. (2019). Mass die-off of Saiga antelopes,

Kazakhstan, 2015. *Emerging Infectious Diseases*, *25*(6), 1169. https://doi.org/10.3201/
eid2506.180990.

75. Kock, R.A., Orynbayev, M.B., Sultankulova, K.T., Strochkov, V.M.,
Omarova, Z.D., Shalgynbayev, E.K., ...Parida, S. (2015), Detection and Genetic
Characterization of Lineage IV Peste Des Petits Ruminant Virus in Kazakhstan.
Transboundary and Emerging Diseases, *62*: 470-479. https://doi.org/10.1111/
tbed.12398.

76. Pruvot, M., Fine, A. E., Hollinger, C., Strindberg, S., Damdinjav, B.,
Buuveibaatar, B....Shiilegdamba, E. (2020). Outbreak of Peste des Petits
Ruminants among Critically Endangered Mongolian Saiga and Other Wild
Ungulates, Mongolia, 2016–2017. *Emerging Infectious Diseases*, *26*(1), 51-62. https://
doi.org/10.3201/eid2601.181998.

77. Richard Kock, personal communication.

78. IUCN SSC Antelope Specialist Group. (2018). *Saiga tatarica*. *The IUCN Red
List of Threatened Species* 2018: e.T19832A50194357. https://dx.doi.org/10.2305/
IUCN.UK.2018-2.RLTS.T19832A50194357.en.

79. IUCN SSC Antelope Specialist Group. (2023). *Saiga tatarica*. *The IUCN Red
List of Threatened Species* 2023: e.T19832A233712210. https://dx.doi.org/10.2305/
IUCN.UK.2023-1.RLTS.T19832A233712210.en.

80. Office International des Epizooties. (2014). Final report of the 82nd General
Session, May 2014: resolution 24, pp 164–166. OIE, Paris. https://www.woah.org/
app/uploads/2021/03/a-fr-2014-public.pdf.

81. FAO and OIE. (2016). PPR Global Eradication Programme. Retrieved from
https://openknowledge.fao.org/items/03711496-4bad-4219-ab57-9b5244e976a5.

82. Ibid.

5. TASMANIAN TROUBLES

1. The thylacine in captivity: the historical thylacine films – film 5, *The Thylacine
Museum*. Retrieved from http://www.naturalworlds.org/thylacine/captivity/films/
flv/film_5.htm.

2. Knight, C.R. (1903). The Tasmanian wolf I. *The Century Illustrated Monthly
Magazine*, *66* (New series 44), 112-114. https://archive.org/details/sim_centur
y-illustrated-monthly-magazine_1903-05_66_1/page/112/mode/2up?q=tiger.

3. Beniuk, D. (2014, May 8). Last thylacine bit cameraman on buttocks,
Australian Geographic. Retrieved from https://www.australiangeographic.com.au/
news/2014/05/last-thylacine-bit-cameraman-on-buttocks/.

4. Owen, D. (2003). *The Tragic Tale of the Tasmanian Tiger*, Allen & Unwin.

5. Tasmanian Tiger. *Wildlife Management*, Department of Natural Resources
and Environment Tasmania. Retrieved from https://nre.tas.gov.au/
wildlife-management/fauna-of-tasmania/mammals/carnivorous-marsupials-an
d-bandicoots/tasmanian-tiger.

6.   Paddle, R. (2000). *The Last Tasmanian Tiger: The History and Extinction of the Thylacine*. Cambridge, UK: Cambridge University Press.

7.   The thylacine in art: aboriginal rock art. The *Thylacine Museum*. Retrieved from http://www.naturalworlds.org/thylacine/art/rockart/image_02.htm.

8.   Ibid.

9.   Paddle, R. (2012). The thylacine's last straw: epidemic disease in a recent mammalian extinction. *Australian Zoologist*, 36(1), 75-92. https://doi.org/10.7882/AZ.2012.008.

10.   Owen, D. (2003). *The Tragic Tale of the Tasmanian Tiger*, Allen & Unwin.

11.   Robert Paddle, personal communication.

12.   Owen, D. (2003). *The Tragic Tale of the Tasmanian Tiger*, Allen & Unwin.

13.   Paddle, R. (2012). The thylacine's last straw: epidemic disease in a recent mammalian extinction. *Australian Zoologist*, 36(1), 75-92. https://doi.org/10.7882/AZ.2012.008.

14.   Robert Paddle, personal communication, unpublished figures.

15.   Paddle, R. (2012). The thylacine's last straw: epidemic disease in a recent mammalian extinction. *Australian Zoologist*, 36(1), 75-92. https://doi.org/10.7882/AZ.2012.008.

16.   Ibid.

17.   Ibid.

18.   Ibid.

19.   Ibid.

20.   Ibid.

21.   Guiler, E.R. (1958, September 15). The thylacine, *The Australian Museum Magazine*, XII (11) 352. Retrieved from https://media.australian.museum/media/dd/Uploads/Documents/30370/AMS_V12-11_lowres.4dff372.pdf.

22.   Paddle, R. (2012). The thylacine's last straw: epidemic disease in a recent mammalian extinction. *Australian Zoologist*, 36(1), 75-92. https://doi.org/10.7882/AZ.2012.008.

23.   Ibid.

24.   Paddle, R. (2014). Scientists and the Construction of the Thylacine's Extinction, chapter in Lang, R. (Ed.) (2014). *The Tasmanian Tiger: Extinct or Extant*. Sydney: Strange Nation Publishing.

25.   McCallum, H., Foufopoulos, J., Grogan, L.F. (2024). Infectious disease as a driver of declines and extinctions. *Cambridge Prisms: Extinction*, 2 e2. https://dx.doi.org/10.1017/ext.2024.1.

26.   Prowse, T.A.A., Johnson, C.N., Lacy, R.C., Bradshaw, C.J.A., Pollak, J.P., Watts, M.J. & Brook, B.W. (2013). No need for disease: testing extinction hypotheses for the thylacine using multi-species metamodels. *Journal of Animal Ecology*, 82, 355-364. https://doi.org/10.1111/1365-2656.12029.

27.   Bob Paddle, personal communication.

28.   Paddle, R. (2012). The thylacine's last straw: epidemic disease in a recent mammalian extinction. *Australian Zoologist*, 36(1), 75-92. https://doi.org/10.7882/AZ.2012.008.

29.   Ibid.

30.   Robert Paddle, personal communication.

31.   Ibid.

32.   Paddle, R. (2000). *The Last Tasmanian Tiger: The History and Extinction of the Thylacine*. Cambridge, UK: Cambridge University Press.

33.   Biography of Theodore Thomson Flynn. *University of Tasmania Library*. Retrieved from https://www.utas.edu.au/library/exhibitions/flynn_and_flynn/ttFlynn.html.

34.   Ibid.

35.   Paddle, R. (2000). *The Last Tasmanian Tiger: The History and Extinction of the Thylacine*. Cambridge, UK: Cambridge University Press.

36.   Owen, D. (2003). *The Tragic Tale of the Tasmanian Tiger*, Allen & Unwin.

37.   Ibid.

38.   Paddle, R. (2000). The Last Tasmanian Tiger: The History and Extinction of the Thylacine. Cambridge, UK: Cambridge University Press.

39.   Ibid.

40.   Paddle, R. N., & Medlock, K. M. (2023). The discovery of the remains of the last Tasmanian tiger (*Thylacinus cynocephalus*). *Australian Zoologist*, 43(1), 97-108. https://doi.org/10.7882/AZ.2023.017.

41.   Brook, B. W., Sleightholme, S. R., Campbell, C. R., Jarić, I., & Buettel, J. C. (2023). Resolving when (and where) the Thylacine went extinct. *Science of The Total Environment*, *877*, 162878. https://dx.doi.org/10.1016/j.scitotenv.2023.162878.

42.   Smith, S.J. (1981). The Tasmanian tiger – 1980: a report on an investigation of the current status of thylacine *Thylacinus cynocephalus*. Funded by the World Wildlife Fund Australia, published by Tasmania: Department of Primary Industries, Parks, Water and Environment. *TROVE*. Retrieved from https://nla.gov.au/nla.obj-2939261753/view.

43.   Burbidge, A.A. & Woinarski, J. (2016.) *Thylacinus cynocephalus*. The IUCN Red List of Threatened Species 2016, e.T21866A21949291. https://dx.doi.org/10.2305/IUCN.UK.2016-2.RLTS.T21866A21949291.en.

44.   Reserve Listing. Tasmania Parks and Wildlife Service. Retrieved from https://parks.tas.gov.au/about-us/managing-our-parks-and-reserves/reserve-listing.

45.   Met Office. (2012, October 16). Top ten coldest recorded temperatures in the UK. [Weblog]. Retrieved from https://blog.metoffice.gov.uk/2012/10/16/top-ten-coldest-recorded-temperatures-in-the-uk/.

46.   Weymouth, A. (2014, July 21). Was this the last wild wolf of Britain? *The Guardian*. Retrieved from https://www.theguardian.com/science/animal-magic/2014/jul/21/last-wolf.

47.   Owen, D. (2003). *The Tragic Tale of the Tasmanian Tiger*, Allen & Unwin.

48.   Ibid.

49.   Hamish McCallum, personal communication.

50.   Bob Paddle, personal communication.

51.   Ibid.

52.   Owen, D. (2003). *The Tragic Tale of the Tasmanian Tiger*, Allen & Unwin.

53.   Pask, A. De-extinction of the thylacine. Chapter in Holmes B., Linnard, G. (2023). *Thylacine: The History, Ecology and Loss of the Tasmanian Tiger*, CSIRO Publishing.

54.   Owen, D. (2003). *The Tragic Tale of the Tasmanian Tiger*, Allen & Unwin.

55.   Threatened Species List – Vertebrate Animals. Department of Natural Resources and Environment Tasmania. Retrieved from https://nre.tas.gov.au/conservation/threatened-species-and-communities/lists-of-threatened-species/threatened-species-vertebrates.

56.   Woinarski, J. & Burbidge, A.A. (2016.) *Perameles gunnii*. *The IUCN Red List of Threatened Species 2016*: e.T16572A21966027. Retrieved from https://dx.doi.org/10.2305/IUCN.UK.2016-2.RLTS.T16572A21966027.en.

57.   National Film and Sound Archive of Australia (NFSA). (2021, September 7) Tasmanian tiger: video footage of last-known thylacine remastered in 4K colour – video, *The Guardian*. N.B. The headline says it's the last-known animal but the elderly female that succeeded it was the last. Retrieved from https://www.theguardian.com/australia-news/video/2021/sep/07/video-footage-of-last-known-tasmanian-tiger-remastered-in-colour-video.

6. BLACK FEET AND BLACK DEATH

1.   Audubon, J.W. (1840/1846). Black-Footed Ferret. *National Gallery of Art*. Retrieved from https://www.nga.gov/collection/art-object-page.39764.html.

2.   Kimberly Fraser, personal communication.

3.   McClure, N. (2017, December). The Black-footed Ferret – Points West Online. Buffalo Bill Center of the West. Retrieved from https://centerofthewest.org/2017/12/01/points-west-black-footed-ferret/.

4.   Esch, K. L., Beauvais, D. G. P., & Keinath, D. A. (2005). *Species conservation assessment for black footed ferret* (Mustela nigripes) *in Wyoming*. US Department of the Interior Bureau of Land Management, Wyoming State Office, Cheyenne, WY.

5.   McClure, N. (2017, December). The Black-footed Ferret – Points West Online. Buffalo Bill Center of the West. Retrieved from https://centerofthewest.org/2017/12/01/points-west-black-footed-ferret/.

6.   Coues, E. (1877). *Fur-Bearing Animals: A Monograph of North American Mustelidae*. Government Printing Office, Washington, D.C. via McClure, N. (2017, December). The Black-footed Ferret – Points West Online. Buffalo Bill Center of the West. Retrieved from https://centerofthewest.org/2017/12/01/point s-west-black-footed-ferret/.

7.   Anderson, E., Forrest, S.C., Clark, T.W., & Richardson, L. (1986). Paleobiology, biogeography, and systematics of the black-footed ferret, *Mustela nigripes* (Audubon and Bachman), 1851. In: The Black-Footed Ferret, *Great Basin Naturalist Memoirs 8*, 11–62 via McClure, N. (2017, December). The Black-footed Ferret – Points West Online. Buffalo Bill Center of the West. Retrieved from https://centerofthewest.org/2017/12/01/points-west-black-footed-ferret/.

8.   Abbott, R.C. and Rocke, T.E. (2012). Plague, Circular 1372, USGS National Wildlife Health Center. Retrieved via https://pubs.usgs.gov/circ/1372/pdf/C1372_Plague.pdf.

9.   Madison, M. (2018, September 25). USFWS coordinator Pete Gober interview– Black footed ferret recovery, a conservation success story. [YouTube video]. Free Nature Interviews. Retrieved from https://www.youtube.com/watch?app=deskt op&v=427e8VoLcMg.

10.   Slobodchikoff, C.N., Paseka, A. & Verdolin, J.L. (2009). Prairie dog alarm calls encode labels about predator colors. *Animal Cognition, 12*, 435–439. https://doi.org/10.1007/s10071-008-0203-y.

11.   Slobodchikoff, C. (2012). *Chasing Dr Dolittle: Learning the Language of Animals*. St Martin's Press.

12.   Wade, S. 8 surprising prairie dog facts, WWF. Retrieved from https://www.worldwildlife.org/stories/8-surprising-prairie-dog-facts.

13.   McClure, N. (2017, December). The Black-footed Ferret – Points West Online. Buffalo Bill Center of the West. Retrieved from https://centerofthewest.org/2017/12/01/points-west-black-footed-ferret/.

14.   Miller, B., Reading, R.P., Forrest, S. (1996). *Prairie Night: Black-Footed Ferrets and the Recovery of Endangered Species*. Smithsonian Institution Press, p254 via McClure, N. (2017, December). The Black-footed Ferret – Points West Online. Buffalo Bill Center of the West. Retrieved from https://centerofthewest.org/2017/12/01/points-west-black-footed-ferret/.

15.   Cassola, F. (2016). *Cynomys ludovicianus* (errata version published in 2017). *The IUCN Red List of Threatened Species*: e.T6091A115080297. https://dx.doi.org/10.2305/IUCN.UK.2016-3.RLTS.T6091A22261137.en.

16.   Sylvatic Plague Vaccine and Management of Prairie Dogs. (August 2012). Fact Sheet 2012–3087, USGS National Wildlife Health Center. Retrieved from https://pubs.usgs.gov/fs/2012/3087/pdf/FS2012_3087_web-lr_th.pdf.

17.   Abbott, R.C. and Rocke, T.E. (2012). Plague, Circular 1372, USGS National Wildlife Health Center. Retrieved via https://pubs.usgs.gov/circ/1372/pdf/C1372_Plague.pdf.

18.   Ibid.

19.   Perry, R. D., & Fetherston, J. D. (1997). *Yersinia pestis* – etiologic agent of plague. *Clinical Microbiology Reviews, 10*(1), 35-66.

20.   Abbott, R.C. and Rocke, T.E. (2012). Plague, Circular 1372, USGS National Wildlife Health Center. Retrieved via https://pubs.usgs.gov/circ/1372/pdf/C1372_Plague.pdf.

21.   Ibid.

22.   Ibid.

23.   Ibid.

24.   Ibid.

25.   Ibid.

26.   WHO. Facts about Plague. (October 2017). Retrieved from https://cdn. who.int/media/images/default-source/infographics/plague-february-2017-jpg. jpg?sfvrsn=af65c05c_0.

27.   Plague Maps and Statistics (May 2024). *CDC*. Retrieved from https://www. cdc.gov/plague/maps-statistics/.

28.   Abbott, R.C. and Rocke, T.E. (2012). Plague, Circular 1372, USGS National Wildlife Health Center. Retrieved via https://pubs.usgs.gov/circ/1372/pdf/C1372_Plague.pdf.

29.   Ibid.

30.   Ibid.

31.   Merriam Webster Dictionary. Retrieved from https://www.merriam-webster. com/dictionary/sylvatic.

32.   Abbott, R.C. and Rocke, T.E. (2012). Plague, Circular 1372, USGS National Wildlife Health Center. Retrieved via https://pubs.usgs.gov/circ/1372/pdf/C1372_Plague.pdf.

33.   Ibid.

34.   Kimberly Fraser, personal communication.

35.   Cassola, F. (2016). *Cynomys ludovicianus* (errata version published in 2017). *The IUCN Red List of Threatened Species 2016*, e.T6091A115080297. https://dx.doi. org/10.2305/IUCN.UK.2016-3.RLTS.T6091A22261137.en.

36.   Biggins, D. E., & Godbey, J. L. (2003). Challenges to reestablishment of free-ranging populations of black-footed ferrets. *Comptes Rendus. Biologies*, 326(S1), 104-111. https://doi.org/10.1016/S1631-0691(03)00046-5.

37.   Kimberly Fraser, personal communication.

38.   Esch, K. L., Beauvais, D. G. P., & Keinath, D. A. (2005). Species conservation assessment for black footed ferret (*Mustela nigripes*) in Wyoming.

US Department of the Interior Bureau of Land Management, Wyoming State Office, Cheyenne, WY.

39.   Kimberly Fraser, personal communication.

40.   Durham, M. (1981, November 6). Rare Black-Footed Ferret Found in Wyoming. *Department of the Interior News Release, Fish and Wildlife Service.* Retrieved from https://www.fws.gov/sites/default/files/documents/historic-news-releases/1981/19811106B.PDF.

41.   Kimberly Fraser, personal communication.

42.   Carpenter, J. W., Appel, M. J., Erickson, R. C., & Novilla, M. N. (1976). Fatal vaccine-induced canine distemper virus infection in black-footed ferrets. *Journal of the American Veterinary Medical Association, 169*(9), 961–964.

43.   Durham, M. (1981, November 6). Rare Black-Footed Ferret Found in Wyoming. *Department of the Interior News Release, Fish and Wildlife Service.* Retrieved from https://www.fws.gov/sites/default/files/documents/historic-news-releases/1981/19811106B.PDF.

44.   Durham, M. (1988, October 28). As the ferrets turn: endangered mammal's story a soap opera of misfortune and triumph. *Department of the Interior News Release, Fish and Wildlife Service.* Retrieved from https://www.fws.gov/sites/default/files/documents/historic-news-releases/1988/19881028B.PDF.

45.   Ibid.

46.   Kimberly Fraser, personal communication.

47.   Durham, M. (1981, November 6). Rare Black-Footed Ferret Found in Wyoming. *Department of the Interior News Release, Fish and Wildlife Service.* Retrieved from https://www.fws.gov/sites/default/files/documents/historic-news-releases/1981/19811106B.PDF.

48.   Durham, M. (1988, October 28). As the ferrets turn: endangered mammal's story a soap opera of misfortune and triumph. Department of the Interior News Release, Fish and Wildlife Service. Retrieved from https://www.fws.gov/sites/default/files/documents/historic-news-releases/1988/19881028B.PDF.

49.   Ibid.

50.   Esch, K. L., Beauvais, D. G. P., & Keinath, D. A. (2005). Species conservation assessment for black footed ferret (*Mustela nigripes*) in Wyoming. US Department of the Interior Bureau of Land Management, Wyoming State Office, Cheyenne, WY.

51.   Abbott, R.C. and Rocke, T.E. (2012). Plague, Circular 1372, USGS National Wildlife Health Center. Retrieved via https://pubs.usgs.gov/circ/1372/pdf/C1372_Plague.pdf.

52.   Esch, K. L., Beauvais, D. G. P., & Keinath, D. A. (2005). Species conservation assessment for black footed ferret (*Mustela nigripes*) in Wyoming. US Department of the Interior Bureau of Land Management, Wyoming State Office, Cheyenne, WY.

53. Abbott, R.C. and Rocke, T.E. (2012). Plague, Circular 1372, USGS National Wildlife Health Center. Retrieved via https://pubs.usgs.gov/circ/1372/pdf/C1372_Plague.pdf.

54. Ibid.

55. Fritts, R. (2020). Plague on the Prairie: The Fight to Save Black-Footed Ferrets from the West's Most Insidious Disease (Master's dissertation). Retrieved from https://dspace.mit.edu/bitstream/handle/1721.1/128984/1227040739-MIT.pdf?sequence=1&isAllowed=y.

56. Abbott, R.C. and Rocke, T.E. (2012). Plague, Circular 1372, USGS National Wildlife Health Center. Retrieved via https://pubs.usgs.gov/circ/1372/pdf/C1372_Plague.pdf.

57. Fritts, R. (2020). Plague on the Prairie: The Fight to Save Black-Footed Ferrets from the West's Most Insidious Disease (Master's dissertation). Retrieved from https://dspace.mit.edu/bitstream/handle/1721.1/128984/1227040739-MIT.pdf?sequence=1&isAllowed=y.

58. Thorne, E. T., & Williams, E. S. (1988). Disease and endangered species: the black-footed ferret as a recent example. *Conservation Biology*, *2*(1), 66-74. https://doi.org/10.1111/j.1523-1739.1988.tb00336.x.

59. National Wildlife Health Center. (2021, August 24). Development of SARS-CoV-2 vaccine to support black-footed ferret conservation. Retrieved from https://www.usgs.gov/centers/nwhc/science/development-sars-cov-2-vaccine-support-black-footed-ferret-conservation.

60. Leon, A. E., Garelle, D., Hartwig, A., Falendysz, E. A., Ip, H. S., Lankton, J. S., ... & Rocke, T. E. (2022). Immunogenicity, safety, and anti-viral efficacy of a subunit SARS-CoV-2 vaccine candidate in captive black-footed ferrets (*Mustela nigripes*) and their susceptibility to viral challenge. *Viruses*, *14*(10), 2188. https://doi.org/10.3390/v14102188.

61. The Black-footed Ferret Project. *revive & restore*. Retrieved from https://reviverestore.org/projects/black-footed-ferret-old/.

62. McCallum, H., Foufopoulos, J., & Grogan, L. F. (2024). Infectious disease as a driver of declines and extinctions. *Cambridge Prisms: Extinction*, *2*, e2. https://doi.org/10.1017/ext.2024.1.

63. Kimberly Fraser, personal communication.

64. Szuszwalak, J. (2024, April 17). U.S. Fish and Wildlife Service and Partners Announce Innovative Cloning Advancements for Black-footed Ferret Conservation. US Fish and Wildlife Service. Retrieved from https://www.fws.gov/press-release/2024-04/innovative-cloning-advancements-black-footed-ferret-conservation.

65. U.S. Fish and Wildlife Service. Recovery plan for the black-footed ferret (*Mustela nigripes*). (Second Revision). (2013, November). Retrieved from https://www.blackfootedferret.org/uploads/1/2/7/7/127791157/usfws_bff_2nd_rev._final_recovery_plan.pdf.

7. SEALS GO TO THE DOGS

1. Land area (sq. km) – Kenya. (2021). *World Bank Group*. Retrieved from https://data.worldbank.org/indicator/AG.LND.TOTL.K2?locations=KE.

2. Woodroffe, R. & Sillero-Zubiri, C. (2020). *Lycaon pictus* (amended version of 2012 assessment). *The IUCN Red List of Threatened Species*, e.T12436A166502262. https://dx.doi.org/10.2305/IUCN.UK.2020-1.RLTS.T12436A166502262. en. Supplementary information at https://www.iucnredlist.org/species/pdf/166502262/attachment.

3. Rise of the Phoenix Pack: Mpala's African Wild Dogs Return. (2020, January 17). *Mpala*. Retrieved from https://mpala.org/rise-of-the-wild-dog-phoenix-pack/.

4. Woodroffe, R. (2021). Modified live distemper vaccines carry low mortality risk for captive African wild dogs, *Lycaon pictus*. *Journal of Zoo and Wildlife Medicine*, *52*(1), 176-184.

5. Drexler, J. F., Corman, V. M., Müller, M. A., Maganga, G. D., Vallo, P., Binger, T., ... & Drosten, C. (2012). Bats host major mammalian paramyxoviruses. *Nature Communications*, *3*(1), 796. https://doi.org/10.1038/ncomms1796.

6. Ibid.

7. Uhl, E. W., Kelderhouse, C., Buikstra, J., Blick, J. P., Bolon, B., & Hogan, R. J. (2019). New world origin of canine distemper: Interdisciplinary insights. *International Journal of Paleopathology*, *24*, 266-278. https://doi.org/10.1016/j.ijpp.2018.12.007.

8. Karki, M., Rajak, K. K., & Singh, R. P.(2022). Canine morbillivirus (CDV): a review on current status, emergence and the diagnostics. *VirusDisease*, *33*(3), 309-321. https://doi.org/10.1007/s13337-022-00779-7.

9. Heide-Jorgensen, M. P., Harkonen, T., Dietz, R., & Thompson, P. M. (1992). Retrospective of the 1988 European seal epizootic. *Diseases of Aquatic Organisms*, *13*(1), 37-62. https://doi.org/10.3354/dao013037.

10. Kennedy, S. (1990). A review of the 1988 European seal morbillivirus epizootic. *The Veterinary Record*, *127*(23), 563–567. https://pubmed.ncbi.nlm.nih.gov/2288059/.

11. Lowry, L. (2016). *Phoca vitulina. The IUCN Red List of Threatened Species*, e.T17013A45229114. https://dx.doi.org/10.2305/IUCN.UK.2016-1.RLTS. T17013A45229114.en.

12. Harwood, J. (1989). Lessons from the seal epidemic. *New Scientist*. Retrieved from https://www.vliz.be/imisdocs/publications/263615.pdf.

13. Heide-Jorgensen, M. P., Harkonen, T., Dietz, R., & Thompson, P. M. (1992). Retrospective of the 1988 European seal epizootic. *Diseases of Aquatic Organisms*, *13*(1), 37-62. https://doi.org/10.3354/dao013037.

14. BBC News. (1988). Mystery seal disease spreads. (1988, August 10). Retrieved from http://news.bbc.co.uk/onthisday/hi/dates/stories/august/10/newsid_2528000/2528665.stm.

15.   Jensen, T., Van de Bildt, M., Dietz, H. H., Andersen, T. H., Hammer, A. S., Kuiken, T., & Osterhaus, A. (2002). Another phocine distemper outbreak in Europe. *Science*, *297*(5579), 209-209.

16.   Powell, K. (2002, July 12). Virus threatens seals. *Nature*. https://doi.org/10.1038/news020708-15.

17.   Heide-Jorgensen, M. P., Harkonen, T., Dietz, R., & Thompson, P. M. (1992). Retrospective of the 1988 European seal epizootic. *Diseases of Aquatic Organisms*, *13*(1), 37-62. https://doi.org/10.3354/dao013037.

18.   Duignan, P. J., Van Bressem, M. F., Baker, J. D., Barbieri, M., Colegrove, K. M., De Guise, S., ... & Wellehan, J. F. (2014). Phocine distemper virus: current knowledge and future directions. *Viruses*, *6*(12), 5093-5134.

19.   Heide-Jorgensen, M. P., Harkonen, T., Dietz, R., & Thompson, P. M. (1992). Retrospective of the 1988 European seal epizootic. *Diseases of Aquatic Organisms*, *13*(1), 37-62. https://doi.org/10.3354/dao013037.

20.   Osterhaus, A.D., Vedder, E.J. (1988). Identification of virus causing recent seal deaths. *Nature*, *335*(20). https://doi.org/10.1038/335020a0.

21.   Ibid.

22.   BBC News (1988): Mystery seal disease spreads. (1988, August 10). Retrieved from http://news.bbc.co.uk/onthisday/hi/dates/stories/august/10/newsid_2528000/2528665.stm.

23.   Osterhaus, A.D., Vedder E.J. (1988). Identification of virus causing recent seal deaths. *Nature*, *335*, 20. https://doi: .org/10.1038/335020a0.

24.   Heide-Jorgensen, M. P., Harkonen, T., Dietz, R., & Thompson, P. M. (1992). Retrospective of the 1988 European seal epizootic. *Diseases of Aquatic Organisms*, *13*(1), 37-62. https://doi.org/10.3354/dao013037.

25.   Kock, R.A. (2023). Great Viral Plagues of Large Mammals. In D.A. Jessup and R.W. Radcliffe (Eds), Wildlife Disease and Health in Conservation, Johns Hopkins University Press.

26.   Ibid.

27.   Osterhaus, A.D., Vedder, E.J. (1988). Identification of virus causing recent seal deaths. *Nature*, *335*, 20. https://doi.org/10.1038/335020a0.

28.   Albert Osterhaus, personal communication.

29.   Laksono, B. M., de Vries, R. D., Verburgh, R. J., Visser, E. G., de Jong, A., Fraaij, P. L., ... & de Swart, R. L. (2018). Studies into the mechanism of measles-associated immune suppression during a measles outbreak in the Netherlands. *Nature Communications*, *9*(1), 1-10. https://doi.org/10.1038/s41467-018-07515-0.

30.   Duignan, P. J., Van Bressem, M. F., Baker, J. D., Barbieri, M., Colegrove, K. M., De Guise, S., ... & Wellehan, J. F. (2014). Phocine distemper virus: current knowledge and future directions. *Viruses*, *6*(12), 5093-5134. https://doi.org/10.3390/v6125093.

31.   VanWormer, E., Mazet, J. A. K., Hall, A., Gill, V. A., Boveng, P. L., London, J. M., ... & Goldstein, T. (2019). Viral emergence in marine mammals in the North Pacific may be linked to Arctic sea ice reduction. *Scientific Reports*, *9*(1), 1-11.

32.   Duignan, P. J., Van Bressem, M. F., Baker, J. D., Barbieri, M., Colegrove, K. M., De Guise, S., ... & Wellehan, J. F. (2014). Phocine distemper virus: current knowledge and future directions. *Viruses*, *6*(12), 5093-5134. https://doi.org/10.3390/v6125093.

33.   Ibid.

34.   Ibid.

35.   Harvell, C. D., Kim, K., Burkholder, J. M., Colwell, R. R., Epstein, P. R., Grimes, D. J., ... & Vasta, G. R. (1999). Emerging marine diseases – climate links and anthropogenic factors. *Science*, *285*(5433), 1505-1510.

36.   Heide-Jorgensen, M. P., Harkonen, T., Dietz, R., & Thompson, P. M. (1992). Retrospective of the 1988 European seal epizootic. *Diseases of Aquatic Organisms*, *13*(1), 37-62. https://doi.org/10.3354/dao013037.

37.   Ibid.

38.   Groch, K. R., Blazquez, D. N., Marcondes, M. C., Santos, J., Colosio, A., Díaz Delgado, J., & Catão-Dias, J. L. (2021). Cetacean morbillivirus in Humpback whales' exhaled breath. *Transboundary and Emerging Diseases*, *68*(4), 1736-1743. https://doi.org/10.1111/tbed.13883.

39.   Heide-Jorgensen, M. P., Harkonen, T., Dietz, R., & Thompson, P. M. (1992). Retrospective of the 1988 European seal epizootic. *Diseases of Aquatic Organisms*, *13*(1), 37-62. https://doi.org/10.3354/dao013037.

40.   Ibid.

41.   de Swart, R., Ross, P., Vedder, L., Timmerman, H., Heisterkamp, S., Van Loveren, H., ... & Osterhaus, A. (1994). Impairment of immune function in harbor seals (*Phoca vitulina*) feeding on fish from polluted waters. *Ambio*, *23*(2), 155-159. https://doi.org/10.1579/0044-7447-31.1.1.

42.   Ibid.

43.   Heide-Jorgensen, M. P., Harkonen, T., Dietz, R., & Thompson, P. M. (1992). Retrospective of the 1988 European seal epizootic. *Diseases of Aquatic Organisms*, *13*(1), 37-62. https://doi.org/10.3354/dao013037.

44.   Ibid.

45.   Mamaev, L. V., Denikina, N. N., Belikov, S. I., Volchkov, V. E., Visser, I. K. G., Fleming, M., ... & Barrett, T. (1995). Characterisation of morbilliviruses isolated from Lake Baikal seals (*Phoca sibirica*). *Veterinary Microbiology*, *44*(2-4), 251-259.

46.   Harvell, C. D., Kim, K., Burkholder, J. M., Colwell, R. R., Epstein, P. R., Grimes, D. J., ... & Vasta, G. R. (1999). Emerging marine diseases – climate links and anthropogenic factors. *Science*, *285*(5433), 1505-1510.

47.	Grachev, M. A., Kumarev, V. P., Mamaev, L. V., Zorin, V. L., Baranova, L V., Denikina, N. N., ... Sidorov, V. N. (1989). Distemper virus in Baikal seals. *Nature*, *338*, 209–210. https://doi.org/10.1038/338209b0, quoted in Heide-Jørgensen, M. P., Härkönen, T., Dietz, R., & Thompson, P. M. (1992). Retrospective of the 1988 European seal epizootic. *Diseases of Aquatic Organisms*, *13*(1), 37-62. https://doi.org/10.3354/dao013037.

48.	Heide-Jorgensen, M. P., Harkonen, T., Dietz, R., & Thompson, P. M. (1992). Retrospective of the 1988 European seal epizootic. *Diseases of Aquatic Organisms*, *13*(1), 37-62. https://doi.org/10.3354/dao013037.

49.	Ibid.

50.	Laws, R. M., & Taylor, R. J. F. (1957, November). A mass dying of crabeater seals, *Lobodon carcinophagus* (Gray). In *Proceedings of the Zoological Society of London* (*129*(3), 315-324). Blackwell Publishing Ltd. Quoted in Heide-Jørgensen, M. P., Härkönen, T., Dietz, R., & Thompson, P. M. (1992). Retrospective of the 1988 European seal epizootic. *Diseases of Aquatic Organisms*, *13*(1), 37-62. https://doi.org/10.3354/dao013037.

51.	Virus threats to monk seals. (2002, November). *The Monachus Guardian*, *5*(2). Friends of the Monk Seal. Retrieved from https://citeseerx.ist.psu.edu/document?repid=rep1&type=pdf&doi=131222ca612f22331774b7efe7bde9a621e7eae5.

52.	Bengtson, J. L., Boveng, P., Franzen, U., Have, P., Heide-Jørgensen, M. P., & Härkönen, T. J. (1991). Antibodies to canine distemper virus in Antarctic seals. *Marine Mammal Science*, *7*(1), 85-87. https://doi.org/10.1111/j.1748-7692.1991.tb00553.x, quoted in Heide-Jorgensen, M. P., Harkonen, T., Dietz, R., & Thompson, P. M. (1992). Retrospective of the 1988 European seal epizootic. *Diseases of Aquatic Organisms*, *13*(1), 37-62. https://doi.org/10.3354/dao013037.

53.	Forsyth, M. A., Kennedy, S., Wilson, S., Eybatov, T., & Barrett, T. (1998). Canine distemper virus in a Caspian seal. *The Veterinary Record*, *143*(24), 662.

54.	Jo, W. K., Peters, M., Kydyrmanov, A., van de Bildt, M. W., Kuiken, T., Osterhaus, A., & Ludlow, M. (2019). The canine morbillivirus strain associated with an epizootic in Caspian seals provides new insights into the evolutionary history of this virus. *Viruses*, *11*(10), 894. https://doi.org/10.3390/v11100894.

55.	Ibid.

56.	VanWormer, E., Mazet, J. A. K., Hall, A., Gill, V. A., Boveng, P. L., London, J. M., ... & Goldstein, T. (2019). Viral emergence in marine mammals in the North Pacific may be linked to Arctic sea ice reduction. *Scientific Reports*, *9*(1), 1-11.

57.	Ibid.

58.	Jensen, T., Van de Bildt, M., Dietz, H. H., Andersen, T. H., Hammer, A. S., Kuiken, T., & Osterhaus, A. (2002). Another phocine distemper outbreak in Europe. *Science*, *297*(5579), 209-209.

59.	Härkönen, T., Dietz, R., Reijnders, P., Teilmann, J., Harding, K., Hall, A., ... & Thompson, P. (2006). The 1988 and 2002 phocine distemper virus epidemics in European harbour seals. *Diseases of Aquatic Organisms, 68*(2), 115-130. https://doi.org/10.3354/dao068115.

60. Duignan, P. J., Van Bressem, M. F., Baker, J. D., Barbieri, M., Colegrove, K. M., De Guise, S., ... & Wellehan, J. F. (2014). Phocine distemper virus: current knowledge and future directions. *Viruses, 6*(12), 5093-5134. https://doi.org/10.3390/v6125093.

61. Ibid.

62. Goldstein, T., Mazet, J. A., Gill, V. A., Doroff, A. M., Burek, K. A., & Hammond, J. A. (2009). Phocine distemper virus in northern sea otters in the Pacific Ocean, Alaska, USA. *Emerging Infectious Diseases, 15*(6), 925. https://doi.org/10.3201%2Feid1506.090056.

63. Ibid.

64. VanWormer, E., Mazet, J. A. K., Hall, A., Gill, V. A., Boveng, P. L., London, J. M., ... & Goldstein, T. (2019). Viral emergence in marine mammals in the North Pacific may be linked to Arctic sea ice reduction. *Scientific Reports, 9*(1), 1-11.

65. Doroff, A., Burdin, A. & Larson, S. (2021). *Enhydra lutris* (errata version published in 2022). *The IUCN Red List of Threatened Species*, e.T7750A219377647. https://dx.doi.org/10.2305/IUCN.UK.2021-3.RLTS.T7750A219377647.en.

66. Sun, Z., Li, A., Ye, H., Shi, Y., Hu, Z., & Zeng, L. (2010). Natural infection with canine distemper virus in hand-feeding Rhesus monkeys in China. *Veterinary Microbiology, 141*(3-4), 374-378. https://doi.org/10.1016/j.vetmic.2009.09.024.

67. The Curious Case of the Monk Seal – Why 'Monk'? (2013, December 2). *The Monachus Guardian*. Retrieved from https://monachus-guardian.org/wordpress/2013/12/02/the-curious-case-of-the-monk-seal-why-monk/.

68. Jiddou, A. M., Vedder, E., Martina, B., Dedaii, S., Soueilem, M. M., Diop, M., ... & Osterhaus, A. (1997). Memorandum on the mass die-off of monk seals on the Cap Blanc peninsula. *CNROP/SRRC December*. Retrieved from https://www.monachus-guardian.org/library/cnrop-srrc00.pdf.

69. Osterhaus, A., Groen, J., Niesters, H., van de Bildt, M., Martina, B., Vedder, L., ... & Barham, M. E. O. (1997). Morbillivirus in monk seal mass mortality. *Nature, 388*(6645), 838-839.

70. Jiddou, A. M., Vedder, E., Martina, B., Dedaii, S., Soueilem, M. M., Diop, M., ... & Osterhaus, A. (1997). Memorandum on the mass die-off of monk seals on the Cap Blanc peninsula. *CNROP/SRRC December*. Retrieved from https://www.monachus-guardian.org/library/cnrop-srrc00.pdf.

71. Osterhaus, A., Groen, J., Niesters, H., van de Bildt, M., Martina, B., Vedder, L., ... & Barham, M. E. O. (1997). Morbillivirus in monk seal mass mortality. *Nature, 388*(6645), 838-839.

72. Osterhaus, A., van de Bildt, M., Vedder, L., Martina, B., Niesters, H., Vos, J., ... & Barham, M. E. O. (1998). Monk seal mortality: virus or toxin?. *Vaccine, 16*(9-10), 979-981.

73. Karamanlidis, A.A., Dendrinos, P., Fernandez de Larrinoa, P., Kıraç, C.O., Nicolaou, H. & Pires, R. (2023). *Monachus monachus*. *The IUCN Red List of Threatened Species*, e.T13653A238637039. https://dx.doi.org/10.2305/IUCN. UK.2023-1.RLTS.T13653A238637039.en.

74. Osterhaus, A., Groen, J., Niesters, H., van de Bildt, M., Martina, B., Vedder, L., ... & Barham, M. E. O. (1997). Morbillivirus in monk seal mass mortality. *Nature*, *388*(6645), 838-839.

75. Barrett, T., Visser, I. K. G., Mamaev, L., Goatley, L., Van Bressem, M. F., & Osterhaus, A. D. M. E. (1993). Dolphin and porpoise morbilliviruses are genetically distinct from phocine distemper virus. *Virology*, *193*(2), 1010-1012.

76. Van Bressem, M. F., Duignan, P. J., Banyard, A., Barbieri, M., Colegrove, K. M., De Guise, S., ... & Wellehan, J. F. (2014). Cetacean morbillivirus: current knowledge and future directions. *Viruses*, *6*(12), 5145-5181.

77. Ibid.

78. Kennedy, S., Smyth, J. A., Cush, P. F., McCullough, S. J., Allan, G. M., & McQuaid, S. (1988). Viral distemper now found in porpoises. *Nature*, *336*, 21. https://doi.org/10.1038/336021a0.

79. Visser, I. K., Van Bressem, M. F., de Swart, R. L., van de Bildt, M. W., Vos, H. W., van der Heijden, R. W., ... & Osterhaus, A. D. (1993). Characterization of morbilliviruses isolated from dolphins and porpoises in Europe. *Journal of General Virology*, *74*(4), 631-641.

80. Osterhaus, A. D., de Swart, R. L., Vos, H. W., Ross, P. S., Kenter, M. J., & Barrett, T. (1995). Morbillivirus infections of aquatic mammals: newly identified members of the genus. *Veterinary Microbiology*, *44*(2-4), 219-227. https://doi. org/10.1016/0378-1135 https://doi.org/10.1016/0378-1135(95)00015-3.

81. Van Bressem, M. F., Duignan, P. J., Banyard, A., Barbieri, M., Colegrove, K. M., De Guise, S., ... & Wellehan, J. F. (2014). Cetacean morbillivirus: current knowledge and future directions. *Viruses*, *6*(12), 5145-5181. https://doi.org/10.3390/ v6125145.

82. Albert Osterhaus, personal communication.

83. Groch, K. R., Blazquez, D. N., Marcondes, M. C., Santos, J., Colosio, A., Díaz Delgado, J., & Catão-Dias, J. L. (2021). Cetacean morbillivirus in Humpback whales' exhaled breath. *Transboundary and Emerging Diseases*, *68*(4), 1736-1743. https://doi.org/10.1111/tbed.13883.

84. Jiddou, A. M., Vedder, E., Martina, B., Dedaii, S., Soueilem, M. M., Diop, M., ... & Osterhaus, A. (1997). Memorandum on the mass die-off of monk seals on the Cap Blanc peninsula. *CNROP/SRRC December*. Retrieved from https:// www.monachus-guardian.org/library/cnrop-srrc00.pdf.

85. Ibid.

86. MOm: Hellenic Society for the Study and Protection of the Monk Seal. Biology. Retrieved from https://www.mom.gr/biology.

87.   Jiddou, A. M., Vedder, E., Martina, B., Dedaii, S., Soueilem, M. M., Diop, M., ... & Osterhaus, A. (1997). Memorandum on the mass die-off of monk seals on the Cap Blanc peninsula. *CNROP/SRRC December*. Retrieved from https://www.monachus-guardian.org/library/cnrop-srrc00.pdf.

88.   NOAA Fisheries (2022, August 30). Conserving Hawaiian Monk Seals Through Protections and Vaccinations. Retrieved from https://www.fisheries.noaa.gov/feature-story/conserving-hawaiian-monk-seals-through-protections-and-vaccinations.

89.   Robinson, S. J., Barbieri, M. M., Murphy, S., Baker, J. D., Harting, A. L., Craft, M. E., & Littnan, C. L. (2018). Model recommendations meet management reality: implementation and evaluation of a network-informed vaccination effort for endangered Hawaiian monk seals. *Proceedings of the Royal Society B: Biological Sciences*, *285*(1870), 20171899. https://doi.org/10.1098/rspb.2017.1899.

90.   Osterhaus, A., van de Bildt, M., Vedder, L., Martina, B., Niesters, H., Vos, J., ... & Barham, M. E. O. (1998). Monk seal mortality: virus or toxin?. *Vaccine*, *16*(9-10), 979-981. https://doi.org/10.1016/S0264-410X(98)00006-1.

91.   Virus threats to monk seals. (November 2002). *Monachus Guardian 5*(2). Friends of the Monk Seal. Retrieved from https://citeseerx.ist.psu.edu/document?repid=rep1&type=pdf&doi=131222ca612f22331774b7efe7bde9a621e7eae5.

92.   Karamanlidis, A.A., Dendrinos, P., Fernandez de Larrinoa, P., Kıraç, C.O., Nicolaou, H. & Pires, R. (2023). *Monachus monachus. The IUCN Red List of Threatened Species*, e.T13653A238637039. https://dx.doi.org/10.2305/IUCN.UK.2023-1.RLTS.T13653A238637039.en.

93.   Monk Seal Alliance. Retrieved from https://www.monksealalliance.org/en/index.

94.   Monk Seal Alliance Brochure 2021. (2022, March 03). *Monk Seal Alliance*. Retrieved from https://www.monksealalliance.org/en/news/leaflet/brochure-2021-057.

95.   Karamanlidis, A.A., Dendrinos, P., Fernandez de Larrinoa, P., Kıraç, C.O., Nicolaou, H. & Pires, R. (2023). *Monachus monachus. The IUCN Red List of Threatened Species*, e.T13653A238637039. https://dx.doi.org/10.2305/IUCN.UK.2023-1.RLTS.T13653A238637039.en.

96.   Minta, A.A., Ferrari, M., Antoni, S., Portnoy, A., Sbarra, A., Lambert, B. ... & Crowcroft, N.S. (2023). Progress toward measles elimination – worldwide, 2000–2022. *Morbidity and Mortality Weekly Report (MMWR)*, *72*, 1262–1268. http://dx.doi.org/10.15585/mmwr.mm7246a3.

97.   Mpala. Rise of the Phoenix Pack: Mpala's African Wild Dogs Return. (2020, January 17). Retrieved from https://mpala.org/rise-of-the-wild-dog-phoenix-pack/.

98.   Kaldor, Z. (2024, March 03). Technology that Promotes Coexistence: Utilizing Collars to Track African Wild Dogs and Lions. Mpala. Retrieved from

https://mpala.org/technology-that-promotes-coexistence-utilizing-collars-to-tra ck-african-wild-dogs-and-lions/.

99.   Woodroffe, R. (2021). Modified live distemper vaccines carry low mortality risk for captive African wild dogs, *Lycaon pictus*. *Journal of Zoo and Wildlife Medicine*, *52*(1), 176-184

100.   Climate change driving dogs to extinction. (2023, August 22). Retrieved from https://www.zsl.org/news-and-events/news/climate-change-driving-dog s-extinction.

8. A DEVIL OF A PROBLEM

1.   Department of Natural Resources and Environment Tasmania. About DFTD. Retrieved from https://nre.tas.gov.au/conservation/threatened-specie s-and-communities/lists-of-threatened-species/threatened-species-vertebrates/ save-the-tasmanian-devil-program/about-dftd.

2.   Department of Natural Resources and Environment Tasmania. About the Tasmanian Devil. Retrieved from https://nre.tas.gov.au/conservation/ threatened-species-and-communities/lists-of-threatened-species/ threatened-species-vertebrates/save-the-tasmanian-devil-program/ about-the-tasmanian-devil.

3.   Pemberton, D., & Renouf, D. (1993). A field-study of communication and social-behavior of the Tasmanian devil at feeding sites. *Australian Journal of Zoology*, *41*(5), 507-526.

4.   Professor Elizabeth Murchison. *University of Cambridge Department of Veterinary Medicine*. Retrieved from https://www.vet.cam.ac.uk/directory/ murchison.

5.   Tasmanian Devils and the transmissible cancer that threatens their extinction. (2015, October 14). Retrieved from https://www.cam.ac.uk/research/ features/tasmanian-devils-and-the-transmissible-cancer-that-threatens-their-ex tinction.

6.   Murchison, E. P., Schulz-Trieglaff, O. B., Ning, Z., Alexandrov, L. B., Bauer, M. J., Fu, B., ... & Stratton, M. R. (2012). Genome sequencing and analysis of the Tasmanian devil and its transmissible cancer. *Cell*, *148*(4), 780-791.

7.   Murchison, E. P., Tovar, C., Hsu, A., Bender, H. S., Kheradpour, P., Rebbeck, C. A., ... & Papenfuss, A. T. (2010). The Tasmanian devil transcriptome reveals Schwann cell origins of a clonally transmissible cancer. *Science*, *327*(5961), 84-87. https://doi.org/10.1126/science.1180616.

8.   Mulford, R. D. (1967). Experimentation on human beings. *Stan L. Rev.*, *20*, 99.

9.   Elizabeth Murchison, personal communication.

10.   Ibid.

11.   Ibid.

12.   Ibid.

13.   Hawkins, C.E., McCallum, H., Mooney, N., Jones, M. & Holdsworth, M. (2008). *Sarcophilus harrisii. The IUCN Red List of Threatened Species 2008*: e.T40540A10331066. https://dx.doi.org/10.2305/IUCN.UK.2008.RLTS.T40540A10331066.en.

14.   Department of Natural Resources and Environment Tasmania. About the Tasmanian Devil. Retrieved from https://nre.tas.gov.au/conservation/threatened-species-and-communities/lists-of-threatened-species/threatened-species-vertebrates/save-the-tasmanian-devil-program/about-the-tasmanian-devil.

15.   Hamish McCallum, personal communication.

16.   Bradshaw, C.J.A. & Brook, B.W. (2005). Disease and the devil: density-dependent epidemiological processes explain historical population fluctuations in the Tasmanian devil. *Ecography, 28*, 181-190. https://doi.org/10.1111/j.0906-7590.2005.04088.x.

17.   Brüniche-Olsen, A., Jones, M. E., Austin, J. J., Burridge, C. P., & Holland, B. R. (2014). Extensive population decline in the Tasmanian devil predates European settlement and devil facial tumour disease. *Biology Letters, 10*(11). https://doi.org/10.1098/RSBL.2014.0619.

18.   Department of Natural Resources and Environment Tasmania. About DFTD. Retrieved from https://nre.tas.gov.au/conservation/threatened-species-and-communities/lists-of-threatened-species/threatened-species-vertebrates/save-the-tasmanian-devil-program/about-dftd.

19.   Murchison, E. P., Schulz-Trieglaff, O. B., Ning, Z., Alexandrov, L. B., Bauer, M. J., Fu, B., ... & Stratton, M. R. (2012). Genome sequencing and analysis of the Tasmanian devil and its transmissible cancer. *Cell, 148*(4), 780-791.

20.   Elizabeth Murchison, personal communication.

21.   Ibid.

22.   Pye, R. J., Pemberton, D., Tovar, C., Tubio, J. M., Dun, K. A., Fox, S., ... & Woods, G. M. (2016). A second transmissible cancer in Tasmanian devils. *Proceedings of the National Academy of Sciences, 113*(2), 374-379.

23.   Murchison, E. P., Tovar, C., Hsu, A., Bender, H. S., Kheradpour, P., Rebbeck, C. A., ... & Papenfuss, A. T. (2010). The Tasmanian devil transcriptome reveals Schwann cell origins of a clonally transmissible cancer. *Science, 327*(5961), 84-87.

24.   Jones, M. E., Cockburn, A., Hamede, R., Hawkins, C., Hesterman, H., Lachish, S., ... Pemberton, D. (2008). Life-history change in disease-ravaged Tasmanian devil populations. *Proceedings of the National Academy of Sciences of the United States of America, 105*, 10023–10027. https://doi.org/10.1073/pnas.0711236105.

25.   Hogg, C. J., Lee, A. V., Srb, C., & Hibbard, C. (2017). Metapopulation management of an endangered species with limited genetic diversity in the presence of disease: the Tasmanian devil *Sarcophilus harrisii. International Zoo Yearbook, 51*(1), 137-153. https://doi.org/10.1111/izy.12144.

26.   Hogg, C. J., Ivy, J. A., Srb, C., Hockley, J., Lees, C., Hibbard, C., & Jones, M. (2015). Influence of genetic provenance and birth origin on productivity of the Tasmanian devil insurance population. *Conservation Genetics*, *16*, 1465-1473. https://doi.org/10.1007/s10592-015-0754-9.

27.   Hogg, C. J., Lee, A. V., Srb, C., & Hibbard, C. (2017). Metapopulation management of an endangered species with limited genetic diversity in the presence of disease: the Tasmanian devil *Sarcophilus harrisii*. *International Zoo Yearbook*, *51*(1), 137-153. https://doi.org/10.1111/izy.12144.

28.   Bittel, J. (2020, October 05). Tasmanian devils return to mainland Australia for first time in 3,000 years. *National Geographic*. Retrieved from https://www.nationalgeographic.com/animals/article/tasmanian-devils-return-to-mainland-australia.

29.   Scoleri, V. P., Johnson, C. N., Vertigan, P., & Jones, M. E. (2020). Conservation trade-offs: Island introduction of a threatened predator suppresses invasive mesopredators but eliminates a seabird colony. *Biological Conservation*, *248*, 108635. https://doi.org/10.1016/j.biocon.2020.108635.

30.   Lu, D. (2021, June 21). Tasmanian devils wipe out thousands of penguins on tiny Australian island. *The Guardian*. Retrieved from https://www.theguardian.com/environment/2021/jun/21/tasmanian-devils-wipe-out-thousands-of-penguins-maria-island-australia.

31.   Pye, R., Darby, J., Flies, A. S., Fox, S., Carver, S., Elmer, J., ... & Lyons, A. B. (2021). Post-release immune responses of Tasmanian devils vaccinated with an experimental devil facial tumour disease vaccine. *Wildlife Research*, *48*(8), 701-712.

32.   Wait, L. F., Peck, S., Fox, S., & Power, M. L. (2017). A review of parasites in the Tasmanian devil (*Sarcophilus harrisii*). *Biodiversity and Conservation*, *26*, 509-526.

33.   Ibid.

34.   Save the Tasmanian Devil Program. *Department of Natural Resources and Environment Tasmania*. Retrieved from https://nre.tas.gov.au/conservation/threatened-species-and-communities/lists-of-threatened-species/threatened-species-vertebrates/save-the-tasmanian-devil-program.

35.   Wait, L. F., Peck, S., Fox, S., & Power, M. L. (2017). A review of parasites in the Tasmanian devil (*Sarcophilus harrisii*). *Biodiversity and Conservation*, *26*, 509-526.

36.   Hollings, T., Jones, M., Mooney, N., & McCallum, H. (2016). Disease-induced decline of an apex predator drives invasive dominated states and threatens biodiversity. *Ecology*, *97*(2), 394-405. https://doi.org/10.1890/15-0204.1.

37.   McCallum, H., & Jones, M. (2006). To lose both would look like carelessness: Tasmanian devil facial tumour disease. *PLoS Biology*, *4*(10), e342. https://doi.org/10.1371/journal.pbio.0040342.

38.   Hollings, T., Jones, M., Mooney, N., & McCallum, H. (2013). Wildlife disease ecology in changing landscapes: mesopredator release and toxoplasmosis. *International Journal for Parasitology: Parasites and Wildlife*, *2*, 110-118. https://doi.org/10.1016/j.ijppaw.2013.02.002.

39. Cunningham, C. X., Johnson, C. N., Barmuta, L. A., Hollings, T., Woehler, E. J., & Jones, M. E. (2018). Top carnivore decline has cascading effects on scavengers and carrion persistence. *Proceedings of the Royal Society B, 285*(1892), 20181582. https://doi.org/10.1098/rspb.2018.1582.

40. Ní Leathlobhair, M., Perri, A. R., Irving-Pease, E. K., Witt, K. E., Linderholm, A., Haile, J., ... & Frantz, L. A. (2018). The evolutionary history of dogs in the Americas. *Science, 361*(6397), 81-85. https://doi.org/10.1126/science.aao4776.

41. Baez-Ortega, A., Gori, K., Strakova, A., Allen, J. L., Allum, K. M., Bansse-Issa, L., ... & Murchison, E. P. (2019). Somatic evolution and global expansion of an ancient transmissible cancer lineage. *Science, 365*(6452), eaau9923.

42. Ibid.

43. Ní Leathlobhair, M., Perri, A. R., Irving-Pease, E. K., Witt, K. E., Linderholm, A., Haile, J., ... & Frantz, L. A. (2018). The evolutionary history of dogs in the Americas. *Science, 361*(6397), 81-85. https://doi.org/10.1126/science.aao4776.

44. Ibid.

45. Murchison, E.P. (2021). Rising incidence of canine transmissible venereal tumours in the UK. *The Veterinary Record, 189*(12), 472-474. https://doi.org/10.1002/vetr.1299.

46. Metzger, M. J., Villalba, A., Carballal, M. J., Iglesias, D., Sherry, J., Reinisch, C., Muttray, A. F., Baldwin, S. A., & Goff, S. P. (2016). Widespread transmission of independent cancer lineages within multiple bivalve species. *Nature*, 534(7609), 705– 709. https://doi.org/10.1038/nature18599.

47. Michnowska, A., Hart, S. F. M., Smolarz, K., Hallmann, A., & Metzger, M. J. (2022). Horizontal transmission of disseminated neoplasia in the widespread clam *Macoma balthica* from the Southern Baltic Sea. *Molecular Ecology, 31*, 3128–3136. https://doi.org/10.1111/mec.16464.

48. Skazina, M., Odintsova, N., Maiorova, M., Ivanova, A., Väinölä, R., & Strelkov, P. (2021). First description of a widespread *Mytilus trossulus*-derived bivalve transmissible cancer lineage in *M. trossulus* itself. *Scientific Reports, 11*(1), 5809.

49. Yonemitsu, M. A., Giersch, R. M., Polo-Prieto, M., Hammel, M., Simon, A., Cremonte, F. ... & Metzger, M. J. (2019). A single clonal lineage of transmissible cancer identified in two marine mussel species in South America and Europe. *eLife*, 8. https://doi.org/10.7554/eLife.47788.

50. Skazina, M., Odintsova, N., Maiorova, M., Ivanova, A., Väinölä, R., & Strelkov, P. (2021). First description of a widespread *Mytilus trossulus*-derived bivalve transmissible cancer lineage in M. trossulus itself. *Scientific Reports, 11*(1), 5809.

51. Dujon, A. M., Bramwell, G., Roche, B., Thomas, F., & Ujvari, B. (2021). Transmissible cancers in mammals and bivalves: How many examples are there? *BioEssays, 43*, e2000222. https://doi.org/10.1002/bies.202000222.

52. McCallum, H., Jones, M. (2006). To lose both would look like carelessness: Tasmanian devil facial tumour disease. *PLoS Biology 4*(10), e342. https://doi.org/10.1371/journal.pbio.0040342.

9. FUNGUS AND FROGS

1. Can we save frogs from a deadly fungus? (2022, October 17). *The Royal Institution*. [YouTube video]. Retrieved from https://www.youtube.com/watch?v=WpOl_yc8n6Q.

2. IUCN Red List. Retrieved from https://www.iucnredlist.org/.

3. Stuart, S. N., Chanson, J. S., Cox, N. A., Young, B. E., Rodrigues, A. S., Fischman, D. L., & Waller, R. W. (2004). Status and trends of amphibian declines and extinctions worldwide. *Science, 306*(5702), 1783-1786.

4. Blaustein, A. R. (1990). Declining amphibian populations: a global phenomenon?. *Trends in Ecology and Evolution, 5*, 203-204.

5. Laurance, W.F., McDonald, K.R. and Speare, R. (1996). Epidemic Disease and the Catastrophic Decline of Australian Rain Forest Frogs. *Conservation Biology, 10*, 406-413. https://doi.org/10.1046/j.1523-1739.1996.10020406.x.

6. Andrew Cunningham, personal communication.

7. Berger, L., Speare, R., Daszak, P., Green, D. E., Cunningham, A. A., Goggin, C. L., ... & Parkes, H. (1998). Chytridiomycosis causes amphibian mortality associated with population declines in the rain forests of Australia and Central America. *Proceedings of the National Academy of Sciences, 95*(15), 9031-9036. https://doi.org/10.1073/pnas.95.15.9031.

8. Ibid.

9. Gascon, C., Collins, J. P., Moore, R. D., Church, D. R., McKay, J. E., Mendelson, J. R., III. Amphibian conservation action plan. IUCN/SSC Amphibian Specialist Group, Gland, 2007.

10. IUCN SSC Amphibian Specialist Group. (2022). *Taudactylus acutirostris. The IUCN Red List of Threatened Species*, e.T21529A78447380. https://dx.doi.org/10.2305/IUCN.UK.2022-2.RLTS.T21529A78447380.en.

11. IUCN SSC Amphibian Specialist Group. (2023). *Litoria rheocola. The IUCN Red List of Threatened Species*, e.T12153A78437571. https://dx.doi.org/10.2305/IUCN.UK.2023-1.RLTS.T12153A78437571.en.

12. IUCN SSC Amphibian Specialist Group. (2022). *Litoria nannotis. The IUCN Red List of Threatened Species* 2022: e.T12148A78436726. https://dx.doi.org/10.2305/IUCN.UK.2022-2.RLTS.T12148A78436726.en.

13. Berger, L., Speare, R., Daszak, P., Green, D. E., Cunningham, A. A., Goggin, C. L., ... & Parkes, H. (1998). Chytridiomycosis causes amphibian mortality associated with population declines in the rain forests of Australia and Central America. *Proceedings of the National Academy of Sciences, 95*(15), 9031-9036. https://doi.org/10.1073/pnas.95.15.9031.

14.   Ibid.

15.   Kaiser, J. (1998). Fungus may drive frog genocide. *Science*, *281*(5373), 23. https://doi.org/10.1126/science.281.5373.23a.

16.   Andrew Cunningham, personal communication.

17.   Stokstad, E. (2019, March 28). This fungus has wiped out more species than any other disease. *Science*. Retrieved from https://doi.org/10.1126/science.aax5025.

18.   Scheele, B. C., Pasmans, F., Skerratt, L. F., Berger, L., Martel, A. N., Beukema, W., ... & Canessa, S. (2019). Amphibian fungal panzootic causes catastrophic and ongoing loss of biodiversity. *Science*, *363*(6434), 1459-1463. https://doi.org/10.1126/science.aav0379.

19.   Fisher, M. C., Henk, D. A., Briggs, C. J., Brownstein, J. S., Madoff, L. C., McCraw, S. L., & Gurr, S. J. (2012). Emerging fungal threats to animal, plant and ecosystem health. *Nature*, *484*(7393), 186-194. https://doi.org/10.1038/nature10947.

20.   Ibid.

21.   Andrew Cunningham, personal communication.

22.   Scheele, B. C., Pasmans, F., Skerratt, L. F., Berger, L., Martel, A. N., Beukema, W., ... & Canessa, S. (2019). Amphibian fungal panzootic causes catastrophic and ongoing loss of biodiversity. *Science*, *363*(6434), 1459-1463. https://doi.org/10.1126/science.aav0379.

23.   O'Hanlon, S. J., Rieux, A., Farrer, R. A., Rosa, G. M., Waldman, B., Bataille, A., ... & Fisher, M. C. (2018). Recent Asian origin of chytrid fungi causing global amphibian declines. *Science*, *360*(6389), 621-627. https://doi.org/10.1126/science.aar1965.

24.   Ibid.

25.   Ibid.

26.   Weldon, C., Du Preez, L. H., Hyatt, A. D., Muller, R., & Speare, R. (2004). Origin of the amphibian chytrid fungus. *Emerging Infectious Diseases*, *10*(12), 2100-2105. https://doi.org/10.3201/eid1012.030804.

27.   Ibid.

28.   Downie, R. (2022, August 30). Croaking Science: The International Trade in Reptiles and Amphibians. *Froglife*. Retrieved from https://www.froglife.org/2022/08/30/croaking-science-the-international-trade-in-reptiles-and-amphibians-2/

29.   Ibid.

30.   Altherr, S. and Lambert, K. (2020). The rush for the rare: reptiles and amphibians in the European pet trade. *Animals 10*(11), 2085. https://doi.org/10.3390/ani10112085.

31.   Walker, S. F., Bosch, J., James, T. Y., Litvintseva, A. P., Valls, J. A. O., Piña, S., ... & Fisher, M. C. (2008). Invasive pathogens threaten species recovery programs. *Current Biology*, *18*(18), R853-R854.

32.   Fisher, M. C., & Garner, T. W. (2020). Chytrid fungi and global amphibian declines. *Nature Reviews Microbiology*, *18*(6), 332-343.

33.   Bosch, J., Sanchez-Tomé, E., Fernández-Loras, A., Oliver, J. A., Fisher, M. C., & Garner, T. W. (2015). Successful elimination of a lethal wildlife infectious disease in nature. *Biology Letters*, *11*(11), 20150874.

34.   Garner, T. W., Schmidt, B. R., Martel, A., Pasmans, F., Muths, E., Cunningham, A. A., ... & Bosch, J. (2016). Mitigating amphibian chytridiomycoses in nature. *Philosophical Transactions of the Royal Society B: Biological Sciences*, *371*(1709), 20160207.

35.   Andrew Cunningham, personal communication.

36.   Mountain Chicken Conservation. *ZSL*. Retrieved from https://www.zsl.org/what-we-do/projects/mountain-chicken-frog-conservation.

37.   Ibid.

38.   Ibid.

39.   Saving the Mountain Chicken: Breeding. (2014, December 21). *ZSL – Zoological Society of London*. [YouTube video]. Retrieved from https://youtu.be/oA2znXVTNWI.

40.   Mountain Chicken Conservation. *ZSL*. Retrieved from https://www.zsl.org/what-we-do/projects/mountain-chicken-frog-conservation.

41.   Hudson, M. A., Young, R. P., D'Urban Jackson, J., Orozco-terWengel, P., Martin, L., James, A., ... & Cunningham, A. A. (2016). Dynamics and genetics of a disease-driven species decline to near extinction: lessons for conservation. *Scientific Reports*, *6*, 30772. https://doi.org/10.1038/srep30772.

42.   Ibid.

43.   Ibid.

44.   IUCN SSC Amphibian Specialist Group. (2017). *Leptodactylus fallax*. *The IUCN Red List of Threatened Species*: e.T57125A3055585. https://dx.doi.org/10.2305/IUCN.UK.2017-3.RLTS.T57125A3055585.en.

45.   Hudson, M. (2017). Conserving the mountain chicken frog, a model for chytridiomycosis mitigation in nature. *Solitaire 28*. Retrieved from https://books.google.co.uk/books?hl=en&lr=&id=eCNFDwAAQBAJ&oi=fnd&pg=PA26&dq=heat+treatment+mountain+chicken+chytrid&ots=cbtVVmtQn4&sig=BAxCgbEHmaekxin_L9Oqt4Et2_I#v=onepage&q=heat%20treatment%20mountain%20chicken%20chytrid&f=false.

46.   Hudson, Michael. (2016). Conservation Management of the Mountain Chicken Frog. Doctor of Philosophy (PhD) thesis, University of Kent. Retrieved from http://kar.kent.ac.uk/57950.

47.   Garner, T. W., Schmidt, B. R., Martel, A., Pasmans, F., Muths, E., Cunningham, A. A., ... & Bosch, J. (2016). Mitigating amphibian chytridiomycoses in nature. *Philosophical Transactions of the Royal Society B: Biological Sciences*, *371*(1709), 20160207.

48.   Dagnano, M. (2018, November 9). It's getting hot hot hot! Controlling the chytrid fungus. [Weblog]. Mountain Chicken Recovery Programme. Retrieved from https://www.mountainchicken.org/blog/its-getting-hot-hot-ho t-controlling-the-chytrid-fungus/, now accessible via https://web.archive.org/ web/20240223163113/https://www.mountainchicken.org/blog/its-getting-hot-ho t-hot-controlling-the-chytrid-fungus/ (worth a visit for the photo of Kermit in a bath, if nothing else).

49.   ZSL. Mountain Chicken Conservation. Retrieved from https://www.zsl.org/ what-we-do/projects/mountain-chicken-frog-conservation.

50.   Andrew Cunningham, personal communication.

51.   Hudson, M. A., Young, R. P., D'Urban Jackson, J., Orozco-terWengel, P., Martin, L., James, A., ... & Cunningham, A. A. (2016). Dynamics and genetics of a disease-driven species decline to near extinction: lessons for conservation. *Scientific Reports*, *6*, 30772. https://doi.org/10.1038/srep30772.

52.   IUCN SSC Amphibian Specialist Group. (2017). *Leptodactylus fallax. The IUCN Red List of Threatened Species* 2017: e.T57125A3055585. https://dx.doi. org/10.2305/IUCN.UK.2017-3.RLTS.T57125A3055585.en.

53.   Dominica. Mountain Chicken Recovery Programme. Retrieved from https:// www.mountainchicken.org/conservation/dominica/, now accessible via https:// web.archive.org/web/20230930122419/https://www.mountainchicken.org/ conservation/dominica/.

54.   Mountain Chicken Conservation. *ZSL*. Retrieved from https://www.zsl.org/ what-we-do/projects/mountain-chicken-frog-conservation.

55.   Garner, T. W., Schmidt, B. R., Martel, A., Pasmans, F., Muths, E., Cunningham, A. A., ... & Bosch, J. (2016). Mitigating amphibian chytridiomycoses in nature. *Philosophical Transactions of the Royal Society B: Biological Sciences*, *371*(1709), 20160207. http://dx.doi.org/10.1098/rstb.2016.0207.

56.   Bletz, M. C., Loudon, A. H., Becker, M. H., Bell, S. C., Woodhams, D. C., Minbiole, K. P., & Harris, R. N. (2013). Mitigating amphibian chytridiomycosis with bioaugmentation: characteristics of effective probiotics and strategies for their selection and use. *Ecology Letters*, *16*(6), 807-820. https://doi.org/10.1111/ele.12099.

57.   Garner, T. W., Schmidt, B. R., Martel, A., Pasmans, F., Muths, E., Cunningham, A. A., ... & Bosch, J. (2016). Mitigating amphibian chytridiomycoses in nature. *Philosophical Transactions of the Royal Society B: Biological Sciences*, *371*(1709), 20160207. http://dx.doi.org/10.1098/rstb.2016.0207.

58.   Ibid.

59.   Zippel, K., Johnson, K., Gagliardo, R., Gibson, R., McFadden, M., Browne, R., ... & Townsend, E. (2011). The Amphibian Ark: a global community for ex situ conservation of amphibians. *Herpetological Conservation and Biology*, *6*(3), 340-352. https://herpconbio.org/Volume_6/Issue_3/Zippel_etal_2011.pdf.

60.   Garner, T. W., Schmidt, B. R., Martel, A., Pasmans, F., Muths, E., Cunningham, A. A., ... & Bosch, J. (2016). Mitigating amphibian

chytridiomycoses in nature. *Philosophical Transactions of the Royal Society B: Biological Sciences, 371*(1709), 20160207. http://dx.doi.org/10.1098/rstb.2016.0207.

61.   Waddle, A. (2024, June 26). Our 'frog saunas' could help save endangered species from the devastating chytrid fungus. *The Conversation.* Retrieved from https://theconversation.com/our-frog-saunas-could-help-save-endangered-s pecies-from-the-devastating-chytrid-fungus-231605.

62.   Holden, W. M., Reinert, L. K., Hanlon, S. M., Parris, M. J., & Rollins-Smith, L. A. (2015). Development of antimicrobial peptide defenses of southern leopard frogs, *Rana sphenocephala*, against the pathogenic chytrid fungus, *Batrachochytrium dendrobatidis. Developmental & Comparative Immunology, 48*(1), 65-75. https://doi.org/10.1016/j.dci.2014.09.003.

63.   Holden, W. M., Hanlon, S. M., Woodhams, D. C., Chappell, T. M., Wells, H. L., Glisson, S. M., ... & Rollins-Smith, L. A. (2015). Skin bacteria provide early protection for newly metamorphosed southern leopard frogs (*Rana sphenocephala*) against the frog-killing fungus, *Batrachochytrium dendrobatidis. Biological Conservation, 187*, 91-102. https://doi.org/10.1016/j.biocon.2015.04.007.

64.   Andrew Cunningham, personal communication.

65.   Crump, M. (2024). *Frog Day: A Story of 24 Hours and 24 Amphibian Lives*, The University of Chicago Press.

66.   Borzée, A.., Kielgast, J., Wren, S., Angulo, A., Chen, S., Magellan, K., ... & Bishop, P. J. (2021). Using the 2020 global pandemic as a springboard to highlight the need for amphibian conservation in eastern Asia. *Biological Conservation, 255*, 108973. https://doi.org/10.1016/j.biocon.2021.108973.

67.   Fisher, M. C., Henk, D. A., Briggs, C. J., Brownstein, J. S., Madoff, L. C., McCraw, S. L., & Gurr, S. J. (2012). Emerging fungal threats to animal, plant and ecosystem health. *Nature, 484*(7393), 186-194. https://doi.org/10.1038/nature10947.

68.   Hudson, M. A., Young, R. P., D'Urban Jackson, J., Orozco-terWengel, P., Martin, L., James, A., ... & Cunningham, A. A. (2016). Dynamics and genetics of a disease-driven species decline to near extinction: lessons for conservation. *Scientific Reports, 6*, 30772. https://doi.org/10.1038/srep30772.

69.   Andrew Cunningham, personal communication.

70.   Fisher, M. C., Henk, D. A., Briggs, C. J., Brownstein, J. S., Madoff, L. C., McCraw, S. L., & Gurr, S. J. (2012). Emerging fungal threats to animal, plant and ecosystem health. *Nature, 484*(7393), 186-194. https://doi.org/10.1038/nature10947.

71.   Andrew Cunningham, personal communication.

72.   Spitzen-van der Sluijs, A., Martel, A., Asselberghs, J., Bales, E.K.,Beukema, W., Bletz, M. C., ... & Lötters, S. (2016). Expanding distribution of lethal amphibian fungus *Batrachochytrium salamandrivorans* in Europe. *Emerging Infectious Diseases, 22*(7), 1286. https://doi.org/10.3201/eid2207.160109.

73. Martel, A., Blooi, M., Adriaensen, C., Van Rooij, P., Beukema, W., Fisher, M. C., ... & Pasmans, F. (2014). Recent introduction of a chytrid fungus endangers Western Palearctic salamanders. *Science, 346*(6209), 630-631. https://dx.doi.org/10.1126/science.1258268.

74. Ibid.

75. A North American strategic plan to prevent and control invasions of the lethal salamander pathogen *Batrachochytrium salamandrivorans*. (2022). *North American Bsal Task Force*. Retrieved from https://www.salamanderfungus.org/wp-content/uploads/sites/2/2022/03/Bsal-Strategic-Plan_March-2022_FINAL.pdf.

76. Miller, D.L., Carter, E.D., Hardman, R.H., Gray, M.J. (2023). Ranavirosis and Chytridiomycosis: the Impact on Amphibian Species. In D.A. Jessup and R.W. Radcliffe (Eds.), *Wildlife Disease and Health in Conservation*, Johns Hopkins University Press.

77. North American Bsal Task Force. A North American strategic plan to prevent and control invasions of the lethal salamander pathogen *Batrachochytrium salamandrivorans*. (2022). Retrieved from https://www.salamanderfungus.org/wp-content/uploads/sites/2/2022/03/Bsal-Strategic-Plan_March-2022_FINAL.pdf.

78. North American Bsal Task Force. A North American strategic plan to prevent and control invasions of the lethal salamander pathogen *Batrachochytrium salamandrivorans*. (2022). Retrieved from https://www.salamanderfungus.org/wp-content/uploads/sites/2/2022/03/Bsal-Strategic-Plan_March-2022_FINAL.pdf.

79. Spitzen-van der Sluijs, A., Martel, A., Asselberghs, J., Bales, E. K., Beukema, W., Bletz, M. C., ... & Lötters, S. (2016).Expanding distribution of lethal amphibian fungus *Batrachochytrium salamandrivorans* in Europe. *Emerging Infectious Diseases, 22*(7), 1286. https://doi.org/10.3201/eid2207.160109.

80. What is Bsal? *North American Bsal Task Force*. Retrieved from https://www.salamanderfungus.org/about-bsal/.

81. Grear, D. A., Mosher, B. A., Richgels, K. L., & Grant, E. H. (2021). Evaluation of regulatory action and surveillance as preventive risk-mitigation to an emerging global amphibian pathogen *Batrachochytrium salamandrivorans* (Bsal). *Biological Conservation, 260*, 109222. https://doi.org/10.1016/j.biocon.2021.109222.

82. North American Bsal Task Force. A North American strategic plan to prevent and control invasions of the lethal salamander pathogen *Batrachochytrium salamandrivorans*. (2022). Retrieved from https://www.salamanderfungus.org/wp-content/uploads/sites/2/2022/03/Bsal-Strategic-Plan_March-2022_FINAL.pdf.

83. Ibid.

84. Ibid.

85. Ibid.

86.   Hoyt, J. R., Kilpatrick, A. M., & Langwig, K. E. (2021). Ecology and impacts of white-nose syndrome on bats. *Nature Reviews Microbiology*, *19*(3), 196-210. https://doi.org/10.1038/s41579-020-00493-5.

87.   Ibid.

88.   White-Nose Syndrome Response Team. What is White-nose Syndrome? Retrieved from https://www.whitenosesyndrome.org/static-page/what-is-white-nose-syndrome.

89.   Hoyt, J. R., Kilpatrick, A. M., & Langwig, K. E. (2021). Ecology and impacts of white-nose syndrome on bats. *Nature Reviews Microbiology*, *19*(3), 196-210. https://doi.org/10.1038/s41579-020-00493-5.

90.   Ibid..

91.   Ibid.

92.   What is White-nose Syndrome? *White-Nose Syndrome Response Team*. Retrieved from https://www.whitenosesyndrome.org/static-page/what-is-white-nose-syndrome.

93.   Hoyt, J. R., Kilpatrick, A. M., & Langwig, K. E. (2021). Ecology and impacts of white-nose syndrome on bats. *Nature Reviews Microbiology*, *19*(3), 196-210. https://doi.org/10.1038/s41579-020-00493-5.

94.   Ibid.

95.   Ibid.

96.   Bats Affected by WNS. *White-Nose Syndrome Response Team*. Retrieved from https://www.whitenosesyndrome.org/static-page/bats-affected-by-wns.

97.   Frick, W. F., Pollock, J. F., Hicks, A. C., Langwig, K. E., Reynolds, D. S., Turner, G. G., ... & Kunz, T. H. (2010). An emerging disease causes regional population collapse of a common North American bat species. *Science*, *329*(5992), 679-682. https://doi.org/10.1126/science.1188594.

98.   U.S. Fish and Wildlife Service. Northern Long-eared Bat Final Rule News Release. (2022, November 29). Retrieved from https://www.fws.gov/media/northern-long-eared-bat-final-rule-news-release.

99.   Ibid.

100.   U.S. Fish and Wildlife Service. Tricolored Bat. Retrieved from https://www.fws.gov/species/tricolored-bat-perimyotis-subflavus.

101.   Hoyt, J. R., Kilpatrick, A. M., & Langwig, K. E. (2021). Ecology and impacts of white-nose syndrome on bats. *Nature Reviews Microbiology*, *19*(3), 196-210. https://doi.org/10.1038/s41579-020-00493-5.

102.   The Fungus. *White-Nose Syndrome Response Team*. Retrieved from https://www.whitenosesyndrome.org/static-page/the-fungus.

103.   Palmer, J. M., Drees, K. P., Foster, J. T., & Lindner, D. L. (2018). Extreme sensitivity to ultraviolet light in the fungal pathogen causing white-nose syndrome of bats. *Nature Communications*, *9*(1), 35. https://doi.org/10.1038/s41467-017-02441-z.

104.   Northern Long-eared Bat. *U.S. Fish and Wildlife Service.* Retrieved from https://www.fws.gov/species/northern-long-eared-bat-myotis-septentrionalis.

105.   Plowright, R. K., Ahmed, A. N., Coulson, T., Crowther, T. W., Ejotre, I., Faust, C. L., ... & Keeley, A. T. H. (2024). Ecological countermeasures to prevent pathogen spillover and subsequent pandemics. *Nature Communications, 15*(1), 2577. https://doi.org/10.1038/s41467-024-46151-9.

106.   Decontamination Information. *White-Nose Syndrome Response Team.* Retrieved from https://www.whitenosesyndrome.org/static-page/decontamination-information.

107.   Hill, K. (2017, March 29). Caving community on alert as threat of white-nose syndrome poses risk to Australian bat populations. ABC. Retrieved from https://www.abc.net.au/news/2017-03-30/caving-community-on-alert-with-threat-of-white-nose-syndrome/8397970.

108.   Holz, P., Hufschmid, J., Boardman, W. S., Cassey, P., Firestone, S., Lumsden, L. F., ... & Stevenson, M. (2019). Does the fungus causing white-nose syndrome pose a significant risk to Australian bats?. *Wildlife Research, 46*(8), 657-668. https://doi.org/10.1071/WR18194.

109.   Turbill, C. & Welbergen, J.A. (2020, January 20). Australia's threatened bats need protection from a silent killer: white-nose syndrome. *The Conversation.* Retrieved from https://theconversation.com/australias-threatened-bats-need-protection-from-a-silent-killer-white-nose-syndrome-129186.

10. CITY CATCHES

1.   Soulsbury, C. D., Iossa, G., Baker, P. J., Cole, N. C., Funk, S. M., & Harris, S. (2007). The impact of sarcoptic mange *Sarcoptes scabiei* on the British fox *Vulpes vulpes* population. *Mammal Review, 37*(4), 278-296. https://doi.org/10.1111/j.1365-2907.2007.00100.x.

2.   Escobar, L. E., Carver, S., Cross, P. C., Rossi, L., Almberg, E. S., Yabsley, M. J., ... & Astorga, F. (2022). Sarcoptic mange: An emerging panzootic in wildlife. *Transboundary and Emerging Diseases, 69*(3), 927-942. https://doi.org/10.1111/tbed.14082.

3.   Mange. *Cornell Wildlife Health Lab.* Retrieved from https://cwhl.vet.cornell.edu/disease/mange#collapse19.

4.   Escobar, L. E., Carver, S., Cross, P. C., Rossi, L., Almberg, E. S., Yabsley, M. J., ... & Astorga, F. (2022). Sarcoptic mange: An emerging panzootic in wildlife. *Transboundary and Emerging Diseases, 69*(3), 927-942. https://doi.org/10.1111/tbed.14082.

5.   Mange. *Cornell Wildlife Health Lab.* Retrieved from https://cwhl.vet.cornell.edu/disease/mange#collapse19.

6.   Rohde, K. (2001). Parasitism. *Encyclopedia of Biodiversity*, S.A. Levin (Ed.) p463-484. https://doi.org/10.1016/B0-12-226865-2/00217-0.

7.   Emily Almberg, personal communication.

8.   Ibid.

9.   Ibid.

10.   Scott, D. M., Baker, R., Tomlinson, A., Berg, M. J., Charman, N., & Tolhurst, B. A. (2020). Spatial distribution of sarcoptic mange (*Sarcoptes scabiei*) in urban foxes (*Vulpes vulpes*) in Great Britain as determined by citizen science. *Urban Ecosystems, 23*(5), 1127-1140. https://doi.org/10.1007/s11252-020-00985-5.

11.   Soulsbury, C. D., Iossa, G., Baker, P. J., Cole, N. C., Funk, S. M., & Harris, S. (2007). The impact of sarcoptic mange *Sarcoptes scabiei* on the British fox *Vulpes vulpes* population. *Mammal Review, 37*(4), 278-296. https://doi.org/10.1111/j.1365-2907.2007.00100.x.

12.   Ibid.

13.   Ibid.

14.   Ibid.

15.   Downs, A. M. R., Harvey, I., & Kennedy, C. T. C. (1999). The epidemiology of head lice and scabies in the UK. *Epidemiology & Infection, 122*(3), 471-477.

16.   Rasmussen J. E. (1994). Scabies. *Pediatrics in Review, 15*(3), 110–114. https://doi.org/10.1542/pir.15-3-110.

17.   Beeton, N. J., Carver, S., & Forbes, L. K. (2019). A model for the treatment of environmentally transmitted sarcoptic mange in bare-nosed wombats (*Vombatus ursinus*). *Journal of Theoretical Biology, 462*, 466-474. https://doi.org/10.1016/j.jtbi.2018.11.033.

18.   Urban foxes, your questions answered. *Birmingham & Black Country Wildlife Trust.* Retrieved from https://www.bbcwildlife.org.uk/urban-fox.

19.   Soulsbury, C. D., Iossa, G., Baker, P. J., Cole, N. C., Funk, S. M., & Harris, S. (2007). The impact of sarcoptic mange *Sarcoptes scabiei* on the British fox *Vulpes vulpes* population. *Mammal Review, 37*(4), 278-296. https://doi.org/10.1111/j.1365-2907.2007.00100.x.

20.   Walz, G.H. (1809). *Natur und Behandlung der Schaaf-Raude.* Stuttgart, Württemberg: bei Johann Friedrich Steinkopf.

21.   Escobar, L. E., Carver, S., Cross, P. C., Rossi, L., Almberg, E. S., Yabsley, M. J., ... & Astorga, F. (2022). Sarcoptic mange: An emerging panzootic in wildlife. *Transboundary and Emerging Diseases, 69*(3), 927-942. https://doi.org/10.1111/tbed.14082.

22.   Emily Almberg, personal communication.

23.   Andriantsoanirina, V., Fang, F., Ariey, F., Izri, A., Foulet, F., Botterel, F., ... & Durand, R. (2016). Are humans the initial source of canine mange? *Parasites & Vectors, 9*, 177. https://doi.org/10.1186/s13071-016-1456-y.

24.   Ibid.

25.   Janet Foley, personal communication.

26. Soulsbury, C. D., Iossa, G., Baker, P. J., Cole, N. C., Funk, S. M., & Harris, S. (2007). The impact of sarcoptic mange *Sarcoptes scabiei* on the British fox *Vulpes vulpes* population. *Mammal Review*, *37*(4), 278-296. https://doi.org/10.1111/j.1365-2907.2007.00100.x.

27. Lindström, E. R., Andrén, H., Angelstam, P., Cederlund, G., Hörnfeldt, B., Jäderberg, L., ... & Swenson, J. E. (1994). Disease reveals the predator: sarcoptic mange, red fox predation, and prey populations. *Ecology*, *75*(4), 1042-1049. https://doi.org/10.2307/1939428.

28. Monk, J. D., Smith, J. A., Donadío, E., Perrig, P. L., Crego, R. D., Fileni, M., ... & Middleton, A. D. (2022). Cascading effects of a disease outbreak in a remote protected area. *Ecology Letters*, *25*(5), 1152-1163. https://doi.org/10.1111/ele.13983.

29. Escobar, L. E., Carver, S., Cross, P. C., Rossi, L., Almberg, E. S., Yabsley, M. J., ... & Astorga, F. (2022). Sarcoptic mange: An emerging panzootic in wildlife. *Transboundary and Emerging Diseases*, *69*(3), 927-942. https://doi.org/10.1111/tbed.14082.

30. Ibid.

31. Old, J. M., Sengupta, C., Narayan, E., & Wolfenden, J. (2018). Sarcoptic mange in wombats – A review and future research directions. *Transboundary and Emerging Diseases*, *65*(2), 399-407. https://doi.org/10.1111/tbed.12770.

32. Escobar, L. E., Carver, S., Cross, P. C., Rossi, L., Almberg, E. S., Yabsley, M. J., ... & Astorga, F. (2022). Sarcoptic mange: An emerging panzootic in wildlife. *Transboundary and Emerging Diseases*, *69*(3), 927-942. https://doi.org/10.1111/tbed.14082.

33. Ibid.

34. Ibid.

35. Ibid.

36. Ibid.

37. Bandi, K. M., & Saikumar, C. (2013). Sarcoptic mange: a zoonotic ectoparasitic skin disease. *Journal of Clinical and Diagnostic Research* *7*(1), 156–157. https://doi.org/10.7860/JCDR/2012/4839.2694.

38. Gakuya, F., Rossi, L., Ombui, J., Maingi, N., Muchemi, G., Ogara, W., ... & Alasaad, S. (2011). The curse of the prey: *Sarcoptes* mite molecular analysis reveals potential prey-to-predator parasitic infestation in wild animals from Masai Mara, Kenya. *Parasites & Vectors*, *4*, 1-7. https://doi.org/10.1186/1756-3305-4-193.

39. Escobar, L. E., Carver, S., Cross, P. C., Rossi, L., Almberg, E. S., Yabsley, M. J., ... & Astorga, F. (2022). Sarcoptic mange: An emerging panzootic in wildlife. *Transboundary and Emerging Diseases*, *69*(3), 927-942. https://doi.org/10.1111/tbed.14082.

40. San Joaquin Kit Fox. *U.S. Fish & Wildlife Service*. Retrieved from https://www.fws.gov/rivers/species/san-joaquin-kit-fox-vulpes-macrotis-mutica.

41. Ibid.

42. Ibid.

43. Ibid.

44. Ibid.

45. QuickFacts, Bakersfield city, California. (2023, July 1). *United States Census Bureau*. Retrieved from https://www.census.gov/quickfacts/fact/table/ bakersfieldcitycalifornia/PST045222.

46. Janet Foley, personal communication.

47. Riley, S. P., Bromley, C., Poppenga, R. H., Uzal, F. A., Whited, L., & Sauvajot, R. M. (2007). Anticoagulant exposure and notoedric mange in bobcats and mountain lions in urban southern California. *The Journal of Wildlife Management, 71*(6), 1874-1884. https://doi.org/10.2193/2005-615.

48. Saito, M. U., & Sonoda, Y. (2017). Symptomatic raccoon dogs and sarcoptic mange along an urban gradient. *EcoHealth, 14*(2), 318-328. https://doi.org/10.1007/ s10393-017-1233-1.

49. Cummings, C. R., Hernandez, S. M., Murray, M., Ellison, T., Adams, H. C., Cooper, R. E., ... & Navara, K. J. (2020). Effects of an anthropogenic diet on indicators of physiological challenge and immunity of white ibis nestlings raised in captivity. *Ecology and Evolution, 10*(15), 8416-8428. https://doi.org/10.1002/ece3.6548.

50. Riley, S. P., Bromley, C., Poppenga, R. H., Uzal, F. A., Whited, L., & Sauvajot, R. M. (2007). Anticoagulant exposure and notoedric mange in bobcats and mountain lions in urban southern California. *The Journal of Wildlife Management, 71*(6), 1874-1884. https://doi.org/10.2193/2005-615.

51. Ibid.

52. Ibid.

53. Ibid.

54. Ibid.

55. Ibid.

56. Almberg, E.S. (2015). *The invasion, dynamics, and impacts of infectious disease in Yellowstone's wolf population* (Doctoral dissertation). Retrieved from https://etda. libraries.psu.edu/files/final_submissions/10628.

57. Ibid.

58. Almberg, E. S., Cross, P. C., Dobson, A. P., Smith, D. W., & Hudson, P. J. (2012). Parasite invasion following host reintroduction: a case study of Yellowstone's wolves. *Philosophical Transactions of the Royal Society B: Biological Sciences, 367*(1604), 2840-2851. https://doi.org/10.1098/rstb.2011.0369.

59. Mange. *Cornell Wildlife Health Lab*. Retrieved from https://cwhl.vet.cornell. edu/disease/mange#collapse21.

60. Janet Foley, personal communication.

61. Beeton, N. J., Carver, S., & Forbes, L. K. (2019). A model for the treatment of environmentally transmitted sarcoptic mange in bare-nosed wombats (*Vombatus*

ursinus). *Journal of Theoretical Biology, 462*, 466-474. https://doi.org/10.1016/j.jtbi.2018.11.033.

62.   Old, J. M., Sengupta, C., Narayan, E., & Wolfenden, J. (2018). Sarcoptic mange in wombats – A review and future research directions. *Transboundary and Emerging Diseases, 65*(2), 399-407. https://doi.org/10.1111/tbed.12770.

63.   Bevan, C. (2018, September 12). Success in treating wombat mange. *WPSA*. Retrieved from https://static1.squarespace.com/static/5d0991e19ea6e200018bdb37/t/5d1d9f04ac21b90001bb988b/1562222346339/WPSA-landowner-mange-treatment.pdf.

64.   Skerratt, L., Death, C., Hufschmid, J., Carver, S., Meredith, A. (2021). Guidelines for the treatment of Australian wildlife with sarcoptic mange, Part 2 – Literature review. NESP Threatened Species Recovery Hub. Retrieved from https://www.nespthreatenedspecies.edu.au/media/bx5pmfxu/1-4-4-guidelines-for-treatment-of-australian-wildlife-with-sarcoptic-mange-report-part-2-lit-review_v4.pdf.

65.   Ibid.

66.   Old, J. M., Sengupta, C., Narayan, E., & Wolfenden, J. (2018). Sarcoptic mange in wombats – A review and future research directions. *Transboundary and Emerging Diseases, 65*(2), 399-407. https://doi.org/10.1111/tbed.12770.

67.   WPSA. Mange treatment using pole method. (2021, June 4). [YouTube video]. Retrieved from https://youtu.be/czwkuib-Efw.

68.   Old, J. M., Sengupta, C., Narayan, E., & Wolfenden, J. (2018). Sarcoptic mange in wombats – A review and future research directions. *Transboundary and Emerging Diseases, 65*(2), 399-407. https://doi.org/10.1111/tbed.12770.

69.   How to make a burrow flap. *WPSA*. Retrieved from https://static1.squarespace.com/static/5d0991e19ea6e200018bdb37/t/663acbdbc88e0f5db3e78529/1715129318597/How+to+make+a+Burrow+Flap+new.pdf.

70.   Rowe, M. L., Whiteley, P. L., & Carver, S. (2019). The treatment of sarcoptic mange in wildlife: a systematic review. *Parasites & Vectors, 12*, 1-14. https://doi.org/10.1186/s13071-019-3340-z.

71.   Skerratt, L., Death, C., Hufschmid, J., Carver, S., Meredith, A. (2021). Guidelines for the treatment of Australian wildlife with sarcoptic mange, Part 2 – Literature review. NESP Threatened Species Recovery Hub. Retrieved from https://www.nespthreatenedspecies.edu.au/media/bx5pmfxu/1-4-4-guidelines-for-treatment-of-australian-wildlife-with-sarcoptic-mange-report-part-2-lit-review_v4.pdf.

72.   Escobar, L. E., Carver, S., Cross, P. C., Rossi, L., Almberg, E. S., Yabsley, M. J., ... & Astorga, F. (2022). Sarcoptic mange: An emerging panzootic in wildlife. *Transboundary and Emerging Diseases, 69*(3), 927-942. https://doi.org/10.1111/tbed.14082.

73.   Foley, P., Foley, J., Rudd, J., Clifford, D., Westall, T., & Cypher, B. (2023). Spatio-temporal and transmission dynamics of sarcoptic mange in an endangered

New World kit fox. *PLoS ONE, 18*(2), e0280283. https://doi.org/10.1371/journal.pone.0280283.

74. Bandi, K. M., & Saikumar, C. (2013). Sarcoptic mange: a zoonotic ectoparasitic skin disease. *Journal of Clinical and Diagnostic Research 7*(1), 156–157. https://doi.org/10.7860/JCDR/2012/4839.2694.

75. Pisano, S., Ryser-Degiorgis, M., Rossi, L., Peano, A., Keckeis, K., & Roosje, P. (2019). Sarcoptic Mange of Fox Origin in Multiple Farm Animals and Scabies in Humans, Switzerland, 2018. *Emerging Infectious Diseases, 25*(6), 1235-1238. https://doi.org/10.3201/eid2506.181891.

76. Ibid.

77. Ibid.

78. Emily Almberg, personal communication

11. WARMING AND WESTERNING

1. Roehrig, J. T. (2013). West Nile virus in the United States – a historical perspective. *Viruses, 5*(12), 3088-3108. https://doi.org/10.3390/v5123088.

2. Ibid.

3. Lanciotti, R. S., Roehrig, J. T., Deubel, V., Smith, J., Parker, M., Steele, K., ... & Gubler, D. J. (1999). Origin of the West Nile virus responsible for an outbreak of encephalitis in the north-eastern United States. *Science, 286*(5448), 2333-2337. https://doi.org/10.1126/science.286.5448.2333.

4. Centers for Disease Control and Prevention (CDC). (1999). Outbreak of West Nile-like viral encephalitis – New York, 1999. *Morbidity and Mortality Weekly Report (MMWR), 48*(38), 845-849. https://www.cdc.gov/mmwr/preview/mmwrhtml/mm4838a1.htm.

5. Roehrig, J. T. (2013). West Nile virus in the United States – a historical perspective. *Viruses, 5*(12), 3088-3108. https://doi.org/10.3390/v5123088.

6. Nash, D., Mostashari, F., Fine, A., Miller, J., O'Leary, D., Murray, K., ... & Layton, M. (2001). The outbreak of West Nile virus infection in the New York City area in 1999. *New England Journal of Medicine, 344*(24), 1807-1814. https://doi.org/10.1056/NEJM200106143442401.

7. Roehrig, J. T. (2013). West Nile virus in the United States – a historical perspective. *Viruses, 5*(12), 3088-3108. https://doi.org/10.3390/v5123088.

8. CDC. Data and Maps for St. Louis Encephalitis. (2024, June 18). Retrieved from https://www.cdc.gov/sle/data-maps/index.html.

9. Roehrig, J. T. (2013). West Nile virus in the United States – a historical perspective. *Viruses, 5*(12), 3088-3108. https://doi.org/10.3390/v5123088.

10. Ibid.

11. Bhatia, M. (2022, September 2022). In a hopeful first, American crows survive West Nile virus. *Cornell Lab, All About Birds*. Retrieved from https://www.allaboutbirds.org/news/in-a-hopeful-first-american-crows-survive-west-nile-virus/.

12.   Roehrig, J. T. (2013). West Nile virus in the United States – a historical perspective. *Viruses*, *5*(12), 3088-3108. https://doi.org/10.3390/v5123088.

13.   Ibid.

14.   Ibid.

15.   Briese, T., Jia, X. Y., Huang, C., Grady, L. J., & Lipkin, W. L. (1999). Identification of a Kunjin/West Nile-like flavivirus in brains of patients with New York encephalitis. *The Lancet*, *354*(9186), 1261-1262. https://doi.org/10.1016/S0140-6736(99)04576-6.

16.   Lanciotti, R. S., Roehrig, J. T., Deubel, V., Smith, J., Parker, M., Steele, K., ... & Gubler, D. J. (1999). Origin of the West Nile virus responsible for an outbreak of encephalitis in the north-eastern United States. *Science*, *286*(5448), 2333-2337. https://doi.org/10.1126/science.286.5448.2333.

17.   Gavier-Widén, D., Duff, J.P, & Meredith, A. (2012). *Infectious diseases of wild mammals and birds in Europe*, Chichester, UK: Wiley-Blackwell.

18.   Roehrig, J. T. (2013). West Nile virus in the United States – a historical perspective. *Viruses*, *5*(12), 3088-3108. https://doi.org/10.3390/v5123088.

19.   Gavier-Widén, D., Duff, J.P, & Meredith, A. (2012). *Infectious diseases of wild mammals and birds in Europe*, Chichester, UK: Wiley-Blackwell.

20.   Roehrig, J. T. (2013). West Nile virus in the United States – a historical perspective. *Viruses*, *5*(12), 3088-3108. https://doi.org/10.3390/v5123088.

21.   Ibid.

22.   Gavier-Widén, D., Duff, J.P, & Meredith, A. (2012). *Infectious diseases of wild mammals and birds in Europe, Chichester*, UK: Wiley-Blackwell.

23.   Roehrig, J. T. (2013). West Nile virus in the United States – a historical perspective. *Viruses*, *5*(12), 3088-3108. https://doi.org/10.3390/v5123088.

24.   Malkinson, M., Banet, C., Weisman, Y., Pokamunski, S., King, R., Drouet, M. T., & Deubel, V. (2002). Introduction of West Nile virus in the Middle East by migrating white storks. *Emerging Infectious Diseases*, *8*(4), 392–397. https://doi.org/10.3201/eid0804.010217.

25.   Ibid.

26.   Gavier-Widén, D., Duff, J.P, & Meredith, A. (2012). *Infectious diseases of wild mammals and birds in Europe*, Chichester, UK: Wiley-Blackwell.

27.   Ibid.

28.   Ibid.

29.   Ibid.

30.   Ibid.

31.   Roehrig, J. T. (2013). West Nile virus in the United States – a historical perspective. *Viruses*, *5*(12), 3088-3108. https://doi.org/10.3390/v5123088.

32.   Malkinson, M., Banet, C., Machany, S., Weisman, Y., Frommer, A., Bock, R., ... & Lachmi, B. (1998). Virus encephalomyelitis of geese: some properties of the viral isolate. *Israel Journal of Veterinary Medicine*, *53*, 44.

33. Roehrig, J. T. (2013). West Nile virus in the United States – a historical perspective. *Viruses*, *5*(12), 3088-3108. https://doi.org/10.3390/v5123088.

34. Rompoti, S. (2024, March 27). The Top USA Ports – The Largest and Busiest Ports in America. *Thomas*. Retrieved from https://www.thomasnet.com/insights/top-usa-ports/.

35. Roehrig, J. T. (2013). West Nile virus in the United States – a historical perspective. *Viruses*, *5*(12), 3088-3108. https://doi.org/10.3390/v5123088.

36. Rappole, J. H., Derrickson, S. R., & Hubalek, Z. (2000). Migratory birds and spread of West Nile virus in the Western Hemisphere. *Emerging Infectious Diseases*, *6*(4), 319. https://doi.org/10.3201%2Feid0604.000401.

37. Roehrig, J. T. (2013). West Nile virus in the United States – a historical perspective. *Viruses*, *5*(12), 3088-3108. https://doi.org/10.3390/v5123088.

38. LaDeau, S. L., Kilpatrick, A. M., & Marra, P. P. (2007). West Nile virus emergence and large-scale declines of North American bird populations. *Nature*, *447*(7145), 710-713. https://doi.org/10.1038/nature05829.

39. American Crow. *Cornell Lab, All About Birds*. Retrieved from https://www.allaboutbirds.org/guide/american_crow/overview.

40. Shannon LaDeau, personal communication.

41. Nemeth, N.M. and Oesterle, P.T. (2014). West Nile Virus and Avian Conservation. *International Zoo Yearbook*, *48*, 101-115. https://doi.org/10.1111/izy.12031.

42. LaDeau, S. L., Kilpatrick, A. M., & Marra, P. P. (2007). West Nile virus emergence and large-scale declines of North American bird populations. *Nature*, *447*(7145), 710-713. https://doi.org/10.1038/nature05829.

43. CDC Species of Dead Birds in Which West Nile Virus Has Been Detected, United States, 1999–2016. *Centers for Disease Control and Prevention*. Retrieved from https://www.cdc.gov/westnile/dead-birds/index.html#https://www.cdc.gov/west-nile-virus/media/files/Bird_species_West_Nile_virus_U.S..pdf.

44. Bhatia, M. (2022, September 2022). In a hopeful first, American crows survive West Nile virus. *Cornell Lab, All About Birds*. Retrieved from https://www.allaboutbirds.org/news/in-a-hopeful-first-american-crows-survive-west-nile-virus/.

45. Nemeth, N.M. and Oesterle, P.T. (2014). West Nile Virus and Avian Conservation. *International Zoo Yearbook*, *48*, 101-115. https://doi.org/10.1111/izy.12031.

46. Jiménez de Oya, N., Escribano-Romero, E., Blázquez, A. B., Martín-Acebes, M. A., & Saiz, J. C. (2019). Current Progress of Avian Vaccines Against West Nile Virus. *Vaccines*, *7*(4), 126. https://doi.org/10.3390/vaccines7040126.

47. Nemeth, N.M. and Oesterle, P.T. (2014). West Nile Virus and Avian Conservation. *International Zoo Yearbook*, *48*, 101-115. https://doi.org/10.1111/izy.12031.

48. American Crow Identification. *Cornell Lab, All About Birds*. Retrieved from https://www.allaboutbirds.org/guide/American_Crow/id.

49.   McLean, R. G. (2006). West Nile virus in North American birds. *USDA National Wildlife Research Center – Staff Publications 425*. Retrieved from https://digitalcommons.unl.edu/icwdm_usdanwrc/425.

50.   Shannon LaDeau, personal communication.

51.   Gavier-Widén, D., Duff, J.P, & Meredith, A. (2012). *Infectious diseases of wild mammals and birds in Europe*, Chichester, UK: Wiley-Blackwell.

52.   Malkinson, M., Banet, C., Weisman, Y., Pokamunski, S., King, R., Drouet, M. T., & Deubel, V. (2002). Introduction of West Nile virus in the Middle East by migrating white storks. *Emerging Infectious Diseases, 8*(4), 392–397. https://doi.org/10.3201/eid0804.010217.

53.   Nelms, B. M., Macedo, P. A., Kothera, L., Savage, H. M., & Reisen, W. K. (2013). Overwintering biology of Culex (Diptera: Culicidae) mosquitoes in the Sacramento valley of California. *Journal of Medical Entomology, 50*(4), 773-790.

54.   Hinton, M. G., Reisen, W. K., Wheeler, S. S., & Townsend, A. K. (2015). West Nile virus activity in a winter roost of American crows (*Corvus brachyrhynchos*): is bird-to-bird transmission important in persistence and amplification? *Journal of Medical Entomology, 52*(4), 683-692. https://doi.org/10.1093/jme/tjv040.

55.   Ibid.

56.   Cornell Lab, All About Birds. American Crow. Retrieved from https://www.allaboutbirds.org/guide/american_crow/overview.

57.   LaDeau, S. L., Calder, C. A., Doran, P. J., & Marra, P. P. (2011). West Nile virus impacts in American crow populations are associated with human land use and climate. *Ecological Research, 26*, 909-916. https://doi.org/10.1007/s11284-010-0725-z.

58.   Shannon LaDeau, personal communication.

59.   Heidecke, J., Lavarello Schettini, A., & Rocklöv, J. (2023). West Nile virus eco-epidemiology and climate change. *PLoS Climate, 2*(5), e0000129. https://doi.org/10.1371/journal.pclm.0000129.

60.   Petersen, L. R., Brault, A. C., & Nasci, R. S. (2013). West Nile virus: review of the literature. *JAMA, 310*(3), 308–315. https://doi.org/10.1001/jama.2013.8042.

61.   Roehrig, J. T. (2013). West Nile virus in the United States – a historical perspective. *Viruses, 5*(12), 3088-3108. https://doi.org/10.3390/v5123088.

62.   Ibid.

63.   Petersen, L. R., Brault, A. C., & Nasci, R. S. (2013). West Nile virus: review of the literature. *JAMA, 310*(3), 308–315. https://doi.org/10.1001/jama.2013.8042.

64.   Roehrig, J. T. (2013). West Nile virus in the United States – a historical perspective. *Viruses, 5*(12), 3088-3108. https://doi.org/10.3390/v5123088.

65.   Ulbert S. (2019). West Nile virus vaccines – current situation and future directions. *Human Vaccines & Immunotherapeutics, 15*(10), 2337–2342. https://doi.org/10.1080/21645515.2019.1621149.

66.   Montecino-Latorre, D., & Barker, C. M. (2018). Overwintering of West Nile virus in a bird community with a communal crow roost. *Scientific Reports*, *8*(1), 6088. https://doi.org/10.1038/s41598-018-24133-4.

67.   Ibid.

68.   Shannon LaDeau, personal communication.

69.   LaDeau, S. L., Calder, C. A., Doran, P. J., & Marra, P. P. (2011). West Nile virus impacts in American crow populations are associated with human land use and climate. *Ecological Research*, *26*, 909-916. https://doi.org/10.1007/s11284-010-0725-z.

70.   Kramer, L. D., Ciota, A. T., & Kilpatrick, A. M. (2019). Introduction, spread, and establishment of West Nile virus in the Americas. *Journal of Medical Entomology*, *56*(6), 1448-1455. https://doi.org/10.1093/jme/tjz151.

71.   Shocket, M. S., Verwillow, A. B., Numazu, M. G., Slamani, H., Cohen, J. M., El Moustaid, F., ... & Mordecai, E. A. (2020). Transmission of West Nile and five other temperate mosquito-borne viruses peaks at temperatures between 23°C and 26°C. *Elife*, *9*, e58511. https://doi.org/10.7554/eLife.58511.

72.   Ibid.

73.   Gavier-Widén, D., Duff, J.P, & Meredith, A. (2012). *Infectious diseases of wild mammals and birds in Europe*, Chichester, UK: Wiley-Blackwell.

74.   Heidecke, J., Lavarello Schettini, A., & Rocklöv, J. (2023). West Nile virus eco-epidemiology and climate change. *PLoS Climate*, *2*(5), e0000129. https://doi.org/10.1371/journal.pclm.0000129.

75.   Paull, S. H., Horton, D. E., Ashfaq, M., Rastogi, D., Kramer, L. D., Diffenbaugh, N. S., & Kilpatrick, A. M. (2017). Drought and immunity determine the intensity of West Nile virus epidemics and climate change impacts. *Proceedings of the Royal Society B: Biological Sciences*, *284*(1848), 20162078. https://doi.org/10.1098/rspb.2016.2078.

76.   Ibid.

77.   Heidecke, J., Lavarello Schettini, A., & Rocklöv, J. (2023). West Nile virus eco-epidemiology and climate change. *PLoS Climate*, *2*(5), e0000129. https://doi.org/10.1371/journal.pclm.0000129.

78.   Historical data by year – West Nile virus seasonal surveillance. *European Centre for Disease Prevention and Control*. Retrieved from https://www.ecdc.europa.eu/en/west-nile-fever/surveillance-and-disease-data/historical.

79.   Farooq, Z., Sjödin, H., Semenza, J. C., Tozan, Y., Sewe, M. O., Wallin, J., & Rocklöv, J. (2023). European projections of West Nile virus transmission under climate change scenarios. *One Health*, *16*, 100509. https://doi.org/10.1016/j.onehlt.2023.100509.

80.   Ulbert S. (2019). West Nile virus vaccines – current situation and future directions. *Human Vaccines & Immunotherapeutics*, *15*(10), 2337–2342. https://doi.org/10.1080/21645515.2019.1621149.

81.   Ewing, D. A., Purse, B. V., Cobbold, C. A., & White, S. M. (2021). A novel approach for predicting risk of vector-borne disease establishment in marginal temperate environments under climate change: West Nile virus in the UK. *Journal of the Royal Society Interface*, *18*(178), 20210049. https://doi.org/10.1098/rsif.2021.0049.

82.   Akinsulie, O. C., Adesola, R. O., Bakre, A., Adebowale, O. O., Adeleke, R., Ogunleye, S. C., & Oladapo, I. P. (2023). Usutu virus: An emerging flavivirus with potential threat to public health in Africa: Nigeria as a case study. *Frontiers in Veterinary Science*, *10*, 1115501. https://doi.org/10.3389/fvets.2023.1115501.

83.   Folly, A. J., Lawson, B., Lean, F. Z., McCracken, F., Spiro, S., John, S. K., ... & McElhinney, L. M. (2020). Detection of Usutu virus infection in wild birds in the United Kingdom, 2020. *Eurosurveillance*, *25*(41), 2001732. https://doi.org/10.2807/1560-7917.ES.2020.25.41.2001732.

84.   Kilpatrick, A. M., Dupuis, A. P., Chang, G. J., & Kramer, L. D. (2010). DNA vaccination of American robins (*Turdus migratorius*) against West Nile virus. *Vector-borne and Zoonotic Diseases*, *10*(4), 377–380. https://doi.org/10.1089/vbz.2009.0029.

85.   Jiménez de Oya, N., Escribano-Romero, E., Blázquez, A.-B., Martín-Acebes, M.A., Saiz, J.-C. (2019). Current Progress of Avian Vaccines Against West Nile Virus. *Vaccines*, *7*, 126. https://doi.org/10.3390/vaccines7040126.

86.   Chang, G. J. J., Davis, B. S., Stringfield, C., & Lutz, C. (2007). Prospective immunization of the endangered California condor (*Gymnogyps californianus*) protects this species from lethal West Nile virus infection. *Vaccine*, *25*(12), 2325-2330. https://doi.org/10.1016/j.vaccine.2006.11.056.

87.   Shannon LaDeau, personal communication.

88.   Hopf, C., Bunting, E., Clark, A., & Childs-Sanford, S. (2022). Survival and release of 5 American Crows (*Corvus brachyrhynchos*) naturally infected with West Nile virus. *Journal of Avian Medicine and Surgery*, *36*(1), 85-91. https://doi.org/10.1647/20-00112.

89.   Bhatia, M. (2022, September 2022). In a hopeful first, American crows survive West Nile virus. *Cornell Lab, All About Birds*. Retrieved from https://www.allaboutbirds.org/news/in-a-hopeful-first-american-crows-survive-west-nile-virus/.

90.   Hopf, C., Bunting, E., Clark, A., & Childs-Sanford, S. (2022). Survival and release of 5 American Crows (*Corvus brachyrhynchos*) naturally infected with West Nile virus. *Journal of Avian Medicine and Surgery*, *36*(1), 85-91. https://doi.org/10.1647/20-00112.

91.   Kilpatrick, A. M., Gluzberg, Y., Burgett, J. & Daszak, P. (2004): Quantitative risk assessment of the pathways by which West Nile virus could reach Hawaii. *EcoHealth 1*, 205–209. https://doi.org/10.1007/s10393-004-0086-6.

12. MONKEY MIX-UP

1.   de Melo, F.R., Boubli, J.P., Mittermeier, R.A., Jerusalinsky, L., Tabacow, F.P., Ferraz, D.S. & Talebi, M. (2021). *Brachyteles hypoxanthus* (amended version of 2019 assessment). *The IUCN Red List of Threatened Species*, e.T2994A191693399. https://dx.doi.org/10.2305/IUCN.UK.2021-1.RLTS.T2994A191693399.en.

2.   Edkins, T. *Brachyteles hypoxanthus* northern muriqui. *Animal Diversity Web, University of Michigan Museum of Zoology*. Retrieved from https://animaldiversity. org/accounts/Brachyteles_hypoxanthus//.

3.   Karen Strier, personal communication.

4.   Mahon, E. (2023, October 12). Conservation, community, and a love for big monkeys: Karen Strier celebrates 40-year study of Northern muriqui. *University of Wisconsin-Madison News*. Retrieved from https://news.wisc.edu/conservatio n-community-and-a-love-for-big-monkeys-karen-strier-celebrates-40-year-stud y-of-northern-muriqui/.

5.   Yellow fever. (2023, May 31). *World Health Organization*. Retrieved from https:// www.who.int/news-room/fact-sheets/detail/yellow-fever.

6.   Possamai, C. B., Rodrigues de Melo, F., Mendes, S. L., & Strier, K. B. (2022). Demographic changes in an Atlantic Forest primate community following a yellow fever outbreak. *American Journal of Primatology*, *84*, e23425. https://doi. org/10.1002/ajp.23425.

7.   Ibid.

8.   Chippaux, J.-P., & Chippaux, A. (2018). Yellow fever in Africa and the Americas: a historical and epidemiological perspective. *Journal of Venomous Animals and Toxins including Tropical Diseases*, *24*, 20. https://doi.org/10.1186/ s40409-018-0162-y.

9.   Stokes, A., Bauert, J. H., & Hudson, N. P. (1997). Experimental transmission of yellow fever virus to laboratory animals. *International Journal of Infectious Diseases*, *2*(1), 54-59.

10.   Chippaux, J.-P., & Chippaux, A. (2018). Yellow fever in Africa and the Americas: a historical and epidemiological perspective. *Journal of Venomous Animals and Toxins including Tropical Diseases*, *24*, 20. https://doi.org/10.1186/ s40409-018-0162-y.

11.   Stokes, A., Bauert, J. H., & Hudson, N. P. (1997). Experimental transmission of yellow fever virus to laboratory animals. *International Journal of Infectious Diseases*, *2*(1), 54-59.

12.   Yellow fever. (2023, May 31). *World Health Organization*. Retrieved from https://www.who.int/news-room/fact-sheets/detail/yellow-fever.

13.   Chippaux, J.-P., & Chippaux, A. (2018). Yellow fever in Africa and the Americas: a historical and epidemiological perspective. *Journal of Venomous Animals and Toxins Including Tropical Diseases*, *24*, 20. https://doi.org/10.1186/ s40409-018-0162-y.

14.   Ibid.

15. Ibid.

16. Etymologia: *Aedes aegypti*. (2016). *Emerging Infectious Diseases*, *22*(10), 1807. https://doi.org/10.3201/eid2210.ET2210.

17. Ibid.

18. Yellow fever. (2023, May 31). *World Health Organization*. Retrieved from https://www.who.int/news-room/fact-sheets/detail/yellow-fever.

19. *Aedes aegypti* – Factsheet for experts. *Centre for Disease Prevention and Control*. Retrieved from https://www.ecdc.europa.eu/en/disease-vectors/facts/mosquito-factsheets/aedes-aegypti.

20. Ibid.

21. Chippaux, J.-P., & Chippaux, A. (2018). Yellow fever in Africa and the Americas: a historical and epidemiological perspective. *Journal of Venomous Animals and Toxins including Tropical Diseases*, *24*, 20. https://doi.org/10.1186/s40409-018-0162-y.

22. Marr, J. S., & Cathey, J. T. (2013). The 1802 Saint-Domingue yellow fever epidemic and the Louisiana Purchase. *Journal of Public Health Management and Practice*, *19*(1), 77-82. https://doi.org/10.1097/PHH.0b013e318252eea8.

23. Chippaux, J.-P., & Chippaux, A.. (2018). Yellow fever in Africa and the Americas: a historical and epidemiological perspective. *Journal of Venomous Animals and Toxins including Tropical Diseases*, *24*, 20. https://doi.org/10.1186/s40409-018-0162-y.

24. Tomlinson, W., & Hodgson, R. S. (2005). Centennial year of yellow fever eradication in New Orleans and the United States, 1905-2005. *The Journal of the Louisiana State Medical Society : Official Organ of the Louisiana State Medical Society*, *157*(4), 216–217.

25. Chippaux, J.-P., & Chippaux, A. (2018). Yellow fever in Africa and the Americas: a historical and epidemiological perspective. *Journal of Venomous Animals and Toxins including Tropical Diseases*, *24*, 20. https://doi.org/10.1186/s40409-018-0162-y.

26. Ibid.

27. Dietz, J. M., Hankerson, S. J., Alexandre, B. R., Henry, M. D., Martins, A. F., Ferraz, L. P., & Ruiz-Miranda, C. R. (2019). Yellow fever in Brazil threatens successful recovery of endangered golden lion tamarins. *Scientific Reports*, *9*(1), 12926. https://doi.org/10.1038/s41598-019-49199-6.

28. Chippaux, J.-P., & Chippaux, A. (2018). Yellow fever in Africa and the Americas: a historical and epidemiological perspective. *Journal of Venomous Animals and Toxins including Tropical Diseases*, *24*, 20. https://doi.org/10.1186/s40409-018-0162-y.

29. Ibid.

30. Ibid..

31. Ibid.

32. Ibid.

33. Possas, C., Lourenço-de-Oliveira, R., Tauil, P. L., Pinheiro, F. D. P., Pissinatti, A., Cunha, R. V. D., ... & Homma, A. (2018). Yellow fever outbreak in Brazil: the puzzle of rapid viral spread and challenges for immunisation. *Memórias do Instituto Oswaldo Cruz, 113*(10), e180278. https://doi.org/10.1590/0074-02760180278.

34. Jerusalinsky, L., Bicca-Marques, J.C., Neves, L.G., Alves, S.L., Ingberman, B., Buss, G., ... & Cortes-Ortíz, L. (2021). *Alouatta guariba* (amended version of 2020 assessment). *The IUCN Red List of Threatened Species*, e.T39916A190417874. https://dx.doi.org/10.2305/IUCN.UK.2021-1.RLTS.T39916A190417874.en.

35. Cordeiro, J. & Strier, K.B. (2017, March 21). Brown Howler & Muriqui Monkeys. [YouTube video]. *UW Madison-Campus Connection*. Retrieved from https://youtu.be/esYsS3S-i9w.

36. Mendes, F. D., & Ades, C. (2004). Vocal sequential exchanges and intragroup spacing in the Northern Muriqui *Brachyteles arachnoides hypoxanthus. Anais da Academia Brasileira de Ciências, 76*, 399-404. https://doi.org/10.1590/S0001-37652004000200032.

37. Strier, K. B., Tabacow, F. P., de Possamai, C. B., Ferreira, A. I., Nery, M. S., de Melo, F. R., & Mendes, S. L. (2019). Status of the northern muriqui (*Brachyteles hypoxanthus*) in the time of yellow fever. *Primates, 60*, 21-28. https://doi.org/10.1007/s10329-018-0701-8.

38. Strier, K.B. (2021). The limits of resilience. *Primates, 62*, 861–868. https://doi.org/10.1007/s10329-021-00953-3.

39. Tyrrell, K.A. (2017, March 21). Yellow fever killing thousands of monkeys in Brazil. *University of Wisconsin-Madison News*. Retrieved from https://news.wisc.edu/yellow-fever-killing-thousands-of-monkeys-in-brazil/.

40. Strier, K. B., Tabacow, F. P., de Possamai, C. B., Ferreira, A. I., Nery, M. S., de Melo, F. R., & Mendes, S. L. (2019). Status of the northern muriqui (*Brachyteles hypoxanthus*) in the time of yellow fever. *Primates, 60*, 21-28. https://doi.org/10.1007/s10329-018-0701-8.

41. Possamai, C. B., Rodrigues de Melo, F., Mendes, S. L., & Strier, K. B. (2022). Demographic changes in an Atlantic Forest primate community following a yellow fever outbreak. *American Journal of Primatology, 84*, e23425. https://doi.org/10.1002/ajp.23425.

42. Hance, J. (2021, August 10). Meet the kitten-sized, clown-faced monkey that's leaping toward extinction. Retrieved from https://news.mongabay.com/2021/08/meet-the-kitten-sized-clown-faced-monkey-thats-leaping-toward-extinction/.

43. Possamai, C. B., Rodrigues de Melo, F., Mendes, S. L., & Strier, K. B. (2022). Demographic changes in an Atlantic Forest primate community following a yellow fever outbreak. *American Journal of Primatology, 84*, e23425. https://doi.org/10.1002/ajp.23425.

44. Hance, J. (2021, August 10). Meet the kitten-sized, clown-faced monkey that's leaping toward extinction. Retrieved from https://news.mongabay.com/2021/08/meet-the-kitten-sized-clown-faced-monkey-thats-leaping-toward-extinction/.

45. Possamai, C. B., Rodrigues de Melo, F., Mendes, S. L., & Strier, K. B. (2022). Demographic changes in an Atlantic Forest primate community following a yellow fever outbreak. *American Journal of Primatology*, *84*, e23425. https://doi.org/10.1002/ajp.23425.

46. Dietz, J. M., Hankerson, S. J., Alexandre, B. R., Henry, M. D., Martins, A. F., Ferraz, L. P., & Ruiz-Miranda, C. R. (2019). Yellow fever in Brazil threatens successful recovery of endangered golden lion tamarins. *Scientific Reports*, *9*(1), 12926. https://doi.org/10.1038/s41598-019-49199-6.

47. Ibid.

48. Bicca Marques, J. C., Calegaro Marques, C., Rylands, A., Strier, K. B., Mittermeier, R., De Almeida, M. A., ... & Teixeira, D. S. (2017). Yellow fever threatens Atlantic Forest primates. *Science Advances*, *3*, e1600946/tab-e. eLetter at https://www.science.org/doi/10.1126/sciadv.1600946.

49. Collias, N., & Southwick, C. (1952). A field study of population density and social organization in howling monkeys. *Proceedings of the American Philosophical Society*, *96*(2), 143-156.

50. Ibid.

51. Ibid.

52. Bicca-Marques, J. C. (2009). Outbreak of yellow fever affects howler monkeys in southern Brazil. *Oryx*, *43*(2).

53. Possas, C., Lourenço-de-Oliveira, R., Tauil, P. L., Pinheiro, F. de P., Pissinatti, A., Cunha, R. V. da ., ... & Homma, A. (2018). Yellow fever outbreak in Brazil: the puzzle of rapid viral spread and challenges for immunisation. *Memórias Do Instituto Oswaldo Cruz*, *113*(10), e180278. https://doi.org/10.1590/0074-02760180278.

54. Tyrrell, K.A. (2017, March 21). Yellow fever killing thousands of monkeys in Brazil. *University of Wisconsin-Madison News*. Retrieved from https://news.wisc.edu/yellow-fever-killing-thousands-of-monkeys-in-brazil/.

55. Possas, C., Lourenço-de-Oliveira, R., Tauil, P. L., Pinheiro, F. de P., Pissinatti, A., Cunha, R. V. da ., ... & Homma, A. (2018). Yellow fever outbreak in Brazil: the puzzle of rapid viral spread and challenges for immunisation. *Memórias Do Instituto Oswaldo Cruz*, *113*(10), e180278. https://doi.org/10.1590/0074-02760180278.

56. Ibid.

57. Tyrrell, K.A. (2017, March 21). Yellow fever killing thousands of monkeys in Brazil. *University of Wisconsin-Madison News*. Retrieved from https://news.wisc.edu/yellow-fever-killing-thousands-of-monkeys-in-brazil/.

58.	Possas, C., Lourenço-de-Oliveira, R., Tauil, P. L., Pinheiro, F. de P., Pissinatti, A., Cunha, R. V. da ., ... & Homma, A. (2018). Yellow fever outbreak in Brazil: the puzzle of rapid viral spread and challenges for immunisation. *Memórias Do Instituto Oswaldo Cruz, 113*(10), e180278. https://doi.org/10.1590/0074-02760180278.

59.	Ibid.

60.	Ibid.

61.	Ibid.

62.	Ibid.

63.	Chapman, C. A., Gillespie, T. R., & Goldberg, T. L. (2005). Primates and the ecology of their infectious diseases: how will anthropogenic change affect host-parasite interactions? *Evolutionary Anthropology: Issues, News, and Reviews, 14*(4), 134-144.

64.	Possas, C., Lourenço-de-Oliveira, R., Tauil, P. L., Pinheiro, F. D. P., Pissinatti, A., Cunha, R. V. D., ... & Homma, A. (2018). Yellow fever outbreak in Brazil: the puzzle of rapid viral spread and challenges for immunisation. *Memórias do Instituto Oswaldo Cruz, 113*(10), e180278. https://doi.org/10.1590/0074-02760180278.

65.	Ferreira-de-Lima, V. H., Câmara, D. C. P., Honório, N. A., & Lima-Camara, T. N. (2020). The Asian tiger mosquito in Brazil: Observations on biology and ecological interactions since its first detection in 1986. *Acta Tropica, 205*, 105386. https://doi.org/10.1016/j.actatropica.2020.105386.

66.	Possas, C., Lourenço-de-Oliveira, R., Tauil, P. L., Pinheiro, F. D. P., Pissinatti, A., Cunha, R. V. D., ... & Homma, A. (2018). Yellow fever outbreak in Brazil: the puzzle of rapid viral spread and challenges for immunisation. *Memórias do Instituto Oswaldo Cruz, 113*(10), e180278. https://doi.org/10.1590/0074-02760180278.

67.	Strier, K. B. (2018). Primate social behavior. *American Journal of Physical Anthropology, 165*(4), 801-812. https://doi.org/10.1002/ajpa.23369.

68.	World Health Organization. WHO, UNICEF, and Gavi, the Vaccine Alliance pass the mid-point of the global 10-year strategy to eliminate yellow fever epidemics. (2023, November 8). Retrieved from https://www.who.int/news/item/08-11-2023-who--unicef--and-gavi--the-vaccine-alliance-pass-the-mid-point-of-the-global-10-year-strategy-to-eliminate-yellow-fever-epidemics.

69.	MercoPress South Atlantic News Agency. Massive vaccination, 23.8 million, against yellow fever in Sao Paulo and Rio do Janeiro. (2018, January 26). Retrieved from https://en.mercopress.com/2018/01/26/massive-vaccination-23.8-million-against-yellow-fever-in-sao-paulo-and-rio-do-janeiro.

70.	Possas, C., Lourenço-de-Oliveira, R., Tauil, P. L., Pinheiro, F. de P., Pissinatti, A., Cunha, R. V. da , ... & Homma, A. (2018). Yellow fever outbreak in Brazil: the puzzle of rapid viral spread and challenges for immunisation. *Memórias Do Instituto Oswaldo Cruz, 113*(10), e180278. https://doi.org/10.1590/0074-02760180278.

71. Bicca Marques, J. C., Calegaro Marques, C., Rylands, A., Strier, K. B., Mittermeier, R., De Almeida, M. A., ... & Teixeira, D. S. (2017). Yellow fever threatens Atlantic Forest primates. *Science Advances, 3*, e1600946/tab-e. eLetter at https://www.science.org/doi/10.1126/sciadv.1600946.

72. Larson, C. (2023, February 1). Race to vaccinate rare wild monkeys gives hope for survival. *AP.* Retrieved from https://apnews.com/article/science-brazil-climate-and-environment-health-animals-8914729b25c06e1fb2f9e5eaec9d7981.

73. Larson, C. (2023, August 1). Once nearing extinction, Brazil's golden monkeys have rebounded from yellow fever, scientists say. *AP.* Retrieved from https://apnews.com/article/endangered-monkeys-golden-lion-tamarin-brazil-2a9dd25048e5df6da5ab9d5e5e9a5006.

74. Bicca Marques, J. C., Calegaro Marques, C., Rylands, A., Strier, K. B., Mittermeier, R., De Almeida, M. A., ... & Teixeira, D. S. (2017). Yellow fever threatens Atlantic Forest primates. *Science Advances, 3*, e1600946/tab-e. eLetter at https://www.science.org/doi/10.1126/sciadv.1600946.

75. Bicca-Marques, J. C., & de Freitas, D. S. (2010). The role of monkeys, mosquitoes, and humans in the occurrence of a yellow fever outbreak in a fragmented landscape in south Brazil: protecting howler monkeys is a matter of public health. *Tropical Conservation Science, 3*(1), 78-89. https://doi.org/10.1177/194008291000300107.

76. Possas, C., Lourenço-de-Oliveira, R., Tauil, P. L., Pinheiro, F. D. P., Pissinatti, A., Cunha, R. V. D., ... & Homma, A. (2018). Yellow fever outbreak in Brazil: the puzzle of rapid viral spread and challenges for immunisation. *Memórias do Instituto Oswaldo Cruz, 113*(10), e180278. https://doi.org/10.1590/0074-02760180278.

77. Bicca-Marques, J. C., & de Freitas, D. S. (2010). The role of monkeys, mosquitoes, and humans in the occurrence of a yellow fever outbreak in a fragmented landscape in south Brazil: protecting howler monkeys is a matter of public health. *Tropical Conservation Science, 3*(1), 78-89. https://doi.org/10.1177/194008291000300107.

78. Karen Gilardi, personal communication.

79. Strier, K. B. (1986). *The behavior and ecology of the woolly spider monkey, or muriqui (Brachyteles arachnoides E. Geoffroy 1806).* (Doctoral Dissertation). Harvard University. Retrieved from https://www.proquest.com/openview/41cd87861aa3c72e99212a23d06b5474/1.

80. Strier, K. B. (1994). Myth of the typical primate. *American Journal of Physical Anthropology, 37*(S19), 233-271. https://doi.org/10.1002/ajpa.1330370609.

81. Edkins, T. *Brachyteles hypoxanthus* northern muriqui. *Animal Diversity Web, University of Michigan Museum of Zoology.* Retrieved from https://animaldiversity.org/accounts/Brachyteles_hypoxanthus/.

82. Bicca-Marques, J. C., Chaves, Ó. M., & Hass, G. P. (2020). Howler monkey tolerance to habitat shrinking: Lifetime warranty or death sentence? *American Journal of Primatology, 82*(4), e23089. https://doi.org/10.1002/ajp.23089.

83.	Possas, C., Lourenço-de-Oliveira, R., Tauil, P. L., Pinheiro, F. de P., Pissinatti, A., Cunha, R. V. da, ... & Homma, A. (2018). Yellow fever outbreak in Brazil: the puzzle of rapid viral spread and challenges for immunisation. *Memórias Do Instituto Oswaldo Cruz, 113*(10), e180278. https://doi.org/10.1590/0074-02760180278.

84.	Possamai, C. B., Rodrigues de Melo, F., Mendes, S. L., & Strier, K. B. (2022). Demographic changes in an Atlantic Forest primate community following a yellow fever outbreak. *American Journal of Primatology, 84*, e23425. https://doi.org/10.1002/ajp.23425.

85.	Strier, K. B. (2018). Primate social behavior. *American Journal of Physical Anthropology, 165*(4), 801-812. https://doi.org/10.1002https://doi.org/10.1002/ajpa.23369.

13. BATS FOR BUSHMEAT

1.	This description is a montage from a couple of videos as I haven't seen gorillas or chimps in the wild. The first is Van Duzer, R. (2023, May 14). Mountain Gorilla Trek in Rwanda. [YouTube video]. Retrieved from https://www.youtube.com/watch?v=kKR27Th-zwc.

2.	Volcanoes Safaris. (2023, April 19). Volcanoes Safaris 25 Year Anniversary Film Teaser. [YouTube video]. Retrieved from https://youtu.be/p4JVqurBM8g?feature=shared.

3.	Dian Fossey Gorilla Fund. Dian Fossey Biography. Retrieved from https://gorillafund.org/who-we-are/dian-fossey/dian-fossey-bio/.

4.	Ibid.

5.	Ibid.

6.	Robbins, M. M., Gray, M., Fawcett, K. A., Nutter, F. B., Uwingeli, P., Mburanumwe, I., ... & Robbins, A. M. (2011). Extreme conservation leads to recovery of the Virunga mountain gorillas. *PloS ONE, 6*(6), e19788. https://doi.org/10.1371/journal.pone.0019788.

7.	Gorilla Doctors. History. Retrieved from https://www.gorilladoctors.org/about-us/history-past-gorilla-doctors/.

8.	Ibid.

9.	Ibid.

10.	Chomel, B. B. (2009). Zoonoses. *Encyclopedia of Microbiology*, 820–829. https://doi.org/10.1016/B978-012373944-5.00213-3.

11.	Leligdowicz, A., Fischer, W. A., Uyeki, T. M., Fletcher, T. E., Adhikari, N. K., Portella, G., ... & Fowler, R. A. (2016). Ebola virus disease and critical illness. *Critical Care, 20*, 1-14.

12.	Gilardi, K.V. (2003) Endangered mountain gorillas, Ebola and One Health, Chapter in *Wildlife Disease and Health in Conservation*, edited by DA Jessup and RW Radcliffe, Johns Hopkins University Press.

13. Report of an International Commission (1978). Ebola haemorrhagic fever in Zaire, 1976. *Bulletin of the World Health Organization*, *56*(2), 271–293.

14. Formenty, P., Boesch, C., Wyers, M., Steiner, C., Donati, F., Dind, F., ... & Guenno, B. L. (1999). Ebola virus outbreak among wild chimpanzees living in a rain forest of Côte d'Ivoire. *The Journal of Infectious Diseases*, *179* (Supplement 1), S120-S126. https://doi.org/10.1086/514296.

15. Ibid.

16. Ibid.

17. Ibid.

18. Leroy, E. M., Kumulungui, B., Pourrut, X., Rouquet, P., Hassanin, A., Yaba, P., ... & Swanepoel, R. (2005). Fruit bats as reservoirs of Ebola virus. *Nature*, *438*(7068), 575-576. https://doi.org/10.1038/438575a.

19. Leendertz, S. A. J., Gogarten, J. F., Düx, A., Calvignac-Spencer, S., & Leendertz, F. H. (2016). Assessing the evidence supporting fruit bats as the primary reservoirs for Ebola viruses. *EcoHealth*, *13*, 18-25. https://doi.org/10.1007/s10393-015-1053-0.

20. Leroy, E. M., Kumulungui, B., Pourrut, X., Rouquet, P., Hassanin, A., Yaba, P., ... & Swanepoel, R. (2005). Fruit bats as reservoirs of Ebola virus. *Nature*, *438*(7068), 575-576. https://doi.org/10.1038/438575a.

21. Kirsten Gilardi, personal communication.

22. Irving, A. T., Ahn, M., Goh, G., Anderson, D. E., & Wang, L. F. (2021). Lessons from the host defences of bats, a unique viral reservoir. *Nature*, *589*(7842), 363-370. https://doi.org/10.1038/s41586-020-03128-0.

23. Banerjee, A., Baker, M. L., Kulcsar, K., Misra, V., Plowright, R., & Mossman, K. (2020). Novel insights into immune systems of bats. *Frontiers in Immunology*, *11*, 26. https://doi.org/10.3389/fimmu.2020.00026.

24. Irving, A. T., Ahn, M., Goh, G., Anderson, D. E., & Wang, L. F. (2021). Lessons from the host defences of bats, a unique viral reservoir. *Nature*, *589*(7842), 363-370. https://doi.org/10.1038/s41586-020-03128-0.

25. Wilkinson, G. S., & South, J. M. (2002). Life history, ecology and longevity in bats. *Aging Cell*, *1*(2), 124-131.

26. Formenty, P., Boesch, C., Wyers, M., Steiner, C., Donati, F., Dind, F., ... & Guenno, B. L. (1999). Ebola virus outbreak among wild chimpanzees living in a rain forest of Côte d'Ivoire. *The Journal of Infectious Diseases*, *179* (Supplement 1), S120-S126. https://doi.org/10.1086/514296.

27. Earth Resources Observation and Science (EROS) Center. (2023, May 16). The Deforestation of the Upper Guinean Forest. *USGS*. Retrieved from https://www.usgs.gov/centers/eros/science/deforestation-upper-guinean-forest.

28. Formenty, P., Boesch, C., Wyers, M., Steiner, C., Donati, F., Dind, F., ... & Guenno, B. L. (1999). Ebola virus outbreak among wild chimpanzees living in a

rain forest of Côte d'Ivoire. *The Journal of Infectious Diseases*, *179* (Supplement 1), S120-S126. https://doi.org/10.1086/514296.

29. Wobeser, G. A. (2006). *Essentials of Disease in Wild Animals*, Blackwell Publishing Professional, p197

30. UN Environment Programme. How new laws could help combat the planetary crisis. (2021, June 24). Retrieved from https://www.unep.org/news-and-stories/story/how-new-laws-could-help-combat-planetary-crisis.

31. Bernstein, A. S., Ando, A. W., Loch-Temzelides, T., Vale, M. M., Li, B. V., Li, H., ... & Dobson, A. P. (2022). The costs and benefits of primary prevention of zoonotic pandemics. *Science Advances*, *8*(5), eabl4183. https://doi.org/10.1126/sciadv.abl4183.

32. Leroy, E., Gonzalez, J.P., Pourrut, X. (2007). Ebolavirus and Other Filoviruses. In J. E. Childs, J. S. Mackenzie and J. A. Richt (Eds). Wildlife and Emerging Zoonotic Diseases: The Biology, Circumstances and Consequences of Cross-Species Transmission. *Current Topics in Microbiology and Immunology*, *315*. https://doi.org/10.1007/978-3-540-70962-6_15.

33. Ibid.

34. Alexander, K. A., Sanderson, C. E., Marathe, M., Lewis, B. L., Rivers, C. M., Shaman, J., ... & Eubank, S. (2015). What factors might have led to the emergence of Ebola in West Africa? *PLoS Neglected Tropical Diseases*, *9*(6), e0003652. https://doi.org/10.1371/journal.pntd.0003652.

35. Formenty, P., Boesch, C., Wyers, M., Steiner, C., Donati, F., Dind, F., ... & Guenno, B. L. (1999). Ebola virus outbreak among wild chimpanzees living in a rain forest of Côte d'Ivoire. *The Journal of Infectious Diseases*, *179* (Supplement 1), S120-S126. https://doi.org/10.1086/514296.

36. Ibid.

37. Jahrling, P. B., Geisbert, T. W., Johnson, E. D., Peters, C. J., Dalgard, D. W., & Hall, W. C. (1990). Preliminary report: isolation of Ebola virus from monkeys imported to USA. *The Lancet*, *335*(8688), 502-505. https://doi.org/10.1016/0140-6736(90)90737-P.

38. Towner, J. S., Amman, B. R., Sealy, T. K., Carroll, S. A., Comer, J. A., Kemp, A. ... & Rollin, P. E. (2009). Isolation of genetically diverse Marburg viruses from Egyptian fruit bats. *PLoS Pathogens*, *5*(7), e1000536. https://doi.org/10.1371/journal.ppat.1000536.

39. Leendertz, S.A.J., Wich, S.A., Ancrenaz, M., Bergl, R.A., Gonder, M.K., Humle, T. & Leendertz, F.H. (2017), Ebola in great apes – current knowledge, possibilities for vaccination, and implications for conservation and human health. *Mammal Review*, *47*(2), 98-111. https://doi.org/10.1111/mam.12082.

40. Georges-Courbot, M. C., Sanchez, A., Lu, C. Y., Baize, S., Leroy, E., Lansout-Soukate, J., ... & Ksiazek, T. G. (1997). Isolation and phylogenetic characterization of Ebola viruses causing different outbreaks in Gabon. *Emerging Infectious Diseases*, *3*(1), 59.

41.   Vidal, J. (2004, June 17). Mostly they died. *The Guardian*. Retrieved from https://www.theguardian.com/education/2004/jun/17/research.highereducation.

42.   Formenty, P., Boesch, C., Wyers, M., Steiner, C., Donati, F., Dind, F., ... & Guenno, B. L. (1999). Ebola virus outbreak among wild chimpanzees living in a rain forest of Côte d'Ivoire. *The Journal of Infectious Diseases, 179* (Supplement 1), S120-S126. https://doi.org/10.1086/514296.

43.   Vidal, J. (2004, June 17). Mostly they died. *The Guardian*. Retrieved from https://www.theguardian.com/education/2004/jun/17/research.highereducation.

44.   Ibid.

45.   Leendertz, S. A. J., Wich, S. A., Ancrenaz, M., Bergl, R. A., Gonder, M. K., Humle, T., & Leendertz, F. H. (2017). Ebola in great apes–current knowledge, possibilities for vaccination, and implications for conservation and human health. *Mammal Review, 47*(2), 98-111.

46.   Georges-Courbot, M. C., Sanchez, A., Lu, C. Y., Baize, S., Leroy, E., Lansout-Soukate, J., ... & Ksiazek, T. G. (1997). Isolation and phylogenetic characterization of Ebola viruses causing different outbreaks in Gabon. *Emerging Infectious Diseases, 3*(1), 59.

47.   Walsh, P.D., Tutin, C.E.G., Baillie, J.E.M., Maisels, F., Stokes, E.J. and Gatti, S. (2008). *Gorilla gorilla ssp. Gorilla. The IUCN Red List of Threatened Species*.

48.   Walsh, P.D., Tutin, C.E.G., Oates, J.F., Baillie, J.E.M., Maisels, F., Stokes, E.J., Gatti, ... & Dunn. A. (2007). *Gorilla gorilla. The IUCN Red List of Threatened Species* 2007: e.T9404A12983966.

49.   Maisels, F., Strindberg, S., Breuer, T., Greer, D., Jeffery, K. & Stokes, E. (2018). *Gorilla gorilla* ssp. *gorilla* (amended version of 2016 assessment). *The IUCN Red List of Threatened Species*, e.T9406A136251508. https://dx.doi.org/10.2305/IUCN.UK.2016-2.RLTS.T9406A136251508.en.

50.   Maisels, F., Strindberg, S., Greer, D., Jeffery, K., Morgan, D.L. & Sanz, C. (2016). *Pan troglodytes* ssp. *troglodytes* (errata version published in 2016). *The IUCN Red List of Threatened Species 2016*: e.T15936A102332276. https://dx.doi.org/10.2305/IUCN.UK.2016-2.RLTS.T15936A17990042.en.

51.   Leendertz, S. A. J., Wich, S. A., Ancrenaz, M., Bergl, R. A., Gonder, M. K., Humle, T., & Leendertz, F. H. (2017). Ebola in great apes – current knowledge, possibilities for vaccination, and implications for conservation and human health. *Mammal Review, 47*(2), 98-111.

52.   Leroy, E. M., Rouquet, P., Formenty, P., Souquiere, S., Kilbourne, A., Froment, J. M., ... & Rollin, P. E. (2004). Multiple Ebola virus transmission events and rapid decline of central African wildlife. *Science, 303*(5656), 387-390.

53.   World Health Organization. (2003). Outbreak(s) of Ebola haemorrhagic fever, Congo and Gabon, October 2001–July 2002. *Weekly Epidemiological Record = Relevé épidémiologique hebdomadaire, 78* (26) 223-228. Retrieved from https://iris.who.int/bitstream/handle/10665/232198/WER7826_223-228.PDF.

54.   Leroy, E. M., Rouquet, P., Formenty, P., Souquiere, S., Kilbourne, A., Froment, J. M., ... & Rollin, P. E. (2004). Multiple Ebola virus transmission events and rapid decline of central African wildlife. *Science*, *303*(5656), 387-390.

55.   Bermejo, M., Rodríguez-Teijeiro, J. D., Illera, G., Barroso, A., Vilà, C., & Walsh, P. D. (2006). Ebola outbreak killed 5000 gorillas. *Science*, *314*(5805), 1564-1564.

56.   Caillaud, D., Levréro, F., Cristescu, R., Gatti, S., Dewas, M., Douadi, M., ... & Ménard, N. (2006). Gorilla susceptibility to Ebola virus: the cost of sociality. *Current Biology*, *16*(13), R489-R491.

57.   Ibid.

58.   Genton, C., Cristescu, R., Gatti, S., Levrero, F., Bigot, E., Caillaud, D., ... & Menard, N. (2012). Recovery potential of a western lowland gorilla population following a major Ebola outbreak: results from a ten year study. *PLoS ONE* 7: e37106.

59.   Ryan, S.J. and Walsh, P.D. (2011). Consequences of non-intervention for infectious disease in African great apes. *PLoS ONE 6*: e29030.

60.   Le Gouar, P. J., Vallet, D., David, L., Bermejo, M., Gatti, S., Levréro, F., ... & Ménard, N. (2009). How Ebola impacts genetics of Western lowland gorilla populations. *PLoS ONE*, *4*(12), e8375. https://doi.org/10.1371/journal.pone.0008375 via Zimmerman, D. M., Hardgrove, E., Sullivan, S., Mitchell, S., Kambale, E., Nziza, J., ... & Lacy, R. C. (2023). Projecting the impact of an ebola virus outbreak on endangered mountain gorillas. *Scientific Reports*, *13*(1), 5675. https://doi.org/10.1038/s41598-023-32432-8.

61.   Zimmerman, D. M., Hardgrove, E., Sullivan, S., Mitchell, S., Kambale, E., Nziza, J., ... & Lacy, R. C. (2023). Projecting the impact of an Ebola virus outbreak on endangered mountain gorillas. *Scientific Reports*, *13*(1), 5675. https://doi.org/10.1038/s41598-023-32432-8.

62.   Leendertz, S. A. J., Wich, S. A., Ancrenaz, M., Bergl, R. A., Gonder, M. K., Humle, T., & Leendertz, F. H. (2017). Ebola in great apes–current knowledge, possibilities for vaccination, and implications for conservation and human health. *Mammal Review*, *47*(2), 98-111.

63.   Ibid.

64.   Reed, P. E., Mulangu, S., Cameron, K. N., Ondzie, A. U., Joly, D., Bermejo, M., ... & Sullivan, N. J. (2014). A new approach for monitoring ebolavirus in wild great apes. *PLoS Neglected Tropical Diseases*, *8*(9), e3143. https://journals.plos.org/plosntds/article?id=10.1371/journal.pntd.0003143.

65.   Goldstein, T., Anthony, S. J., Gbakima, A., Bird, B. H., Bangura, J., Tremeau-Bravard, A., ... & Mazet, J. A. (2018). Discovery of a new ebolavirus (Bombali virus) in molossid bats in Sierra Leone. *Nature Microbiology*, *3*(10), 1084. https://doi.org/10.1038/s41564-018-0227-2.

66.   Leroy, E. M., Rouquet, P., Formenty, P., Souquiere, S., Kilbourne, A., Froment, J. M., ... & Rollin, P. E. (2004). Multiple Ebola virus transmission events and rapid decline of central African wildlife. *Science*, *303*(5656), 387-390.

67.   Leroy, E., Gonzalez, J.P., Pourrut, X. (2007). Ebolavirus and Other Filoviruses. In J. E. Childs, J. S. Mackenzie and J. S. Richt (Eds). Wildlife and Emerging Zoonotic Diseases: The Biology, Circumstances and Consequences of Cross-Species Transmission. *Current Topics in Microbiology and Immunology*, *315*. https://doi.org/10.1007/978-3-540-70962-6_15.

68.   World Health Organization (WHO). Ebola, West Africa, March 2014-2016. https://www.who.int/emergencies/situations/ebola-outbreak-2014-2016-West-Africa.

69.   Redding, D. W., Atkinson, P. M., Cunningham, A. A., Lo Iacono, G., Moses, L. M., Wood, J. L., & Jones, K. E. (2019). Impacts of environmental and socio-economic factors on emergence and epidemic potential of Ebola in Africa. *Nature Communications*, *10*(1), 4531. https://doi.org/10.1038/s41467-019-12499-6.

70.   Leendertz, S. A. J., Wich, S. A., Ancrenaz, M., Bergl, R. A., Gonder, M. K., Humle, T., & Leendertz, F. H. (2017). Ebola in great apes–current knowledge, possibilities for vaccination, and implications for conservation and human health. *Mammal Review*, *47*(2), 98-111.

71.   Marí Saéz, A., Weiss, S., Nowak, K., Lapeyre, V., Zimmermann, F., Düx, A., ... & Leendertz, F. H. (2015). Investigating the zoonotic origin of the West African Ebola epidemic. *EMBO Molecular Medicine*, *7*(1), 17-23.

72.   Zimmerman, D. M., Hardgrove, E., Sullivan, S., Mitchell, S., Kambale, E., Nziza, J., ... & Lacy, R. C. (2023). Projecting the impact of an Ebola virus outbreak on endangered mountain gorillas. *Scientific Reports*, *13*(1), 5675. https://doi.org/10.1038/s41598-023-32432-8.

73.   Strindberg, S., Maisels, F., Williamson, E. A., Blake, S., Stokes, E. J., Aba'a, R., ... & Wilkie, D. S. (2018). Guns, germs, and trees determine density and distribution of gorillas and chimpanzees in Western Equatorial Africa. *Science Advances*, *4*(4), eaar2964. https://doi.org/10.1126/sciadv.aar2964.

74.   Zimmerman, D. M., Hardgrove, E., Sullivan, S., Mitchell, S., Kambale, E., Nziza, J., ... & Lacy, R. C. (2023). Projecting the impact of an Ebola virus outbreak on endangered mountain gorillas. *Scientific Reports*, *13*(1), 5675. https://doi.org/10.1038/s41598-023-32432-8.

75.   Henao-Restrepo, A. M., Camacho, A., Longini, I. M., Watson, C. H., Edmunds, W. J., Egger, M., ... & Kieny, M. P. (2017). Efficacy and effectiveness of an rVSV-vectored vaccine in preventing Ebola virus disease: final results from the Guinea ring vaccination, open-label, cluster-randomised trial (Ebola Ça Suffit!). *The Lancet*, *389*(10068), 505-518. https://doi.org/10.1016/S0140-6736(16)32621-6.

76.   Morelle, R. (2017, March 10). Ebola vaccine shows promise for gorillas and chimps. *BBC News*. Retrieved from https://www.bbc.co.uk/news/science-environment-39229625.

77.   Zimmerman, D. M., Hardgrove, E., Sullivan, S., Mitchell, S., Kambale, E., Nziza, J., ... & Lacy, R. C. (2023). Projecting the impact of an Ebola virus

outbreak on endangered mountain gorillas. *Scientific Reports*, *13*(1), 5675. https://
doi.org/10.1038/s41598-023-32432-8.

78.   Ibid.

79.   Britten, F. (2023, April 23). Champion of the gorillas: the vet fighting to save
Uganda's great apes. *The Guardian*. Retrieved from https://www.theguardian.
com/environment/2023/apr/23/champion-of-the-gorillas-the-vet-fighting-to-
save-ugandas-great-apes.

80.   Gilardi, K.V (2023). *Endangered mountain gorillas, Ebola and One Health*,
Chapter in D. A. Jessup & R. W. Radliffe (Eds). *Wildlife Disease and Health in
Conservation*, Johns Hopkins University Press.

81.   Office for National Statistics. Census 2021, Census Maps, Cornwall.
Retrieved from https://www.ons.gov.uk/census/maps/choropleth/population/
population-density/population-density/persons-per-square-kilometre?la
d=E06000052.

82.   United States Census Bureau. 2020 Census Demographic Data Map Viewer,
Population Density. Retrieved from https://maps.geo.census.gov/ddmv/map.
html.

83.   Office for National Statistics. Census 2021, Census Maps, Tower Hamlets.
Retrieved from https://www.ons.gov.uk/census/maps/choropleth/population/
population-density/population-density/persons-per-square-kilometre?la
d=E09000030.

84.   Hassell, J. M., Zimmerman, D., Cranfield, M. R., Gilardi, K.,
Mudakikwa, A., Ramer, J., ... & Lowenstine, L. J. (2017). Morbidity and
mortality in infant mountain gorillas (*Gorilla beringei beringei*): A 46-year
retrospective review. *American Journal of Primatology*, *79*(10), e22686. https://
doi.org/10.1002/ajp.22686.

85.   Mazet, J. A., Genovese, B. N., Harris, L. A., Cranfield, M., Noheri, J. B.,
Kinani, J. F., ... & Gilardi, K. V. (2020). Human respiratory syncytial virus
detected in mountain gorilla respiratory outbreaks. *EcoHealth*, *17*, 449-460.
https://doi.org/10.1007/s10393-020-01506-8.

86.   Spelman, L. H., Gilardi, K. V., Lukasik-Braum, M., Kinani, J. F.,
Nyirakaragire, E., Lowenstine, L. J., & Cranfield, M. R. (2013). Respiratory
disease in mountain gorillas (*Gorilla beringei beringei*) in Rwanda, 1990–2010:
outbreaks, clinical course, and medical management. *Journal of Zoo and Wildlife
Medicine*, *44*(4), 1027-1035. https://doi.org/10.1638/2013-0014R.1.

87.   Hickey, J.R., Basabose, A., Gilardi, K.V., Greer, D., Nampindo, S.,
Robbins, M.M. & Stoinski, T.S. (2020). *Gorilla beringei* ssp. *beringei* (amended
version of 2018 assessment). *The IUCN Red List of Threatened Species*:
e.T39999A176396749. https://dx.doi.org/10.2305/IUCN.UK.2020-3.RLTS.
T39999A176396749.en.

88.   International Gorilla Conservation Programme. Gorilla Families. Retrieved
from https://igcp.org/families/.

89. Our Mountain and Grauer's Gorilla Patients. Gorilla Doctors. Retrieved from https://www.gorilladoctors.org/saving-lives/our-mountain-and-grauers-gorill a-patients/.

90. Gilardi, K.V (2023), *Endangered mountain gorillas, Ebola and One Health*, Chapter in D. A. Jessup & R. W. Radliffe (Eds). *Wildlife Disease and Health in Conservation*, Johns Hopkins University Press.

91. Gilardi, K., & Uwingeli, P. (2022). Keep mountain gorillas free from pandemic virus. *Nature*, *602*(7896), 211. https://doi.org/10.1038/d41586-022-00331-z.

92. Hickey, J.R., Basabose, A., Gilardi, K.V., Greer, D., Nampindo, S., Robbins, M.M. & Stoinski, T.S. (2020). *Gorilla beringei* ssp. *beringei* (amended version of 2018 assessment). *The IUCN Red List of Threatened Species*: e.T39999A176396749. https://dx.doi.org/10.2305/IUCN.UK.2020-3.RLTS.T39999A176396749.en.

93. Mazet, J. A., Genovese, B. N., Harris, L. A., Cranfield, M., Noheri, J. B., Kinani, J. F., ... & Gilardi, K. V. (2020). Human respiratory syncytial virus detected in mountain gorilla respiratory outbreaks. *EcoHealth*, *17*, 449-460. https://doi.org/10.1007/s10393-020-01506-8.

94. Mutu, K. (2024). Wildlife veterinarian Gladys Kalema-Zikusoka on integrating healthcare for human and nonhuman communities. *Earth Island Journal*. Retrieved from https://ctph.org/keeping-gorillas-and-people-healthy/.

95. Gilardi, K.V (2023), *Endangered mountain gorillas, Ebola and One Health*, Chapter in D. A. Jessup & R. W. Radliffe (Eds). *Wildlife Disease and Health in Conservation*, Johns Hopkins University Press.

96. Mazet, J. A., Genovese, B. N., Harris, L. A., Cranfield, M., Noheri, J. B., Kinani, J. F., ... & Gilardi, K. V. (2020). Human respiratory syncytial virus detected in mountain gorilla respiratory outbreaks. *EcoHealth*, *17*, 449-460. https://doi.org/10.1007/s10393-020-01506-8.

97. Our Mountain and Grauer's Gorilla Patients. *Gorilla Doctors*. Retrieved from https://www.gorilladoctors.org/saving-lives/our-mountain-and-grauers-gorill a-patients/.

98. Robbins, M. M., Gray, M., Fawcett, K. A., Nutter, F. B., Uwingeli, P., Mburanumwe, I., ... & Robbins, A. M. (2011). Extreme conservation leads to recovery of the Virunga mountain gorillas. *PLoS ONE*, *6*(6), e19788. https://doi.org/10.1371/journal.pone.0019788.

99. Ibid.

100. Ibid.

101. Ibid.

102. Ibid.

103. Ibid.

104. Hickey, J.R., Basabose, A., Gilardi, K.V., Greer, D., Nampindo, S., Robbins, M.M. & Stoinski, T.S. (2020). *Gorilla beringei* ssp. *beringei* (amended version of 2018 assessment). *The IUCN Red List of Threatened Species*:

e.T39999A176396749. https://dx.doi.org/10.2305/IUCN.UK.2020-3.RLTS. T39999A176396749.en.

105.   Redding, D. W., Atkinson, P. M., Cunningham, A. A., Lo Iacono, G., Moses, L. M., Wood, J. L., & Jones, K. E. (2019). Impacts of environmental and socio-economic factors on emergence and epidemic potential of Ebola in Africa. *Nature Communications, 10*(1), 4531.

106.   Ibid.

14. BIG FARMA

1.   Falchieri, M., Reid, S.M., Ross, C.S., James, J., Byrne, A.M.P., Zamfir, M., ...& Miles, W. (2022), Shift in HPAI infection dynamics causes significant losses in seabird populations across Great Britain. *Veterinary Record, 191*, 294-296. https://doi.org/10.1002/vetr.2311.

2.   Lycett, S. J., Duchatel, F., & Digard, P. (2019). A brief history of bird flu. *Philosophical Transactions of the Royal Society B, 374*(1775), 20180257.

3.   Types of influenza viruses. (2023). *CDC*. Retrieved from https://www.cdc.gov/flu/about/viruses/types.htm.

4.   Lean, F. Z., Falchieri, M., Furman, N., Tyler, G., Robinson, C., Holmes, P., ... & Núñez, A. (2024). Highly pathogenic avian influenza virus H5N1 infection in skua and gulls in the United Kingdom, 2022. *Veterinary Pathology, 61*(3), 421-431. https://doi.org/10.1177/03009858231217224.

5.   Tremlett, C.J., Morley, N., and Wilson, L.J. (2024). UK seabird colony counts in 2023 following the 2021-22 outbreak of Highly Pathogenic Avian Influenza. *RSPB Research Report 76*. RSPB Centre for Conservation Science, RSPB, The Lodge, Sandy, Bedfordshire, SG19 2DL. https://www.rspb.org.uk/birds-and-wildlife/seabird-surveys-project-report.

6.   Lane, J. V., Jeglinski, J. W., Avery-Gomm, S., Ballstaedt, E., Banyard, A. C., Barychka, T., ... & Votier, S. C. (2024). High pathogenicity avian influenza (H5N1) in Northern Gannets (*Morus bassanus*): Global spread, clinical signs and demographic consequences. *Ibis, 166*(2), 633-650. https://doi.org/10.1111/ibi.13275.

7.   Ibid.

8.   Weston, P. (2022, July 20). 'The scale is hard to grasp': avian flu wreaks devastation on seabirds. *The Guardian*. Retrieved from https://www.theguardian.com/environment/2022/jul/20/avian-flu-h5n1-wreaks-devastation-seabirds-aoe.

9.   Lean, F. Z., Falchieri, M., Furman, N., Tyler, G., Robinson, C., Holmes, P., ... & Núñez, A. (2024). Highly pathogenic avian influenza virus H5N1 infection in skua and gulls in the United Kingdom, 2022. *Veterinary Pathology, 61*(3), 421-431. https://doi.org/10.1177/03009858231217224.

10.   EFSA (European Food Safety Authority), ECDC (European Centre for Disease Prevention and Control), EURL (European Reference Laboratory for Avian Influenza), Adlhoch, C., Fusaro, A., Gonzales, J.L., ...& Baldinelli, F.

(2023). Scientific Report: Avian Influenza Overview September–December 2022. *EFSA Journal, 21*(1) e07786.

11.   Weston, P. (2022, July 20). 'The scale is hard to grasp': avian flu wreaks devastation on seabirds. *The Guardian.* Retrieved from https://www.theguardian.com/environment/2022/jul/20/avian-flu-h5n1-wreaks-devastation-seabirds-aoe.

12.   Tremlett, C.J., Morley, N., and Wilson, L.J. (2024). UK seabird colony counts in 2023 following the 2021–22 outbreak of Highly Pathogenic Avian Influenza. *RSPB Research Report 76.* RSPB Centre for Conservation Science, RSPB, The Lodge, Sandy, Bedfordshire, SG19 2DL.

13.   Northern Gannet. The Wildlife Trusts. Retrieved from https://www.wildlifetrusts.org/wildlife-explorer/birds/seabirds/northern-gannet.

14.   Xu, X., Subbarao, K., Cox, N.J. & Guo, Y. (1999). Genetic characterization of the pathogenic influenza A/goose/Guangdong/1/96 (H5N1) virus: Similarity of its hemagglutinin gene to those of H5N1 viruses from the 1997 outbreaks in Hong Kong. *Virology, 261,* 15–19.

15.   Sanjuán, R., & Domingo-Calap, P. (2016). Mechanisms of viral mutation. *Cellular and Molecular Life Sciences : CMLS, 73*(23), 4433–4448. https://doi.org/10.1007/s00018-016-2299-6.

16.   Lycett, S. J., Duchatel, F., & Digard, P. (2019). A brief history of bird flu. *Philosophical Transactions of the Royal Society B, 374*(1775), 20180257.

17.   Richard Kock, personal communication.

18.   The Intravenous Pathogenicity Index Test for Avian Influenza. *Animal & Plant Health Agency.* Retrieved from https://science.vla.gov.uk/fluglobalnet/Documents/english/protocol_IntravenousPathogenicity.pdf.

19.   Questions and answers on avian influenza. *European Centre for Disease Prevention and Control.* Retrieved from https://www.ecdc.europa.eu/en/zoonotic-influenza/facts/faq-avian-influenza.

20.   Lycett, S. J., Duchatel, F., & Digard, P. (2019). A brief history of bird flu. *Philosophical Transactions of the Royal Society B, 374*(1775), 20180257.

21.   Becker, W. B. (1966). The isolation and classification of Tern virus: influenza A-Tern South Africa–1961. *Journal of Hygiene, 64,* 309–320.

22.   Lycett, S. J., Duchatel, F., & Digard, P. (2019). A brief history of bird flu. *Philosophical Transactions of the Royal Society B, 374*(1775), 20180257.

23.   1880-1959 Highlights in the History of Avian Influenza (Bird Flu) Timeline, (2024 April 30). *CDC.* Retrieved from https://www.cdc.gov/flu/avianflu/timeline/avian-timeline-1880-1959.htm.

24.   Becker, W. B. (1966). The isolation and classification of Tern virus: influenza A-Tern South Africa–1961. *Journal of Hygiene, 64,* 309–320.

25.   Richard Kock, personal communication.

26.   Sims, L. D., Domenech, J., Benigno, C., Kahn, S., Kamata, A., Lubroth, J., ... & Roeder, P. (2005). Origin and evolution of highly pathogenic H5N1 avian influenza in Asia. *The Veterinary Record, 157*(6), 159–164. https://doi.org/10.1136/vr.157.6.159.

27.   Duan, L., Campitelli, L., Fan, X. H., Leung, Y. H. C., Vijaykrishna, D., Zhang, J. X., ... & Guan, Y. (2007). Characterization of low-pathogenic H5 subtype influenza viruses from Eurasia: implications for the origin of highly pathogenic H5N1 viruses. *Journal of Virology*, *81*(14), 7529-7539. https://doi.org/10.1128/JVI.00327-07 found via Lycett, S. J., Duchatel, F., & Digard, P. (2019). A brief history of bird flu. *Philosophical Transactions of the Royal Society B*, *374*(1775), 20180257.

28.   1960-1999 Highlights in the History of Avian Influenza (Bird Flu) Timeline. (2024, April 30). *CDC*. Retrieved from https://www.cdc.gov/bird-flu/avian-timeline/1960-1999.html.

29.   H5N1 highly pathogenic avian influenza: Timeline of major events. (2014, December 4). *FAO, OIE, WHO*. Retrieved from https://cdn.who.int/media/docs/default-source/influenza/avian-and-other-zoonotic-influenza/h5n1_avian_influenza_update20141204.pdf.

30.   Sims, L. D., Domenech, J., Benigno, C., Kahn, S., Kamata, A., Lubroth, J., ... & Roeder, P. (2005). Origin and evolution of highly pathogenic H5N1 avian influenza in Asia. *Veterinary Record*, *157*(6), 159-164.

31.   Lycett, S. J., Duchatel, F., & Digard, P. (2019). A brief history of bird flu. *Philosophical Transactions of the Royal Society B*, *374*(1775), 20180257.

32.   Cumulative number of confirmed human cases for avian influenza A(H5N1) reported to WHO, 2003-2021, 15 April 2021. (2021, April 15). *World Health Organization*. Retrieved from https://www.who.int/publications/m/item/cumulative-number-of-confirmed-human-cases-for-avian-influenza-a(h5n1)-reported-to-who-2003-2021-15-april-2021.

33.   Wang, G., Zhan, D., Li, L., Lei, F., Liu, B., Liu, D., ... & Zhu, Q. (2008). H5N1 avian influenza re-emergence of Lake Qinghai: phylogenetic and antigenic analyses of the newly isolated viruses and roles of migratory birds in virus circulation. *Journal of General Virology*, *89*(3), 697-702.

34.   Ibid.

35.   Li, K. S., Guan, Y., Wang, J., Smith, G. J. D., Xu, K. M., Duan, L., ... & Peiris, J. S. M. (2004). Genesis of a highly pathogenic and potentially pandemic H5N1 influenza virus in eastern Asia. *Nature*, *430*(6996), 209-213. https://doi.org/10.1038/nature02746, quoted in Lycett, S. J., Duchatel, F., & Digard, P. (2019). A brief history of bird flu. *Philosophical Transactions of the Royal Society B*, *374*(1775), 20180257.

36.   World Health Organisation. (2014). *H5N1 avian influenza: timeline of major events*. Geneva, Switzerland: WHO. See https://www.who.int/influenza/human_animal_interface/H5N1_cumulative_table_archives/en/, quoted in Lycett, S. J., Duchatel, F., & Digard, P. (2019). A brief history of bird flu. *Philosophical Transactions of the Royal Society B*, *374*(1775), 20180257.

37. Sims, L. D., Domenech, J., Benigno, C., Kahn, S., Kamata, A., Lubroth, J., ... & Roeder, P. (2005). Origin and evolution of highly pathogenic H5N1 avian influenza in Asia. *Veterinary Record*, *157*(6), 159-164.

38. Ibid.

39. Plaza, P. I., Gamarra-Toledo, V., Euguí, J. R., & Lambertucci, S. A. (2024). Recent Changes in Patterns of Mammal Infection with Highly Pathogenic Avian Influenza A(H5N1) Virus Worldwide. *Emerging Infectious Diseases*, *30*(3), 444–452. https://doi.org/10.3201/eid3003.231098.

40. Roberton, S. I., Bell, D. J., Smith, G. J., Nicholls, J. M., Chan, K. H., Nguyen, D. T., ... & Peiris, J. S. (2006). Avian influenza H5N1 in viverrids: implications for wildlife health and conservation. *Proceedings Biological Sciences*, *273*(1595), 1729–1732. https://doi.org/10.1098/rspb.2006.3549.

41. Sims, L. D., Domenech, J., Benigno, C., Kahn, S., Kamata, A., Lubroth, J., ... & Roeder, P. (2005). Origin and evolution of highly pathogenic H5N1 avian influenza in Asia. *Veterinary Record*, *157*(6), 159-164.

42. Chen, H., Smith, G.J.D., Zhang, S.Y., Qin, K., Wang, J., Li, K.S., ... & Guan, Y. (2005) H5N1 virus outbreak in migratory waterfowl. *Nature*, *436*(7048), 191-192. https://doi.org/10.1038/nature03974, quoted in Lycett, S. J., Duchatel, F., & Digard, P. (2019). A brief history of bird flu. *Philosophical Transactions of the Royal Society B*, *374*(1775), 20180257.

43. Wang, G., Zhan, D., Li, L., Lei, F., Liu, B., Liu, D., ... & Zhu, Q. (2008). H5N1 avian influenza re-emergence of Lake Qinghai: phylogenetic and antigenic analyses of the newly isolated viruses and roles of migratory birds in virus circulation. *Journal of General Virology*, *89*(3), 697-702.

44. Olsen, B., Munster, V. J., Wallensten, A., Waldenström, J., Osterhaus, A. D., & Fouchier, R. A. (2006). Global patterns of influenza A virus in wild birds. *Science*, *312*(5772), 384-388.

45. Sims, L. D., Domenech, J., Benigno, C., Kahn, S., Kamata, A., Lubroth, J., ... & Roeder, P. (2005). Origin and evolution of highly pathogenic H5N1 avian influenza in Asia. *The Veterinary Record*, *157*(6), 159–164. https://doi.org/10.1136/vr.157.6.159.

46. Van Borm, S., Thomas, I., Hanquet, G., Lambrecht, B., Boschmans, M., Dupont, G., ... & van den Berg, T. (2005). Highly pathogenic H5N1 influenza virus in smuggled Thai eagles, Belgium. *Emerging Infectious Diseases*, *11*(5), 702–705. https://doi.org/10.3201/eid1105.050211.

47. Sims, L. D., Domenech, J., Benigno, C., Kahn, S., Kamata, A., Lubroth, J., ... & Roeder, P. (2005). Origin and evolution of highly pathogenic H5N1 avian influenza in Asia. *The Veterinary Record*, *157*(6), 159–164. https://doi.org/10.1136/vr.157.6.159.

48. Gavier-Widén, D., Duff, J.P, & Meredith, A., (2012), *Infectious diseases of wild mammals and birds in Europe*, Chichester, UK: Wiley-Blackwell.

49. Ibid.

50. Ibid.

51.   Lycett, S. J., Duchatel, F., & Digard, P. (2019). A brief history of bird flu. *Philosophical Transactions of the Royal Society B*, *374*(1775), 20180257.

52.   Ibid..

53.   Kleyheeg, E., Slaterus, R., Bodewes, R., Rijks, J. M., Spierenburg, M. A., Beerens, N., ... & van der Jeugd, H. P. (2017). Deaths among wild birds during highly pathogenic avian influenza A (H5N8) virus outbreak, the Netherlands. *Emerging Infectious Diseases*, *23*(12), 2050. https://doi.org/10.3201/eid2312.171086 found via Lycett, S. J., Duchatel, F., & Digard, P. (2019). A brief history of bird flu. *Philosophical Transactions of the Royal Society B*, *374*(1775), 20180257.

54.   Lycett, S. J., Duchatel, F., & Digard, P. (2019). A brief history of bird flu. *Philosophical Transactions of the Royal Society B*, *374*(1775), 20180257.

55.   Caliendo, V., Leijten, L., van de Bildt, M. W., Poen, M. J., Kok, A., Bestebroer, T., ... & Kuiken, T. (2022). Long-term protective effect of serial infections with H5N8 highly pathogenic avian influenza virus in wild ducks. *Journal of Virology*, *96*(18), e01233-22. https://doi.org/10.1128/jvi.01233-22.

56.   Caliendo, V., Kleyheeg, E., Beerens, N., Camphuysen, K. C., Cazemier, R., Elbers, A. R., ... & Rijks, J. M. (2024). Effect of 2020–21 and 2021–22 Highly Pathogenic Avian Influenza H5 Epidemics on Wild Birds, the Netherlands. *Emerging Infectious Diseases*, *30*(1), 50. https://doi.org/10.3201/eid3001.230970.

57.   Valentina Caliendo, personal communication.

58.   Gavier-Widén, D., Duff, J.P, & Meredith, A., (2012), *Infectious diseases of wild mammals and birds in Europe*, Chichester, UK: Wiley-Blackwell.

59.   Ramey, A.M., Avian influenza in wild birds, 'Chapter in D. A. Jessup & R. W. Radcliffe (Eds). *Wildlife Disease and Health in Conservation*, Johns Hopkins University Press.

60.   Avian influenza (bird flu): how to spot and report the disease. *Scottish Government*. Retrieved from https://www.gov.scot/publications/avian-influenza-bird-flu/.

61.   Roberton, S. I., Bell, D. J., Smith, G. J. D., Nicholls, J. M., Chan, K. H., Nguyen, D. T., ... & Peiris, J. S. M. (2006). Avian influenza H5N1 in viverrids: implications for wildlife health and conservation. *Proceedings of the Royal Society B: Biological Sciences*, *273*(1595), 1729-1732. https://doi.org/10.1098/rspb.2006.3549.

62.   Pohlmann, A., King, J., Fusaro, A., Zecchin, B., Banyard, A. C., Brown, I. H., ... & Harder, T. (2022). Has epizootic become enzootic? Evidence for a fundamental change in the infection dynamics of highly pathogenic avian influenza in Europe, 2021. *MBio*, *13*(4), e00609-22.

63.   Highlights in the History of Avian Influenza (Bird Flu) Timeline, (2024 April 30), *CDC*. https://www.cdc.gov/bird-flu/avian-timeline/2010-2019.html.

64.   Barnacle goose turns out to be more flexible than expected: Not all migratory birds still migrate. (2022, October 17). *University of Amsterdam Institute for Biodiversity and Ecosystem Dynamics*. Retrieved from https://ibed.uva.nl/content/news/2022/10/not-all-migratory-birds-still-migrate.html?cb.

65.   2020-2024 Highlights in the History of Avian Influenza (Bird Flu) Timeline. (2024, April 30). *CDC*. Retrieved from https://www.cdc.gov/bird-flu/avian-timeline/2020s.html.

66.   Caliendo, V., Kleyheeg, E., Beerens, N., Camphuysen, K., Cazemier, R., Elbers, A....Rijks, J. M. (2024). Effect of 2020–21 and 2021–22 Highly Pathogenic Avian Influenza H5 Epidemics on Wild Birds, the Netherlands. *Emerging Infectious Diseases*, *30*(1), 50-57. https://doi.org/10.3201/eid3001.230970.

67.   Oliver, I., Roberts, J., Brown, C. S., Byrne, A. M., Mellon, D., Hansen, R. D., ... & Zambon, M. (2022). A case of avian influenza A (H5N1) in England, January 2022. *Eurosurveillance*, *27*(5), 2200061. https://doi.org/10.2807/1560-7917.ES.2022.27.5.2200061.

68.   Grierson, J. (2022, January 7). First UK person to catch H5N1 bird flu strain is named. *The Guardian*. Retrieved from https://www.theguardian.com/world/2022/jan/07/first-uk-person-to-catch-h5n1-bird-flu-strain-is-named.

69.   Two foxes in east of Groningen diagnosed with Avian influenza virus. (2021, June 3). *Dutch Wildlife Health Center (DWHC)*. Retrieved from https://dwhc.nl/en/2021/06/two-foxes-in-east-of-groningen-diagnosed-with-avian-influenza-virus/.

70.   Ibid.

71.   Chestakova, I. V., van der Linden, A., Bellido Martin, B., Caliendo, V., Vuong, O., Thewessen, S., ... Sikkema, R. S. (2023). High number of HPAI H5 virus infections and antibodies in wild carnivores in the Netherlands, 2020–2022. *Emerging Microbes & Infections*, *12*(2). https://doi.org/10.1080/22221751.2023.2270068.

72.   Vreman, S., Kik, M., Germeraad, E., Heutink, R., Harders, F., Spierenburg, M., ... & Beerens, N. (2023). Zoonotic mutation of highly pathogenic avian influenza H5N1 virus identified in the brain of multiple wild carnivore species. *Pathogens*, *12*(2), 168. https://doi.org/10.3390/pathogens12020168.

73.   Confirmed findings of influenza of avian origin in non-avian wildlife. *Department for Environment, Food & Rural Affairs, Animal & Plant Health Agency*. Retrieved from https://www.gov.uk/government/publications/bird-flu-avian-influenza-findings-in-non-avian-wildlife/confirmed-findings-of-influenza-of-avian-origin-in-non-avian-wildlife.

74.   Gavier-Widén, D., Duff, J.P., & Meredith, A. (2012) *Infectious diseases of wild mammals and birds in Europe*, Chichester, UK: Wiley-Blackwell.

75.   Vreman, S., Kik, M., Germeraad, E., Heutink, R., Harders, F., Spierenburg, M., ... & Beerens, N. (2023). Zoonotic mutation of highly pathogenic avian influenza H5N1 virus identified in the brain of multiple wild carnivore species. *Pathogens*, *12*(2), 168. https://doi.org/10.3390/pathogens12020168.

76.   Plaza, P. I., Gamarra-Toledo, V., Euguí, J. R., & Lambertucci, S. A. (2024). Recent changes in patterns of mammal infection with Highly Pathogenic Avian

Influenza A(H5N1) Virus worldwide. *Emerging Infectious Diseases, 30*(3), 444–452. https://doi.org/10.3201/eid3003.231098.

77. Ibid.

78. Ibid..

79. Vreman, S., Kik, M., Germeraad, E., Heutink, R., Harders, F., Spierenburg, M., ... & Beerens, N. (2023). Zoonotic mutation of highly pathogenic avian influenza H5N1 virus identified in the brain of multiple wild carnivore species. *Pathogens, 12*(2), 168. https://doi.org/10.3390/pathogens12020168.

80. Lycett, S. J., Duchatel, F., & Digard, P. (2019). A brief history of bird flu. *Philosophical Transactions of the Royal Society B, 374*(1775), 20180257.

81. Trilla, A., Trilla, G., & Daer, C. (2008). The 1918 'Spanish flu' in Spain. *Clinical Infectious Diseases, 47*(5), 668-673.

82. Lycett, S. J., Duchatel, F., & Digard, P. (2019). A brief history of bird flu. *Philosophical Transactions of the Royal Society B, 374*(1775), 20180257.

83. Richard Kock, personal communication.

84. Coronavirus Tracker. (2024, April 13). Worldometer. Retrieved from https://www.worldometers.info/coronavirus/?utm_campaign=homeAdvegas1.

85. Lycett, S. J., Duchatel, F., & Digard, P. (2019). A brief history of bird flu. *Philosophical Transactions of the Royal Society B, 374*(1775), 20180257.

86. World Health Organization. Avian Influenza A(H5N1) - United States of America. (2024, April 9). Retrieved from https://www.who.int/emergencies/disease-outbreak-news/item/2024-DON512.

87. CDC Newsroom. Highly Pathogenic Avian Influenza A (H5N1) Virus Infection Reported in a Person in the U.S. (2024, April 1). Retrieved from https://www.cdc.gov/media/releases/2024/p0401-avian-flu.html.

88. Fidler, K. (2024, April 2). Bird flu patient marks concerning first in virus spread. *Metro*. Retrieved from https://www.msn.com/en-gb/news/other/bird-flu-patient-marks-concerning-first-in-virus-spread/ar-BB1kVrbA#image=BB1kVAze.

89. CDC Newsroom. Highly Pathogenic Avian Influenza A (H5N1) Virus Infection Reported in a Person in the U.S. (2024, April 1). Retrieved from https://www.cdc.gov/media/releases/2024/p0401-avian-flu.html.

90. Plaza, P. I., Gamarra-Toledo, V., Euguí, J. R., & Lambertucci, S. A. (2024). Recent changes in patterns of mammal infection with Highly Pathogenic Avian Influenza A(H5N1) Virus worldwide. *Emerging Infectious Diseases, 30*(3), 444–452. https://doi.org/10.3201/eid3003.231098.

91. Ibid.

92. Emergence and Evolution of H5N1 Bird Flu. *CDC*. Retrieved from https://www.cdc.gov/flu/pdf/avianflu/bird-flu-origin-graphic.pdf.

93.   Caliendo, V., Kleyheeg, E., Beerens, N., Camphuysen, K., Cazemier, R., Elbers, A....Rijks, J. M. (2024). Effect of 2020–21 and 2021–22 Highly Pathogenic Avian Influenza H5 epidemics on wild birds, the Netherlands. *Emerging Infectious Diseases*, *30*(1), 50-57. https://doi.org/10.3201/eid3001.230970.

94.   Emergence and Evolution of H5N1 Bird Flu. *CDC*. Retrieved from https://www.cdc.gov/flu/pdf/avianflu/bird-flu-origin-graphic.pdf.

95.   Caliendo, V., Lewis, N. S., Pohlmann, A., Baillie, S.R., Banyard, A.C., Beer, M., ... Berhane, Y. (2022). Transatlantic spread of highly pathogenic avian influenza H5N1 by wild birds from Europe to North America in 2021. *Scientific Reports*, *12*, 11729. https://doi.org/10.1038/s41598-022-13447-z.

96.   First report of suspected cases of High Pathogenicity Avian Influenza (HPAI) in the Antarctic Treaty Area. *SCAR*. Retrieved from https://scar.org/scar-news/life-sciences/awhn-news/hpai-antarctic-treaty-area.

97.   Kock, R. A. (2019). Is it time to reflect, not on the 'what' but the 'why' in emerging wildlife disease research? *Journal of Wildlife Diseases*, *55*(1), 1-2.

98.   Alasiri, A., Soltane, R., Hegazy, A., Khalil, A. M., Mahmoud, S. H., Khalil, A. A., ... & Mostafa, A. (2023). Vaccination and antiviral treatment against Avian influenza H5Nx viruses: a harbinger of virus control or evolution. *Vaccines*, *11*(11), 1628. https://doi.org/10.3390/vaccines11111628.

99.   California Condors & HPAI Update. (2024, February 2). *California Condor Recovery Program, U.S. Fish & Wildlife Service*. Retrieved from https://www.fws.gov/program/california-condor-recovery/southwest-california-condor-flock-hpai-information-updates-2023.

100.   Kozlov, M. (2023, May 26). US will vaccinate birds against avian flu for first time – what researchers think. *Nature*, *618*, 220-221.

101.   Laville, S. (2023, May 18). RSPB calls for suspension of game-bird releases over avian flu fears. *The Guardian*. Retrieved from https://www.theguardian.com/world/2023/may/18/rspb-calls-for-suspension-of-game-bird-releases-over-avian-flu-fears.

102.   Vickers, S. H., Raghwani, J., Banyard, A. C., Brown, I. H., Fournié, G., & Hill, S. C. (2024). Utilising citizen science data to rapidly assess potential wild bridging hosts and reservoirs of infection: avian influenza outbreaks in Great Britain. *bioRxiv*, 2024.03.28.587127. (Preprint, not yet peer-reviewed). https://doi.org/10.1101/2024.03.28.587127.

103.   Ibid.

104.   Ramey, A.M., Avian influenza in wild birds, Chapter in D. A. Jessup & R. W. Radcliffe (Eds). *Wildlife Disease and Health in Conservation*, Johns Hopkins University Press.

105.   Scottish Government. Avian influenza (bird flu): how to spot and report the disease. Retrieved from https://www.gov.scot/publications/avian-influenza-bird-flu/.

106.   Department for Environment, Food & Rural Affairs, Animal & Plant Health Agency. Bird flu: how to keep pets safe. (2023, August 14). Retrieved from

https://www.gov.uk/government/publications/bird-flu-how-to-keep-pets-safe/bir
d-flu-how-to-keep-pets-safe.

107.   Wu, T., Perrings, C., Shang, C., Collins, J. P., Daszak, P., Kinzig, A., & Minteer, B. A. (2020). Protection of wetlands as a strategy for reducing the spread of avian influenza from migratory waterfowl. *Ambio, 49*, 939-949. https://doi.org/10.1007/s13280-019-01238-2.

108.   Knief, U., Bregnballe, T., Alfarwi, I., Ballmann, M. Z., Brenninkmeijer, A., Bzoma, S., ... & Courtens, W. (2024). Highly pathogenic avian influenza causes mass mortality in Sandwich Tern *Thalasseus sandvicensis* breeding colonies across north-western Europe. *Bird Conservation International, 34*, e6.

109.   Valentina Caliendo, personal communication.

110.   Dias, M. P., Martin, R., Pearmain, E. J., Burfield, I. J., Small, C., Phillips, R. A., ... & Croxall, J. P. (2019). Threats to seabirds: a global assessment. *Biological Conservation, 237*, 525-537. https://doi.org/10.1016/j.biocon.2019.06.033.

111.   Camphuysen, C.J. & Gear, S.C. (2022). Great Skuas and Northern Gannets on Foula, summer 2022 - an unprecedented, H5N1 related massacre. NIOZ Report 2022-02, NIOZ Royal Netherlands Institute for Sea Research: Texel. 66 pp. https://doi.org/10.25850/nioz/7b.b.gd.

112.   Avian flu update: NatureScot advises public landings to stop on 23 islands. (2022, July 21). *NatureScot*. Retrieved from https://www.nature.scot/avian-fl
u-update-naturescot-advises-public-landings-stop-23-islands.

113.   Ibid.

114.   Lane, J. V., Jeglinski, J. W., Avery-Gomm, S., Ballstaedt, E., Banyard, A. C., Barychka, T., ... & Votier, S. C. (2024). High pathogenicity avian influenza (H5N1) in Northern Gannets (*Morus bassanus*): Global spread, clinical signs and demographic consequences. *Ibis, 166*(2), 633-650. https://doi.org/10.1111/ibi.13275.

115.   Gannet. *British Trust for Ornithology (BTO) BirdFacts*. Retrieved from https://www.bto.org/understanding-birds/birdfacts/gannet.

116.   BirdLife International. (2018). *Morus bassanus. The IUCN Red List of Threatened Species*, e.T22696657A132587285. https://dx.doi.org/10.2305/IUCN.UK.2018-2.RLTS.T22696657A132587285.en.

117.   Amaral-Rogers, N. (2023, May 4). Black eyes in seabirds indicates bird flu survival. *RSPB*. Retrieved from https://www.rspb.org.uk/media-centre/blac
k-eyes-in-seabirds-indicates-bird-flu-survival.

118.   Lane, J. V., Jeglinski, J. W., Avery-Gomm, S., Ballstaedt, E., Banyard, A. C., Barychka, T., ... & Votier, S. C. (2024). High pathogenicity avian influenza (H5N1) in Northern Gannets (*Morus bassanus*): Global spread, clinical signs and demographic consequences. *Ibis, 166*(2), 633-650. https://doi.org/10.1111/ibi.13275.

119.   Ibid.

120.   Suzy Gold, personal communication.

121.   Ibid.

15. CHANGING OUR STRIPES

1. Dan Rubenstein, personal communication.

2. Marani, M., Katul, G. G., Pan, W. K., & Parolari, A. J. (2021). Intensity and frequency of extreme novel epidemics. *Proceedings of the National Academy of Sciences, 118*(35), e2105482118.

3. Vora, N. M., Hassan, L., Plowright, R. K., Horton, R., Cook, S., Sizer, N., & Bernstein, A. (2024). The Lancet–PPATS Commission on Prevention of Viral Spillover: reducing the risk of pandemics through primary prevention. *The Lancet, 403*(10427), 597-599.

4. Marani, M., Katul, G. G., Pan, W. K., & Parolari, A. J. (2021). Intensity and frequency of extreme novel epidemics. *Proceedings of the National Academy of Sciences, 118*(35), e2105482118.

5. McCallum, H., Foufopoulos, J., & Grogan, L. F. (2024). Infectious disease as a driver of declines and extinctions. *Cambridge Prisms: Extinction, 2*, e2

6. Ibid.

7. Ibid.

8. Both Hippocrates and Franklin are quoted in Foufopoulos, J., Wobeser, G. A., & McCallum, H., (2022). *Infectious Disease Ecology and Conservation*, Oxford, UK: Oxford University Press as well as in Deem, S. L., Karesh, W. B., & Weisman, W. (2001). Putting theory into practice: wildlife health in conservation. *Conservation Biology, 15*(5), 1224-1233, who said that 'Preventing the introduction of new health problems, as opposed to intervening once a situation has already reached the crisis point, can and should be a bigger part of what we do as conservation biologists.'

9. Bernstein, A. S., Ando, A. W., Loch-Temzelides, T., Vale, M. M., Li, B. V., Li, H., ... & Dobson, A. P. (2022). The costs and benefits of primary prevention of zoonotic pandemics. *Science Advances, 8*(5), eabl4183.

10. Ibid.

11. Dobson, A., & Foufopoulos, J. (2001). Emerging infectious pathogens of wildlife. *Philosophical Transactions of the Royal Society of London. Series B: Biological Sciences, 356*(1411), 1001-1012.

12. Jones, I. J., MacDonald, A. J., Hopkins, S. R., Lund, A. J., Liu, Z. Y. C., Fawzi, N. I., ... & Sokolow, S. H. (2020). Improving rural health care reduces illegal logging and conserves carbon in a tropical forest. *Proceedings of the National Academy of Sciences, 117*(45), 28515-28524.

13. Bernstein, A. S., Ando, A. W., Loch-Temzelides, T., Vale, M. M., Li, B. V., Li, H., ... & Dobson, A. P. (2022). The costs and benefits of primary prevention of zoonotic pandemics. *Science Advances, 8*(5), eabl4183.

14. Ibid.

15. Ibid.

16.   Ibid.

17.   He, D., Lin, L., Artzy-Randrup, Y., Demirhan, H., Cowling, B. J., & Stone, L. (2023). Resolving the enigma of Iquitos and Manaus: A modeling analysis of multiple COVID-19 epidemic waves in two Amazonian cities. *Proceedings of the National Academy of Sciences, 120*(10), e2211422120.

18.   Ferreira, S. (2024, January 12). Amazon rainforest: Deforestation rate halved in 2023. BBC News. Retrieved from https://www.bbc.co.uk/news/ world-latin-america-67962297.

19.   Ibid.

20.   Watts, J. & Bedinelli, T. (2023, February 2). How Illegal Mining Caused a Humanitarian Crisis in the Amazon. *YaleEnvironment360*. Retrieved from https://e360.yale.edu/features/brazil-yanomami-mining-malari a-malnutrition-lula.

21.   Nepstad, D., McGrath, D., Stickler, C., Alencar, A., Azevedo, A., Swette, B., ... & Hess, L. (2014). Slowing Amazon deforestation through public policy and interventions in beef and soy supply chains. *Science, 344*(6188), 1118-1123.

22.   Bernstein, A. S., Ando, A. W., Loch-Temzelides, T., Vale, M. M., Li, B. V., Li, H., ... & Dobson, A. P. (2022). The costs and benefits of primary prevention of zoonotic pandemics. *Science Advances, 8*(5), eabl4183.

23.   Nepstad, D., McGrath, D., Stickler, C., Alencar, A., Azevedo, A., Swette, B., ... & Hess, L. (2014). Slowing Amazon deforestation through public policy and interventions in beef and soy supply chains. *Science, 344*(6188), 1118-1123.

24.   de Souza Cunha, F. A. F., Börner, J., Wunder, S., Cosenza, C. A. N., & Lucena, A. F. (2016). The implementation costs of forest conservation policies in Brazil. *Ecological Economics, 130*, 209-220.

25.   Bernstein, A. S., Ando, A. W., Loch-Temzelides, T., Vale, M. M., Li, B. V., Li, H., ... & Dobson, A. P. (2022). The costs and benefits of primary prevention of zoonotic pandemics. *Science Advances, 8*(5), eabl4183.

26.   Al Jazeera. Brazil Amazon deforestation drops in Lula's first month in office. (2023, February 10). Retrieved from https://www.aljazeera.com/news/2023/2/10/ brazil-amazon-deforestation-drops-in-lulas-first-month-in-office.

27.   Muoria, P.K., Muruthi, P., Kariuki, W.K., Hassan, B.A., Mijele, D. and Oguge, N.O. (2007), Anthrax outbreak among Grevy's zebra (*Equus grevyi*) in Samburu, Kenya. *African Journal of Ecology, 45*: 483-489. https://doi. org/10.1111/j.1365-2028.2007.00758.x.

28.   Vora, N. M., Hannah, L., Lieberman, S., Vale, M. M., Plowright, R. K., & Bernstein, A. S. (2022). Want to prevent pandemics? Stop spillovers. *Nature, 605*(7910), 419-422.

29.   Jones, I. J., MacDonald, A. J., Hopkins, S. R., Lund, A. J., Liu, Z. Y. C., Fawzi, N. I., ... & Sokolow, S. H. (2020). Improving rural health care reduces

illegal logging and conserves carbon in a tropical forest. *Proceedings of the National Academy of Sciences, 117*(45), 28515-28524.

30.   Ibid..

31.   Ibid..

32.   Neil Vora, personal communication.

33.   Webb, K., Jennings, J., & Minovi, D. (2018). A community-based approach integrating conservation, livelihoods, and health care in Indonesian Borneo. *The Lancet Planetary Health, 2*, S26.

34.   Foufopoulos, J., Wobeser, G.A., & McCallum, H., (2022). *Infectious Disease Ecology and Conservation*, Oxford, UK: Oxford University Press p196.

35.   Plowright, R. K., Ahmed, A. N., Coulson, T., Crowther, T. W., Ejotre, I., Faust, C. L., ... & Keeley, A. T. (2024). Ecological countermeasures to prevent pathogen spillover and subsequent pandemics. *Nature Communications, 15*(1), 2577. https://doi.org/10.1038/s41467-024-46151-9.

36.   Ibid.

37.   Ibid.

38.   Ibid.

39.   Meltzer, D. G. (1993). Historical survey of disease problems in wildlife populations: Southern Africa mammals. *Journal of Zoo and Wildlife Medicine, 24*, 237-244.

40.   Meltzer, D. G. (1993). Historical survey of disease problems in wildlife populations: Southern Africa mammals. *Journal of Zoo and Wildlife Medicine, 24*, 237-244.

41.   Nasi, R., Taber, A., & Van Vliet, N. (2011). Empty forests, empty stomachs? Bushmeat and livelihoods in the Congo and Amazon Basins. *International Forestry Review, 13*(3), 355-368.

42.   Ripple, W. J., Abernethy, K., Betts, M. G., Chapron, G., Dirzo, R., Galetti, M., ... & Young, H. (2016). Bushmeat hunting and extinction risk to the world's mammals. *Royal Society Open Science, 3*(10), 160498.

43.   Bernstein, A. S., Ando, A. W., Loch-Temzelides, T., Vale, M. M., Li, B. V., Li, H., ... & Dobson, A. P. (2022). The costs and benefits of primary prevention of zoonotic pandemics. *Science Advances, 8*(5), eabl4183.

44.   Ibid.

45.   Dobson, A. P., Pimm, S. L., Hannah, L., Kaufman, L., Ahumada, J. A., Ando, A. W., ... & Vale, M. M. (2020). Ecology and economics for pandemic prevention. *Science, 369*(6502), 379-381.

46.   Ibid.

47.   Bernstein, A. S., Ando, A. W., Loch-Temzelides, T., Vale, M. M., Li, B. V., Li, H., ... & Dobson, A. P. (2022). The costs and benefits of primary prevention of zoonotic pandemics. *Science Advances, 8*(5), eabl4183.

48.   Gary A Wobeser. (2006). *Essentials of Disease in Wild Animals*, Blackwell Publishing Professional, Ames, USA ISBN 0-8138-0589-9 p175.

49.   Bernstein, A. S., Ando, A. W., Loch-Temzelides, T., Vale, M. M., Li, B. V., Li, H., ... & Dobson, A. P. (2022). The costs and benefits of primary prevention of zoonotic pandemics. *Science Advances*, *8*(5), eabl4183.

50.   Ibid.

51.   Ibid.

52.   Ibid.

53.   Ibid.

54.   World Animal Health Organisation (OIE) and CITES agree to collaborate on animal health and welfare issues worldwide to safeguard biodiversity and protect animals, (2015, December 4), *CITES and OIE joint press release*. Retrieved from https://cites.org/eng/node/18857.

55.   Bernstein, A. S., Ando, A. W., Loch-Temzelides, T., Vale, M. M., Li, B. V., Li, H., ... & Dobson, A. P. (2022). The costs and benefits of primary prevention of zoonotic pandemics. *Science Advances*, *8*(5), eabl4183.

56.   Ibid.

57.   Ibid.

58.   Ibid.

59.   Ibid.

60.   Ibid.

61.   Ibid.

62.   2030 Targets, Kunming-Montreal Global Biodiversity Framework, *Convention on Biological Diversity*. Retrieved from https://www.cbd.int/gbf/targets.

63.   Vora, N. M., Hannah, L., Lieberman, S., Vale, M. M., Plowright, R. K., & Bernstein, A. S. (2022). Want to prevent pandemics? Stop spillovers. *Nature*, *605*(7910), 419-422.

64.   Ibid.

65.   Bernstein, A. S., Ando, A. W., Loch-Temzelides, T., Vale, M. M., Li, B. V., Li, H., ... & Dobson, A. P. (2022). The costs and benefits of primary prevention of zoonotic pandemics. *Science Advances*, *8*(5), eabl4183.

66.   Ibid.

67.   Xiao, L., Lu, Z., Li, X., Zhao, X., & Li, B. V. (2021). Why do we need a wildlife consumption ban in China? *Current Biology*, *31*(4), R168-R172, supplementary information.

68.   Xiao, L., Lu, Z., Li, X., Zhao, X., & Li, B. V. (2021). Why do we need a wildlife consumption ban in China? *Current Biology*, *31*(4), R168-R172.

69.   Li, P. J. (2013). Explaining China's wildlife crisis: cultural tradition or politics of development. Chapter in Bekoff, M. (Ed.). *Ignoring Nature No More: The Case*

for Compassionate Conservation. Chicago: University of Chicago Press. https://doi. org/10.7208/9780226925363-029.

70.   Xia, W.; Hughes, J.; Robertson, D.; Jiang, X. (2021). How one pandemic led to another: ASFV, the disruption contributing to SARS-CoV-2 emergence in Wuhan. Preprints, 2021020590. https://doi.org/10.20944/preprints202102.0590.vi. (Not peer-reviewed).

71.   The Wildlife Conservation and Management Act 2013, Revised 2022, Kenya. Retrieved from http://kenyalaw.org:8181/exist/rest//db/kenyalex/ Kenya/Legislation/English/Acts%20and%20Regulations/W/Wildlife%20 Conservation%20and%20Management%20Act%20-%20No.%2047%20of%202013/ docs/WildlifeConservationandManagementAct47of2013.pdf.

72.   Bernstein, A. S., Ando, A. W., Loch-Temzelides, T., Vale, M. M., Li, B. V., Li, H., ... & Dobson, A. P. (2022). The costs and benefits of primary prevention of zoonotic pandemics. *Science Advances*, *8*(5), eabl4183.

73.   Bernstein, A. S., Ando, A. W., Loch-Temzelides, T., Vale, M. M., Li, B. V., Li, H., ... & Dobson, A. P. (2022). The costs and benefits of primary prevention of zoonotic pandemics. *Science Advances*, *8*(5), eabl4183. Supplementary material.

74.   Ibid.

75.   Kock, R. A. (2019). Is it time to reflect, not on the 'what' but the 'why' in emerging wildlife disease research? *Journal of Wildlife Diseases*, *55*(1), 1-2.

76.   Henderson, R. (2024, May 23). Radical hope: reimagining capitalism for climate crisis. [YouTube video]. Oxford Smith School. Retrieved from https:// www.youtube.com/live/-qKmp4iuKLQ?si=hnvKXfLCRT94DIQs.

77.   Bernstein, A. S., Ando, A. W., Loch-Temzelides, T., Vale, M. M., Li, B. V., Li, H., ... & Dobson, A. P. (2022). The costs and benefits of primary prevention of zoonotic pandemics. *Science Advances*, *8*(5), eabl4183.

78.   Vora, N. M., Hannah, L., Lieberman, S., Vale, M. M., Plowright, R. K., & Bernstein, A. S. (2022). Want to prevent pandemics? Stop spillovers. *Nature*, *605*(7910), 419-422.

79.   Bernstein, A. S., Ando, A. W., Loch-Temzelides, T., Vale, M. M., Li, B. V., Li, H., ... & Dobson, A. P. (2022). The costs and benefits of primary prevention of zoonotic pandemics. *Science Advances*, *8*(5), eabl4183.

80.   Doucleff, M. (2023, February 15). How do pandemics begin? There's a new theory – and a new strategy to thwart them. *Goats and Soda, NPR*. Retrieved from https://www.npr.org/sections/goatsandsoda/2023/02/15/1152892721/ how-to-stop-pandemics.

81.   Neil Vora, personal communication.

82.   Hartmann, T. & Martinez, G. What are natural climate solutions? (2021, September 16). *World Economic Forum*. Retrieved from https://www.weforum. org/agenda/2021/09/what-are-natural-climate-solutions-ncs-alliance/.

83.   Vora, N. M., Hannah, L., Lieberman, S., Vale, M. M., Plowright, R. K., & Bernstein, A. S. (2022). Want to prevent pandemics? Stop spillovers. *Nature*, *605*(7910), 419-422.

84.   Gayle, D. (2022, March 10). Millions suffering in deadly pollution 'sacrifice zones', warns UN expert. *The Guardian*. Retrieved from https://www.theguardian.com/environment/2022/mar/10/millions-suffering-in-deadly-pollution-sacrifice-zones-warns-un-expert.

85.   EU Council votes to criminalise cases "comparable to ecocide". (2024, March 26). *Stop Ecocide International*. Retrieved from https://www.stopecocide.earth/2024/eu-council-votes-to-criminalise-cases-comparable-to-ecocide.

Acknowledgements

A huge thank you to everyone in this list, you helped get this book off the ground and keep it there: my agent Doug Young at PEW Literary for being the earliest believer in this book, my UK editor Connor Stait at Icon Books and my US editor Joe Calamia at the University of Chicago Press for also believing, Patrick Walsh for early discussions, Belinda Low of the Grevy's Zebra Trust and Dan Rubenstein of Princeton University, Rhiannon Morris and Elle-Jay Christodoulou from the publicity team at Icon Books and promotions manager Anne Strother from the University of Chicago Press, Steve Burdett at Icon Books for his comprehensive copy edit, Great Grevy's Rally team members Henry Lolosoli, Mpakuyo Lekulmulahau, Lparasian Lalampaa and Steve Xomo, Steve Mwanga and his team of drivers, especially Edwin for the car-jack, also from the Grevy's Zebra Trust, Peter Lalampaa for his time and his stories, Julius Lekenit and Damaris Lekiluai for all their patient translations, Andrew Letura, Erick, Anita and everybody else who was so kind and welcoming, Martha Fischer and the zookeepers for their fine company, Gathia and the team at Mpala ranch, creative writing tutor Tricia Wastvedt for being an incredibly kind genius and friend, Jane Shemilt for lending me her copy of the disease classic *Man and Microbes* and for many lifts and writing chats, Jon Turney for his listening ear and valuable feedback on the proposal in the early stages, everyone at the co-working space for your company and camaraderie, Ulrike Hartmann-Cadey and Vanessa Spedding for listening patiently and being early readers of the proposal, and Ulrike for helping with a translation about mange in sheep, Ben Hamilton, Mark Ashworth and Jake Hodgewood for reading early drafts and Jed Baron and Alex Barker for wanting to, Kate Long for generous confidence-boosting feedback when I was starting out as a writer, Jheni Osman for putting me in touch with her agent (not to be mine, this time!) and Toria Hare for putting me in touch with Jheni, Sophie Hannah and the clear-headed advice offered through her DreamAuthor Programme, the Society of Authors for generously providing me with an Authors' Foundation grant that gave me a much-needed confidence boost and the time and space to write, all the researchers who gave their time on video calls and reading drafts, it was a pleasure to talk to all of you, Rachael, Alex, Jamie and Chris at CABI, Stuart, Andrew, Jan, Liam and Emily at SGR, Wiley and everyone who provided me with access to their scientific papers, everyone who published open access, the staff at the British Library and the University of Bristol libraries for being uniformly pleasant and helpful, science book club for stimulating discussions and chats, and the Association of British Science Writers for events and support, and the World Federation of Science Journalists for funding the trip to the Arctic, Sharpham House for providing amazing meditation retreats, Bristol Writers' Group for early readings, Lesley Wye and Jen Greenwood for the woods time, Harv, Bridget, Nick and Matt for the Shetland trip, David and Deborah for Einstein, Ulrike, Alice, Liam, Eimear, David, Emma, Andy, Lucy, Rebecca, Marric, Hugh, Matt, Clare, Ian, Suzanne, Catherine, Steve, Maria and Liz for all your friendship and support in Bristol and Nicky, Heather, Beth and Su for the same from afar, and last but very much not least my family Sue, Patrick, Catherine, Mags, Justin, Daniel, Tom and Josh for being with me every step of the way. Thank you.

Index